Models of Delay Differential Equations

Models of Delay Differential Equations

Editors

Francisco Rodríguez
Juan Carlos Cortés López
María Ángeles Castro

MDPI • Basel • Beijing • Wuhan • Barcelona • Belgrade • Manchester • Tokyo • Cluj • Tianjin

Editors
Francisco Rodríguez
University of Alicante
Spain

Juan Carlos Cortés López
Universitat Politècnica de
València
Spain

María Ángeles Castro
University of Alicante
Spain

Editorial Office
MDPI
St. Alban-Anlage 66
4052 Basel, Switzerland

This is a reprint of articles from the Special Issue published online in the open access journal *Mathematics* (ISSN 2227-7390) (available at: https://www.mdpi.com/journal/mathematics/special_issues/delay_differential_equations).

For citation purposes, cite each article independently as indicated on the article page online and as indicated below:

LastName, A.A.; LastName, B.B.; LastName, C.C. Article Title. *Journal Name* **Year**, *Volume Number*, Page Range.

ISBN 978-3-0365-0932-7 (Hbk)
ISBN 978-3-0365-0933-4 (PDF)

© 2021 by the authors. Articles in this book are Open Access and distributed under the Creative Commons Attribution (CC BY) license, which allows users to download, copy and build upon published articles, as long as the author and publisher are properly credited, which ensures maximum dissemination and a wider impact of our publications.

The book as a whole is distributed by MDPI under the terms and conditions of the Creative Commons license CC BY-NC-ND.

Contents

About the Editors vii

Preface to "Models of Delay Differential Equations" ix

Abraham J. Arenas, Gilberto González-Parra, Jhon J. Naranjo, Myladis R. Cogollo and Nicolás De La Espriella
Mathematical Analysis and Numerical Solution of a Model of HIV with a Discrete Time Delay
Reprinted from: *Mathematics* **2021**, *9*, 257, doi:10.3390/math9030257 1

Alexandra Kashchenko
Dependence of Dynamics of a System of Two Coupled Generators with Delayed Feedback on the Sign of Coupling
Reprinted from: *Mathematics* **2020**, *8*, 1790, doi:10.3390/math8101790 23

Amar Debbouche and Vladimir E. Fedorov
A Class of Fractional Degenerate Evolution Equations with Delay
Reprinted from: *Mathematics* **2020**, *8*, 1700, doi:10.3390/math8101700 43

A. S. Hendy and R. H. De Staelen
Theoretical Analysis (Convergence and Stability) of a Difference Approximation for Multiterm Time Fractional Convection Diffusion-Wave Equations with Delay
Reprinted from: *Mathematics* **2020**, *8*, 1696, doi:10.3390/math8101696 51

Akio Matsumoto and Ferenc Szidarovszky
DelayStability of n-Firm Cournot Oligopolies
Reprinted from: *Mathematics* **2020**, *8*, 1615, doi:10.3390/math8091615 71

Luis M. Abia, Óscar Angulo, Juan C. López-Marcos, Miguel A. López-Marcos
The Convergence Analysis of a Numerical Method for a Structured Consumer-Resource Model with Delay in the Resource Evolution Rate
Reprinted from: *Mathematics* **2020**, *8*, 1440, doi:10.3390/math8091440 91

Julia Calatayud, Juan Carlos Cortés, Marc Jornet and Francisco Rodríguez
Mean Square Convergent Non-Standard Numerical Schemes for Linear Random Differential Equations with Delay
Reprinted from: *Mathematics* **2020**, *8*, 1417, doi:10.3390/math8091417 109

Juan Carlos Cortés and Marc Jornet
L^p-solution to the Random Linear Delay Differential Equation with a Stochastic Forcing Term
Reprinted from: *Mathematics* **2020**, *8*, 1013, doi:10.3390/math8061013 127

Ewa Majchrzak and Bohdan Mochnacki
Second-Order Dual Phase Lag Equation. Modeling of Melting and Resolidification of Thin Metal Film Subjected to a Laser Pulse
Reprinted from: *Mathematics* **2020**, *8*, 999, doi:10.3390/math8060999 143

L. Gómez-Valle and J. Martínez-Rodríguez
Two New Strategies for Pricing Freight Options by Means of a Valuation PDE and by Functional Bounds
Reprinted from: *Mathematics* **2020**, *8*, 620, doi:10.3390/math8040620 155

Zhe Yin, Yongguang Yu and Zhenzhen Lu
Stability Analysis of an Age-Structured SEIRS Model with Time Delay
Reprinted from: *Mathematics* **2020**, *8*, 455, doi:10.3390/math8030455 . . . . **167**

Bounded Solutions of Semilinear Time Delay HyperbolicDifferential and Difference Equations
Allaberen Ashyralyev, Deniz Agirseven
Reprinted from: *Mathematics* **2019**, *7*, 1163, doi:10.3390/math7121163 . . . . **185**

María Ángeles Castro, Miguel Antonio García, José Antonio Martín and Francisco Rodríguez
Exact and Nonstandard Finite Difference Schemes for Coupled Linear Delay Differential Systems
Reprinted from: *Mathematics* **2019**, *7*, 1038, doi:10.3390/math7111038 . . . . **223**

About the Editors

Francisco Rodríguez is currently an Associate Professor of Applied Mathematics at the University of Alicante (UA), Alicante, Spain. He develops his research in mathematical modelling in Biology and Engineering in the Dep. of Applied Mathematics and the Multidisciplinary Institute for Environmental Studies (IMEM), UA.

Juan Carlos Cortés López is currently a Full Professor of Applied Mathematics at the Universitat Politècnica de València (UPV), València, Spain. He develops his research in mathematical modelling with uncertainty in the Instituto Universitario de Matemática Multidisciplinar, UPV.

María Ángeles Castro is currently an Associate Professor of Applied Mathematics at the University of Alicante (UA), Alicante, Spain. She develops her research in mathematical modelling and numerical analysis in the Dep. of Applied Mathematics, UA.

Preface to "Models of Delay Differential Equations"

Models of differential equations with delay have pervaded many scientific and technical fields in the last decades. The use of delay differential equations and partial delay differential equations to model problems with the presence of lags or hereditary effects have demonstrated a valuable balance between realism and tractability. Of special interest in recent years is the development and analysis of models with interactions between delay and random effects, through the use of stochastic and random delay differential equations.

In this Special Issue we gather quite a balanced picture of mainstreams topics in the realm of delay differential equations. Indeed, we can find contributions dealing with the construction of exact solutions, numerical methods, dynamical properties, and applications to mathematical modeling of phenomena and processes in biology, economics and engineering, in both deterministic and stochastic settings.

In the paper by Arenas et al. a mathematical model is proposed, based on a set of delay differential equations, that describes intracellular HIV infection. The model considers the time delay between viral entry into a target cell and the production of new virions. The study includes local stability analysis and the design of a non-standard difference scheme that preserves some relevant properties of the continuous mathematical model. In his paper, Kashchenko studies the nonlocal dynamics of a system of delay differential equations with large parameters using the method of steps. This system simulates coupled generators with delayed feedback. The study shows that the dynamics of the system are significantly different in the case of positive coupling and in the case of negative coupling. In the paper by Debbouche and Fedorov, local unique solvability for a class of degenerate fractional differential equations and its application to study initial-boundary value problems for systems of equations with delays is proved. Hendy and Staelen introduce a high order numerical approximation method for convection diffusion wave equations with a multiterm time fractional Caputo operator and a nonlinear fixed time delay. In the paper by Matsumoto and Szidarovszky, the dynamic behavior of n-firm oligopolies is studied, assuming the companies are able to face both implementation and information delays. The analysis includes a classification of stability scenarios depending on the relationship between delays. Continuing within the economical setting, the paper by Abia et al. presents a new numerical method to obtain the solution to a size-structured population model that describes the evolution of a consumer feeding on a dynamical resource that reacts to the environment with a lag-time response. The model is formulated by combining a partial and an ordinary differential equation with delay. Two papers address the numerical and theoretical analysis of linear random delay differential equations. The first one, by Calatayud et al., proposes a mean square convergent non-standard numerical scheme while the second one, by Cortés and Jornet, constructs, rigorously, a solution in the important case that the source term is a stochastic process. In the realm of applications, Majchrzak and Mochnacki, propose a second-order dual phase lag equation to model phase changes associated with heating and cooling of thin metal films. In the paper by Gómez-Valle et al. a partial differential equation for pricing an Asian-style option, termed a freight option, derived from a stochastic delay differential equation is established. This partial differential equation permits attainment of lower and upper bounds for the prime of this type of derivative. The theoretical findings are nicely illustrated using real data from the Baltic Exchange. Zhe Ying et al. study the stability of an age-structured susceptible–exposed—infective–recovered–susceptible (SEIRS) model with time delay. After obtaining one disease-free equilibrium point and one endemic equilibrium point of the model, they establish sufficient conditions in order for the local stability to be guaranteed. In the paper by A. Ashyralyev and D. Agirseven, the initial value problem for a semilinear delay hyperbolic equation in Hilbert spaces with a self-adjoint positive definite operator from a theoretical standpoint is studied. This analysis is complemented with some numerical experiments in the case of semilinear hyperbolic equations with unbounded time delay term, since, in general, it is not possible to obtain the exact solution. The volume finishes with the study of exact and nonstandard finite difference schemes for a class of coupled linear delay differential systems. The study includes the analysis of consistency properties of the new nonstandard schemes and several illustrative examples.

Francisco Rodríguez, Juan Carlos Cortés López, María Ángeles Castro
Editors

Article

Mathematical Analysis and Numerical Solution of a Model of HIV with a Discrete Time Delay

Abraham J. Arenas [1], Gilberto González-Parra [2,*], Jhon J. Naranjo [1], Myladis Cogollo [1] and Nicolás De La Espriella [3]

[1] Departamento de Matemáticas y Estadística, Universidad de Córdoba, Montería 230002, Colombia; aarenas@correo.unicordoba.edu.co (A.J.A.); jjnaranjo2014@gmail.com (J.J.N.); myladiscogollo@correo.unicordoba.edu.co (M.C.)
[2] Department of Mathematics, New Mexico Institute of Mining and Technology, Socorro, NM 87801, USA
[3] Departamento de Física y Electrónica, Universidad de Córdoba, Montería 230002, Colombia; ndelaespriella@correo.unicordoba.edu.co
* Correspondence: gilberto.gonzalezparra@nmt.edu

Abstract: We propose a mathematical model based on a set of delay differential equations that describe intracellular HIV infection. The model includes three different subpopulations of cells and the HIV virus. The mathematical model is formulated in such a way that takes into account the time between viral entry into a target cell and the production of new virions. We study the local stability of the infection-free and endemic equilibrium states. Moreover, by using a suitable Lyapunov functional and the LaSalle invariant principle, it is proved that if the basic reproduction ratio is less than unity, the infection-free equilibrium is globally asymptotically stable. In addition, we designed a non-standard difference scheme that preserves some relevant properties of the continuous mathematical model.

Keywords: HIV infection; mathematical delay model; eclipse phase; NSFD; numerical simulation

Citation: Arenas, A.J.; González-Parra, G.; Naranjo, J.J.; Cogollo, M.; De La Espriella, N. Mathematical Analysis and Numerical Solution of a Model of HIV with a Discrete Time Delay. *Mathematics* **2021**, *9*, 257. http://doi.org/10.3390/math9030257

Academic Editor: Francisco Rodríguez
Received: 31 December 2020
Accepted: 25 January 2021
Published: 28 January 2021

Publisher's Note: MDPI stays neutral with regard to jurisdictional claims in published maps and institutional affiliations.

Copyright: © 2021 by the authors. Licensee MDPI, Basel, Switzerland. This article is an open access article distributed under the terms and conditions of the Creative Commons Attribution (CC BY) license (https://creativecommons.org/licenses/by/4.0/).

1. Introduction

History has recorded that infectious diseases have caused devastation in the human population. Despite the great advances in epidemic control, it was believed that infectious diseases would soon be eradicated, but this has clearly not been the case. Microorganisms adapt and evolve, and consequently, new infectious diseases such as AIDS, Ebola or COVID-19 appear, which cause many deaths. In addition, the genome of some microorganisms can sometimes change slightly and consequently, they can acquire resistance to some drugs [1]. According to the World Health Organization, since its first registration in the 1980s, Human Immunodeficiency Virus (HIV), the causative agent of Acquired Immunodeficiency Syndrome (AIDS), has caused more than 35 million deaths worldwide [2]. The greatest impact of deaths caused by AIDS occurs in underdeveloped or very poor countries, especially in sub-Saharan Africa [2,3].

HIV is an RNA virus that belongs to the retroviridae family, specifically to the lentivirus subfamily, and acts against the immune system, weakening its defense systems against infections and certain types of cancer, which is why the infected person gradually loses its immunodeficiency [4,5]. The HIV replication process is active and dynamic in the sense that when the virus enters the body, the cells that have the CD4+ receptor are infected, most of them are TCD4+ lymphocytes. After entering the cell, the HIV virus can remain latent, replicate in a controlled manner, or undergo massive replication that results in a cytopathic effect for the infected cell. In most lymphocytes the virus is latent, and the infection gradually decreases the amount of these in both the tissues and the blood. This leads the patient to a severe state of cellular immunosuppression, and then a group of microorganisms causes infections. As a consequence, there is a great mortality of people affected by HIV [6].

Epidemiologists conduct scientific experiments, sometimes in controlled settings through self-experimentation, to analyze the spread and possible control strategies of infectious diseases. However, designing such controlled experiments is sometimes impossible due to ethical concerns and the possible collection of erroneous data [1,7,8]. These reasons motivate the possibility of using mathematical models as tools to corroborate the perception of disease transmission, test theories, and suggest better intervention and control strategies.

Recently, there have been a growing literature regarding mathematical models for virus infection within-host [9–19]. These mathematical models include a variety of characteristics related to the viral dynamics. For instance, some models include discrete time delays, cell to cell viral transmissions, and the most well-known virus to cell transmission [10,15,17,19–21]. This article presents a new mathematical model, by means of a set of differential equations with delay, to determine the effect of how to produce viruses by target cells inside the dynamics of viruses. In this case, two types of virus-infected cells are analyzed: the cells in the eclipse phase that are not producing the virus I_E, and the cells that are actively producing the virus I. The cells in the eclipse phase change to the state of virus production at a m rate, and the mortality rate of each cell type is δ_{I_E} and δ_I, respectively. Cells in the eclipse phase may die because they could be recognized as infectious by mediators of innate immunity or due to the accumulation of DNA intermediates in the cell cytoplasm, see [22,23].

Numerous mathematical models represented by means of a system of differential equations, with or without delay, have been discretized by means of the non-standard finite difference method proposed by Ronald Mickens, see [24–34]. Their use is mainly because they are very effective in preserving certain qualitative properties of the original differential equations and the convergence, consistency and stability of their solutions have been demonstrated, see [35–46].

In this study, we also design a non-standard finite difference $(NSFD)$ scheme that allows us to obtain numerical solutions of a set of delayed and ordinary differential equations, which describes the dynamics of HIV infection within-host. First, we apply the techniques designed by Mickens for the construction of the non-standard finite difference $(NSFD)$ scheme to our HIV mathematical model. Secondly, we prove some main properties of the non-standard finite difference $(NSFD)$ scheme, and that agree with qualitative properties on the HIV mathematical model with discrete time delay. One important property that the non-standard finite difference $(NSFD)$ scheme has is that it allows us to guarantee accurate and positive solutions. This is very important when solving inverse problems related to estimation of parameters [13,47–51]. Finally, we perform some numerical simulations that show the advantages regarding accuracy and computational cost.

This paper is organized as follows. In Section 2, we present the mathematical model of HIV within-host with discrete time delay. Section 3 is devoted to the stability mathematical analysis. In Section 4, we construct the $NSFD$ numerical scheme. We include in this section the study of some properties of this numerical scheme such as stability analysis. In Section 5, the numerical simulation results using the constructed $NSFD$ scheme are shown, and the last section is devoted to the conclusions.

2. Mathematical Model of HIV Within-Host with Discrete Time Delay

We construct the mathematical model using a combination of virus facts, hypotheses, and previous proposed mathematical models within-host [9,11,14,52–55]. Despite there has been a growing number of studies related to viral dynamics within-host there are still some aspects that are not well understood [8,9,18,56–58]. Moreover, mathematical models include assumptions that make them more tractable to be able to extract useful information and test different hypotheses [9,55,59,60]. We start noticing that it has been argued that at least in vitro, most HIV-infected cells die before virus production begins [22,61,62]. Virus-producing cells produce virions V at a rate of $N\delta_I$, where N is the average number of infectious virions released by an infected cell during its lifetime. In general, it is accepted that most virions produced by infected cells are not infectious [63–65], and since these virions are not contributing to the infection of new cells, non-infectious virions are not considered in this model. On the other hand, V virions that are infectious can be removed by the immune system from the population of virus-free cells at intrinsic clearance rate C, or they can infect target cells (CD4+ T) at a β rate, with T the concentration of target cells, where Λ is the generation rate of uninfected CD4+ T cells and μ_0 mortality rate of uninfected cells. In the constructed mathematical model there are two classes of infected cells. The first one is the class that includes the cells in the eclipse phase which are not making the virus, I_E. The second class include the cells that are actively producing the virus, I. Cells in the eclipse phase transition to the class I at a rate, m. These cells die at rate δ_{I_E}. The cells in class I then die at rate δ_I. Cells in the eclipse phase die due to the immune systems. Notice that the number of targets cells T are not virus-infected and vary depending on the parameters Λ and the particular death rate for target cells μ_0.

Based on the previous assumptions, we propose a model that describes the intracellular dynamics of HIV and is given by the following system of ordinary differential equations,

$$\begin{aligned}
\frac{dI_E(t)}{dt} &= \beta T(t)V(t) - (m + \delta_{I_E})I_E(t) \\
\frac{dI(t)}{dt} &= mI_E(t) - \delta_I I(t) \\
\frac{dV(t)}{dt} &= N\delta_I I(t) - CV(t) - \beta T(t)V(t) \\
\frac{dT(t)}{dt} &= \Lambda - \beta T(t)V(t) - \mu_0 T(t).
\end{aligned} \quad (1)$$

Notice, that the transmission term $\beta T(t)V(t)$ also appears in the equation for virions because of the assumption that it takes only one virion to infect a target cell [52,66]. The possibility of multiple infected cells is excluded [66]. It also should be noted that during the eclipse phase (the time from viral entry to the active production of viral particles) the infected cells are not producing virions [8,9,16,67,68]. This delay affects the maximum viral load time and the probability that a viral infection will be established [9,68–70], and therefore should be explicitly modeled [9,68]. For this case, let $\Delta > 0$ be the duration of the eclipse phase and $i_E(t,\tau)$ the density of cells at a time t that were infected τ units of time before of t, i.e., are infected cells of age τ. Then $i_E(t,\Delta)$ represents the proportion of cells in the eclipse phase that go into the state of virus production, whereby

$$\frac{dI(t)}{dt} = i_E(t,\Delta) - \delta_I I(t). \quad (2)$$

Since the mortality rate of eclipse cells δ_{I_E} is constant, it is appropriate to assume that $i_E(t,\tau)$ satisfies the equation of McKendrick-Von Foerster age-structured population dynamics model, see [71] that is

$$\frac{\partial i_E(t,\tau)}{\partial t} + \frac{\partial i_E(t,\tau)}{\partial \tau} = -\delta_{I_E} i_E(t,\tau), \quad (3)$$

subject to the boundary condition $i_E(t,0) = \beta T(t)V(t)$. Thus, the solution of the Equation (3) is given by $i_E(t,\tau) = \beta T(t-\tau)V(t-\tau)e^{-\delta_{I_E}\tau}$. Therefore, Equation (2) is given by

$$\frac{dI(t)}{dt} = \beta T(t-\Delta)V(t-\Delta)e^{-\delta_{I_E}\Delta} - \delta_I I(t). \quad (4)$$

The total number of cells in the eclipse phase is given by $I_E(t) = \int_0^\Delta i_E(t,\tau)d\tau$, and by integrating both sides of Equation (3) on the interval $[0,\Delta]$ we have

$$\frac{dI_E(t)}{dt} = \beta T(t)V(t) - \beta T(t-\Delta)V(t-\Delta)e^{-\delta_{I_E}\Delta} - \delta_{I_E} I_E(t). \quad (5)$$

Thus the mathematical model given in (1) takes the form

$$\frac{dI_E(t)}{dt} = \beta T(t)V(t) - \beta T(t-\Delta)V(t-\Delta)e^{-\delta_{I_E}\Delta} - \delta_{I_E}I_E(t)$$
$$\frac{dI(t)}{dt} = \beta T(t-\Delta)V(t-\Delta)e^{-\delta_{I_E}\Delta} - \delta_I I(t) \qquad (6)$$
$$\frac{dV(t)}{dt} = N\delta_I I(t) - CV(t) - \beta T(t)V(t)$$
$$\frac{dT(t)}{dt} = \Lambda - \beta T(t)V(t) - \mu_0 T(t),$$

where Δ is the duration of the eclipse phase, $e^{-\delta_{I_E}\Delta}$ represents the probability that an infected cell will survive a time Δ after viral entry. Notice that Δ is a fixed time delay, and then we have a differential equation with a discrete time delay. For a better understanding of the mathematical model, we can analyze it from the following flow chart shown in Figure 1 [54]:

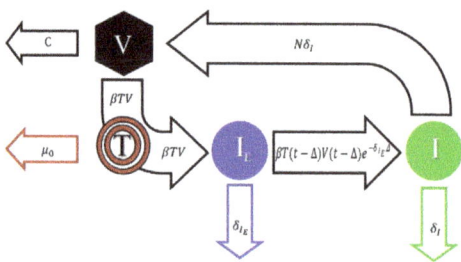

Figure 1. Flow chart of model (6).

In addition, the system (6) satisfies the initial conditions given by

$$T(s) = \xi_1(s), V(s) = \xi_2(s), s \in [-\Delta, 0], \qquad (7)$$

with $\xi_1(s), \xi_2(s)$ positive continuous functions defined from the interval $[-\Delta, 0]$ to $\mathbb{R}^2_+$, and equipped with the norm $\|\xi_{1,2}\| = \sup_{-\Delta \leq s \leq 0} |\xi_{1,2}|$.

3. Properties of the Solutions of the Mathematical Model

Using the fundamental theory of functional differential equations [72,73], it follows that the solution of the system (6) with the initial condition (7) exists for all $t \leq 0$ and is unique.

Next, we will establish different dynamic properties of the solution of the mathematical model described by the system of Equation (6). Since the system (6) represents a biological model, it is important to determine the nature of the solution. Thus, if it is assumed that all the parameters are non-negative as well as the initial conditions $I_E(s), I(s), V(s), T(s)$, for $s \in [-\Delta, 0]$. Therefore, we must guarantee the positivity and boundedness of the solution $(I_E(t), I(t), V(t), T(t))$ of the system (6) at $[0, \infty)$. The following results characterizes these properties.

Theorem 1. *If the initial conditions $I_E(0) = I_{E_0}, I(0) = I_0, V(0) = V_0, T(0) = T_0$ of the mathematical model (6) are positive, then the solutions $(I_E(t), I(t), V(t), T(t))$ of the system (6) are positive for all $t \in [0, \infty)$.*

Proof. Let us start by noting that the solutions of the differential equations of $I_E(t)$ and $I(t)$ given in (6) can be written as

$$I_E(t) = e^{-\delta_{I_E} t} \left[I_{E_0} + \int_{t-\Delta}^{t} \beta T(s) V(s) e^{\delta_{I_E} s} ds \right], \tag{8}$$

$$I(t) = e^{-\delta_I t} \left[I_0 + \int_{0}^{t} \beta T(s-\Delta) V(s-\Delta) e^{(\delta_I - \delta_{I_E})s} ds \right]. \tag{9}$$

Therefore, the positivity of the solutions $T(t)$ and $V(t)$ for all $t > 0$, allows us to guarantee the positivity of $I_E(t)$ and $I(t)$ and thus of system (6). Thus, for $T(t)$ given as in (6) we have that $T(t) > 0$, for all $t \geq 0$. Indeed, suppose that the positivity does not holds, therefore there must be a $t_0 > 0$ such that $T(t_0) = 0$, $\frac{dT(t_0)}{dt} \leq 0$ and $T(t) > 0$ for all $t \in [0, t_0)$, because the initial condition $T_0 > 0$. Thus, $T(t)$ must be negative from some t_0. However, in the interval $[0, t_0)$ the function $T(t)$ is positive, and at point t_0 the derivative at t_0 is non-positive. Thus, from the fourth equation of model (6), it follows that for t_0,

$$\frac{dT(t_0)}{dt} = \Lambda - \beta T(t_0) V(t_0) - \mu_0 T(t_0) = \Lambda > 0,$$

which contradicts that $\frac{dT(t_0)}{dt} \leq 0$. Therefore, we must have $T(t) > 0$, for all $t \geq 0$. Now, we affirm that if $V(t)$ is given by the system (6), it follows that

$$V(t) > 0, \text{ for all } t \geq 0. \tag{10}$$

Indeed, suppose that there exists a $t_1 > 0$ such that $V(t_1) = 0$, $\frac{dV(t_1)}{dt} \leq 0$ and $V(t) > 0$ for all $t \in [0, t_1)$. Then it holds that $I(t) > 0$ for all $t \in [0, t_1)$. Otherwise, there should be a t_2 such that $0 < t_2 < t_1$, $I(t_2) = 0$, $\frac{dI(t_2)}{dt} \leq 0$ and $I(t) > 0$ for all $t \in [0, t_2)$. Thus, from the second equation of system (6), if follows that for $t_2 \in (\Delta, t_1)$ it holds

$$\frac{dI(t_2)}{dt} = \beta T(t_2 - \Delta) V(t_2 - \Delta) e^{-\delta_{I_E} \Delta} - \delta_I I(t_2) = \beta T(t_2 - \Delta) V(t_2 - \Delta) e^{-\delta_{I_E} \Delta}.$$

But, $-\Delta < t_2 - \Delta < t_1 - \Delta < t_1$. From the initial conditions given by (7) and the hypothesis for $V(0) > 0$ it follows that $V(t_2 - \Delta) > 0$. This, contradicts that $\frac{dI(t_2)}{dt} \leq 0$. Thus, $I(t) > 0$, for all $t \in [0, t_1)$. Next, from third equation of system (6) for $t_1 > 0$,

$$\frac{dV(t_1)}{dt} = N \delta_I I(t_1) - CV(t_1) - \beta T(t_1) V(t_1) = N \delta_I I(t_1) > 0.$$

This is a contradiction since $\frac{dV(t_1)}{dt} \leq 0$. Therefore, $V(t) > 0$ for all $t \geq 0$. □

Theorem 2. *The solution $(I_E(t), I(t), V(t), T(t))$ of system (6) is uniformly bounded in $[0, \infty)$.*

Proof. From system (6), one can see that

$$\frac{dI_E(t)}{dt} + \frac{dI(t)}{dt} + \frac{dT(t)}{dt} = \Lambda - \delta_{I_E} I_E(t) - \delta_I I(t) - \mu_0 T(t) \leq \Lambda - M(I_E(t) + I(t) + T(t)),$$

where $M = \min\{\delta_{I_E}, \delta_I, \mu_0\}$. This implies that

$$I_E(t) + I(t) + T(t) \leq e^{-Mt} \left[I_{E_0} + I_0 + T_0 + \int_0^t \Lambda e^{Ms} ds \right] = e^{-Mt} [I_{E_0} + I_0 + T_0] + \frac{\Lambda}{M} \left(1 - e^{-Mt} \right).$$

Thus,
$$\limsup_{t\to\infty}(I_E(t)+I(t)+T(t)) \leq \frac{\Lambda}{M}.$$

Accordingly, $I_E(t), I(t)$ and $T(t)$ are uniformly boundedness. Even more, given $\varepsilon > 0$, there exists $t_1 > 0$ such that for all $t \geq t_1$,

$$I(t) \leq \frac{\Lambda}{M} + \varepsilon.$$

Then, since $\frac{dV(t)}{dt} \leq N\delta_I I(t) - CV(t)$, and $V(t) > 0$ for all $t > t_1$, one obtains that

$$\frac{dV(t)}{dt} + CV(t) \leq N\delta_I\left(\frac{\Lambda}{M}+\varepsilon\right).$$

It follows that

$$V(t) \leq V_0 e^{-Ct} + \left(1-e^{-Ct}\right)\left(\frac{N\delta_I}{C}\right)\left(\frac{\Lambda}{M}+\varepsilon\right).$$

As $t \longrightarrow \infty$, then $V(t) \leq \left(\frac{N\delta_I}{C}\right)\left(\frac{\Lambda}{M}+\varepsilon\right)$. Since $\varepsilon > 0$, $V(t)$ is uniformly boundedness. This completes the proof. □

3.1. Equilibrium Points

The model described by the system of differential Equation (6) has two stationary states, the first one corresponds to the disease-free equilibrium and the second to the endemic equilibrium, which we will denote P^0 and P^* respectively. To determine both states we must calculate the critical points of the system (6) by setting $\frac{dI_E(t)}{dt} = \frac{dI(t)}{dt} = \frac{dV(t)}{dt} = \frac{dT(t)}{dt} = 0$. Thus, we have

$$\begin{aligned} 0 &= \beta TV - \beta TVe^{-\delta_{I_E}\Delta} - \delta_{I_E}I_E \\ 0 &= \beta TVe^{-\delta_{I_E}\Delta} - \delta_I I \\ 0 &= N\delta_I I - CV - \beta TV \\ 0 &= \Lambda - \beta TV - \mu_0 T. \end{aligned} \quad (11)$$

The disease-free equilibrium point of a model are solutions of steady state in the absence of infection. For this case, we must consider $I_E = 0$, $I = 0$, $V = 0$, and $T > 0$, in the system (11). Then P^0 will be of the form $P^0 = (0,0,0,T^0)$, where $T^0 = \frac{\Lambda}{\mu_0}$. Therefore, $P^0 = \left(0,0,0,\frac{\Lambda}{\mu_0}\right)$.

On the other hand, we can determine the basic reproductive number using the next generation matrix methodology. With the terms of infection and viral production in the mathematical model (11), matrices F and V are given by

$$\mathbb{F} = \begin{pmatrix} 0 & 0 & \beta T^0\left(1-e^{-\delta_{I_E}\Delta}\right) \\ 0 & 0 & \beta T^0 e^{-\delta_{I_E}\Delta} \\ 0 & 0 & 0 \end{pmatrix}, \quad \mathbb{V} = \begin{pmatrix} \delta_{I_E} & 0 & 0 \\ 0 & \delta_I & 0 \\ 0 & N\delta_I & C+\beta T^0 \end{pmatrix},$$

where $T^0 = \dfrac{\Lambda}{\mu_0}$, which it is the number of target cells before infection. Thus, the basic reproductive number $\mathcal{R}_0$, is calculated as the spectral radius of the matrix given by

$$\mathbb{F}\mathbb{V}^{-1} = \begin{pmatrix} 0 & 0 & \beta T^0\left(1-e^{-\delta_{I_E}\Delta}\right) \\ 0 & 0 & \beta T^0 e^{-\delta_{I_E}\Delta} \\ 0 & 0 & 0 \end{pmatrix} \begin{pmatrix} \frac{1}{\delta_{I_E}} & 0 & 0 \\ 0 & -\delta_I & 0 \\ 0 & -\frac{N}{C+\beta T^0} & \frac{1}{C+\beta T^0} \end{pmatrix}$$

$$= \begin{pmatrix} 0 & -\dfrac{N\beta T^0\left(1-e^{-\delta_{I_E}\Delta}\right)}{\beta T^0+C} & \dfrac{\beta T^0\left(1-e^{-\delta_{I_E}\Delta}\right)}{\beta T^0+C} \\ 0 & -\dfrac{N\beta T^0 e^{-\delta_{I_E}\Delta}}{\beta T^0+C} & \dfrac{\beta T^0 e^{-\delta_{I_E}\Delta}}{\beta T^0+C} \\ 0 & 0 & 0 \end{pmatrix}.$$

Therefore, the basic reproductive number $\mathcal{R}_0$ is given by

$$\mathcal{R}_0 = \dfrac{N\beta\Lambda e^{-\delta_{I_E}\Delta}}{C\mu_0 + \beta\Lambda}. \tag{12}$$

Now, the endemic equilibrium point of a model is its steady-state solutions in the presence of infection or disease, for which it must be considered $I_E > 0$, $I > 0$, $V > 0$ and $T > 0$ in the system (11.) Then P^* will be of the form $P^* = (I_E^*, I^*, V^*, T^*)$. In this case, from the system (11) the following equalities are obtained

$$I_E^* = \dfrac{\beta T^* V^*(1-e^{-\delta_{I_E}\Delta})}{\delta_{I_E}}. \tag{13}$$

$$I^* = \dfrac{\beta T^* V^* e^{-\delta_{I_E}\Delta}}{\delta_I}. \tag{14}$$

$$I^* = \dfrac{V^*}{N\delta_I}(C+\beta T^*). \tag{15}$$

$$T^* = \dfrac{\Lambda}{\beta V^* + \mu_0}. \tag{16}$$

Replacing (16) in (14) and (15), is obtains

$$I^* = \dfrac{\beta\Lambda V^* e^{-\delta_{I_E}\Delta}}{\delta_I(\beta V^* + \mu_0)} = \dfrac{V^*}{N\delta_I}\left(C + \dfrac{\beta\Lambda}{\beta V^* + \mu_0}\right). \tag{17}$$

Then

$$V^* = \dfrac{N\beta\Lambda e^{-\delta_{I_E}\Delta} - C\mu_0 - \beta\Lambda}{C\beta} = \dfrac{(C\mu_0 + \beta\Lambda)\dfrac{N\beta\Lambda e^{-\delta_{I_E}\Delta}}{C\mu_0 + \beta\Lambda} - 1}{C\beta}. \tag{18}$$

Thus, if $\mathcal{R}_0 > 1$, then $N\beta\Lambda e^{-\delta_{I_E}\Delta} - C\mu_0 - \beta\Lambda > 0$. Hence, we can write V^* as

$$V^* = \dfrac{(C\mu_0 + \beta\Lambda)(\mathcal{R}_0 - 1)}{C\beta}. \tag{19}$$

Next, we replace (18) in (17) to get

$$I^* = \dfrac{N\beta\Lambda e^{-\delta_{I_E}\Delta} - C\mu_0 - \beta\Lambda}{\delta_I \beta e^{\delta_{I_E}\Delta}(Ne^{-\delta_{I_E}\Delta} - 1)} = \dfrac{(C\mu_0 + \beta\Lambda)(\mathcal{R}_0 - 1)}{\delta_I \beta e^{\delta_{I_E}\Delta}(Ne^{-\delta_{I_E}\Delta} - 1)}. \tag{20}$$

Now, substituting (18) in (16) one gets

$$T^* = \frac{\Lambda C}{(C\mu_0 + \beta\Lambda)(\mathcal{R}_0 - 1) + C\mu_0}. \tag{21}$$

Finally, we replace (18) and (21) in (13) to obtain

$$I_E^* = \frac{\left(N\beta\Lambda e^{-\delta_I_E \Delta} - C\mu_0 - \beta\Lambda\right)\left(e^{\delta_{I_E}\Delta} - 1\right)}{\delta_I \beta e^{\delta_{I_E}\Delta}\left(Ne^{-\delta_{I_E}\Delta} - 1\right)} = \frac{(C\mu_0 + \beta\Lambda)(\mathcal{R}_0 - 1)\left(e^{\delta_{I_E}\Delta} - 1\right)}{\delta_I \beta e^{\delta_{I_E}\Delta}\left(Ne^{-\delta_{I_E}\Delta} - 1\right)} \tag{22}$$

Note that $I_E^* > 0$, $I^* > 0$, $V^* > 0$ and $T^* > 0$ if only if $\mathcal{R}_0 = \frac{N\beta\Lambda e^{-\delta_{I_E}\Delta}}{C\mu_0 + \beta\Lambda} > 1$. Thus, $Ne^{-\delta_{I_E}\Delta} > \frac{C\mu_0 + \beta\Lambda}{\beta\Lambda} > 1$.

3.2. Local Stability of the Equilibrium Points

Theorem 3. *The disease-free equilibrium point P^0 of the system (6) is locally asymptotically stable if $\mathcal{R}_0 < 1$.*

Proof. The eigenvalues of the Jacobian matrix of system (6) evaluated at point P^0, are given as the roots of polynomial

$$(-\delta_I - \lambda)(-\mu_0 - \lambda)\left[\lambda^2 + \left(\delta_I + C + \frac{\beta\Lambda}{\mu_0}\right)\lambda + \left(\delta_I C + \frac{\delta_I \beta\Lambda}{\mu_0} - \frac{\beta e^{-\delta_{I_E}\Delta}N\delta_I\Lambda}{\mu_0}\right)\right] = 0.$$

Therefore, the first two eigenvalues of the Jacobian matrix evaluated at P^0 are $\lambda_1 = -\delta_I < 0$ and $\lambda_2 = -\mu_0 < 0$.

Next, since $\mathcal{R}_0 < 1$ then the coefficients of equation

$$\lambda^2 + \left(\delta_I + C + \frac{\beta\Lambda}{\mu_0}\right)\lambda + \frac{\delta_I}{\mu_0}(C\mu_0 + \beta\Lambda)(1 - \mathcal{R}_0) = 0, \tag{23}$$

are positives. Thus, since there is no sign change between its terms, and by Descartes' sign rule it is concluded that Equation (23) does not have positive roots. Now, if λ is replaced by $-\lambda$ in (23) then

$$\lambda^2 - \left(\delta_I + C + \frac{\beta\Lambda}{\mu_0}\right)\lambda + \frac{\delta_I}{\mu_0}(C\mu_0 + \beta\Lambda)(1 - \mathcal{R}_0) = 0. \tag{24}$$

Thus, if $\mathcal{R}_0 < 1$ Equation (24) has two sign changes in its terms, and by Descartes' sign rule it is concluded that there are exactly two negative roots in Equation (23). Therefore, $P^0 = \left(0, 0, 0, \frac{\Lambda}{\mu_0}\right)$ is locally asymptotically stable if $\mathcal{R}_0 < 1$. □

Theorem 4. *The P^* endemic point of the system (6) is locally asymptotically stable if $\mathcal{R}_0 > 1$.*

Proof. We note that $R_1 = Ne^{-\delta_{I_E}\Delta} - 1 > 0$. Thus, the characteristic equation is given by

$$\begin{vmatrix} -\delta_I - \lambda & \dfrac{Ce^{-\delta_{I_E}\Delta}}{R_1} & \dfrac{\beta\Lambda R_1 e^{-\delta_{I_E}\Delta}}{C} - \mu_0 e^{-\delta_{I_E}\Delta} \\ (-\delta_{I_E} - \lambda) \quad N\delta_I & -C - \dfrac{C}{R_1} - \lambda & \mu_0 - \dfrac{\beta\Lambda R_1}{C} \\ 0 & -\dfrac{C}{R_1} & -\dfrac{\beta\Lambda R_1}{C} - \lambda \end{vmatrix} = 0$$

Therefore, the first eigenvalue of the Jacobian matrix will be $\lambda_1 = -\delta_{I_E} < 0$, and the other, are the roots of polynomial

$$\lambda^3 + \left(\delta_I + C + \frac{C}{R_1} + \frac{\beta\Lambda R_1}{C}\right)\lambda^2 + \left(\frac{\delta_I \beta\Lambda R_1}{C} + \beta\Lambda R_1 + \frac{C\mu_0}{R_1}\right)\lambda + \delta_I(C\mu_0 + \beta\Lambda)(\mathcal{R}_0 - 1) = 0. \tag{25}$$

If $\mathcal{R}_0 > 1$ all the coefficients of Equation (25) are positive, i.e., there is no sign change between their terms, and by Descartes's sign rule it is concluded that there are no positive roots. Now if λ is replaced by $-\lambda$ in (25) it gives us that

$$-\lambda^3 + \left(\delta_I + C + \frac{C}{R_1} + \frac{\beta\Lambda R_1}{C}\right)\lambda^2 - \left(\frac{\delta_I \beta\Lambda R_1}{C} + \beta\Lambda R_1 + \frac{C\mu_0}{R_1}\right)\lambda + \delta_I(C\mu_0 + \beta\Lambda)(\mathcal{R}_0 - 1) = 0 \tag{26}$$

Then, if $\mathcal{R}_0 > 1$ the polynomial (26) has three sign changes between its terms, and by Descartes's sign rule it is concluded that there are exactly three negative roots of Equation (25). Thus, P^* is locally asymptotically stable if $\mathcal{R}_0 > 1$. □

3.3. Global Stability Analysis of the Mathematical Model

Since the variable I_E does not appear in the other three equations, without loss of generality we will only consider the following three-dimensional system,

$$\begin{aligned} \frac{dI(t)}{dt} &= \beta T(t-\Delta)V(t-\Delta)e^{-\delta_{I_E}\Delta} - \delta_I I(t) \\ \frac{dV(t)}{dt} &= N\delta_I I(t) - (C + \beta T(t))V(t) \\ \frac{dT(t)}{dt} &= \Lambda - \beta T(t)V(t) - \mu_0 T(t). \end{aligned} \tag{27}$$

To analyze the global stability of the equilibrium points of the system (27), we use the method of the Lyapunov's functions, and we will use the Volterra function

$$G(x) = x - 1 - \ln x \tag{28}$$

for $x > 0$, which is no negative for any $x > 0$ and $G(x) = 0$ if and only if $x = 1$.

Theorem 5. *If $\mathcal{R}_0 < 1$ then the disease-free equilibrium point $P_f = \left(0, 0, \frac{\Lambda}{\mu_0}\right)$ of system (27) is globally asymptotically stable.*

Proof. We define the Lyapunov functional

$$\mathcal{V}(t) = e^{-\delta_{I_E}\Delta} T^0 G\left(\frac{T(t)}{T^0}\right) + \frac{1}{N}V(t) + I(t) + e^{-\delta_{I_E}\Delta}\int_0^\Delta \beta T(t-\theta)V(t-\theta)\,d\theta.$$

Now, calculating the time derivative of $\mathcal{V}(t)$ along the solution of model (27), one gets that

$$\frac{d\mathcal{V}(t)}{dt} = e^{-\delta_{I_E}\Delta}\frac{T(t)-T^0}{T(t)}\frac{dT(t)}{dt} + \frac{1}{N}\frac{dV(t)}{dt} + \frac{dI(t)}{dt} + e^{-\delta_{I_E}\Delta}\frac{d}{dt}\int_0^\Delta \beta T(t-\theta)V(t-\theta)\,d\theta$$
$$= -\mu_0 e^{-\delta_{I_E}\Delta}\frac{(T(t)-T^0)^2}{T(t)} - e^{-\delta_{I_E}\Delta}\beta T(t)V(t) + e^{-\delta_{I_E}\Delta}\beta V(t)T^0 + \delta_I I(t) - \frac{C}{N}V(t) - \frac{\beta T(t)V(t)}{N}$$
$$+ \beta T(t-\Delta)V(t-\Delta)e^{-\delta_{I_E}\Delta} - \delta_I I(t) - \beta T(t-\Delta)V(t-\Delta)e^{-\delta_{I_E}\Delta} - \delta_I I(t) + e^{-\delta_{I_E}\Delta}\beta T(t)V(t)$$
$$\leq -\mu_0 e^{-\delta_{I_E}\Delta}\frac{(T(t)-T^0)^2}{T(t)} + \frac{C}{N}(\mathcal{R}_0 - 1).$$

Thus, $\frac{d\mathcal{V}(t)}{dt} < 0$ when $\mathcal{R}_0 < 1$. Therefore, by Lyapunov–LaSalle Invariance Principle, the infection-free equilibrium E_f is globally asymptotically stable. □

4. Design of a NSFD Scheme for the Mathematical Model

The use of differential equations in the modeling of the transmission of infectious diseases has represented a versatile tool to understand better the dynamics of a variety of infectious diseases [7,59,60,74–76]. Mathematical models based on differential equations have been useful to study how to reduce the burden of infectious diseases. The models allow the determination of optimal controls and estimate the impact of a variety of virus on the disease dynamics [67,75,77]. One advantage of mathematical models is that different simulations can be performed, and this allows us to analyze different main driven factors of epidemics under a variety of complex scenarios [8,11,59]. However, there are no general formulas that allow the obtaining of precise analytical solutions for many differential equation systems. These solutions exist only occasionally and are often difficult to find, so good approximations are necessary that preserve the qualitative properties of said solution, for which numerical methods have been used, see [24,25,31,48–50,78–83].

Discrete epidemic models generated by numerical methods contain additional parameters to those that already exist in differential equations, such as the time and space steps. Variations in these additional parameters can generate solutions to the discrete equations that do not correspond to any solution of the original differential equations, producing fictitious bifurcations, artificial chaos, spurious solutions, and false stable states [24–26,45,83,84]. Therefore, we must choose numerical discrete schemes that guarantee the qualitative properties of the mathematical models. There are several methods that can be used to obtain accurate solutions. For instance, the Richardson extrapolation on uniform and nonuniform grids or NSFD schemes have been used for that end [45,85–90].

Another, important aspect where a robust numerical method plays an important role is when solving inverse problems to estimate the parameters of the model [48,51,56,57,91,92]. Thus, for mathematical models of a variety of virus is of paramount importance to have a robust and efficient numerical method for solving the differential equations [25,49,51]. Usually, when a differential equation is solved numerically a certain tolerance is prescribed and this has an impact in the success in estimating the parameters [48,49,93,94]. In this paper, we deal with a mathematical model that is based on system of differential equations with discrete time delay [17,32,60,72,95–97]. There are different numerical methods to deal with this type of equations, and some are analogous to the ones used for ordinary differential equations but with additional issues [50,98–102]. One particular numerical method that we are interested in is by using NSFD schemes to guarantee some properties of the continuous mathematical model. Some previous works using this methodology have been developed for linear and nonlinear delay differential equations [103–107].

For the construction of a discrete numerical scheme that allows us to efficiently approximate the solutions of the system (6), we use the methodology proposed by Ronald

Mickens, see [24–27]. In that order of ideas, for the discrete approximation of the time derivative of a function $X(t) \in C^1(\mathbb{R})$, we define the non-standard derivative as

$$\frac{d_N X(t)}{dt} = \frac{X(t+h) - X(t)}{\varphi(h)} + \mathcal{O}(h), \qquad (29)$$

where $\varphi(h)$ is a real positive valued function that satisfies $\varphi(h) = h + \mathcal{O}(h^2)$, and N is to denote the non-standard derivative.

Although there is no general algorithm for constructing an *NSFD* schema that approximates the solutions of a given system of differential equations, the following general rules are often useful to correctly design these schemes.

Rule 1. The discrete derivatives in a numerical scheme must be of the same orders as the continuous derivatives that appear in the differential equation.

Rule 2. Discrete derivatives may have non-trivial denominators.

Rule 3. Nonlinear terms that appear in differential equations must have non-local representations.

Rule 4. The numerical solution must preserve all the special conditions that hold for the solutions of the corresponding differential equations.

Rule 5. The scheme must not introduce unnecessary or false solutions, i.e., convergence to false steady states.

Let us denote by I_E^n, I^n, V^n and T^n the approximations of $I_E(nh)$, $I(nh)$, $V(nh)$ and $T(nh)$, respectively, for $n = 0, 1, 2...$, and for h size step in time of the scheme. The value of I_E^{n+1} for $n = 0, 1, \cdots$, is calculated using Equation (8) and a quadrature formula. For this case, we use

$$I_E^{n+1} = e^{-\delta_{I_E} \Delta (n+1)h} \left[I_E^n + \frac{m_1 h}{2} \left(\beta T^{n+1} V^{n+1} e^{\delta_{I_E} \Delta (n+1)h} + \beta T^{n+1-m_1} V^{n+1-m_1} e^{\delta_{I_E} \Delta (n+1-m_1)h} \right) \right], \qquad (30)$$

with $\Delta = m_1 h$.

Next, we make the following non-local approximations of the terms on the right side of the system (27)

$$\begin{cases}
\beta T(t) V(t) & \to \beta T^n V^n \\
-\beta T(t-\Delta) V(t-\Delta) e^{-\delta_{I_E} \Delta} & \to -\beta T^{n-m_1+1} V^{n-m_1} e^{-\delta_{I_E} \Delta} \\
-\delta_{I_E} I_E(t) & \to -\delta_{I_E} I_E^{n+1} \\
-\delta_I I(t) & \to -\delta_I I^{n+1} \\
N \delta_I I(t) & \to N \delta_I I^{n+1} \\
-(C + \beta T(t)) V(t) & \to -(C + \beta T^n) V^{n+1} \\
-\beta T(t) V(t) & \to -\beta T^{n+1} V^n \\
-\mu_0 T(t) & \to -\mu_0 T^{n+1}
\end{cases} \qquad (31)$$

Then, we approximate the derivatives on the left side of the system (27) as follows

$$\begin{cases}
\dfrac{d_N I(t)}{dt} & \to \dfrac{I^{n+1} - I^n}{\varphi(h)} \\
\dfrac{d_N V(t)}{dt} & \to \dfrac{V^{n+1} - V^n}{\varphi(h)} \\
\dfrac{d_N T(t)}{dt} & \to \dfrac{T^{n+1} - T^n}{\varphi(h)}
\end{cases} \qquad (32)$$

Consequently, the system (27) can be discretized as an implicit *NSFD* scheme given by

$$\frac{T^{n+1} - T^n}{\varphi(h)} = \Lambda - \beta T^{n+1} V^n - \mu_0 T^{n+1},$$

$$\frac{I^{n+1} - I^n}{\varphi(h)} = \beta T^{n-m_1+1} V^{n-m_1} e^{-\delta_{I_E} \Delta} - \delta_I I^{n+1}, \quad (33)$$

$$\frac{V^{n+1} - V^n}{\varphi(h)} = N \delta_I I^{n+1} - C V^{n+1} - \beta T^n V^{n+1}.$$

And the explicit form is given by

$$T^{n+1} = \frac{\varphi(h)\Lambda + T^n}{1 + \varphi(h)(\beta V^n + \mu_0)},$$

$$I^{n+1} = \frac{\varphi(h)\beta T^{n-m_1+1} V^{n-m_1} e^{-\delta_{I_E}\Delta} + I^n}{1 + \varphi(h)\delta_I}, \quad (34)$$

$$V^{n+1} = \frac{\varphi(h) N \delta_I I^{n+1} + V^n}{1 + \varphi(h)(C + \beta T^n)},$$

where $m_1 = \frac{\Delta}{h} \in \mathbb{N}$. The initial conditions of scheme (34) are given by

$$T^j = \xi_1^j, \, V^j = \xi_2^j, \, j = -m_1, -m_1 + 1, \cdots, 0.$$

The positivity of scheme (34) is trivially satisfied, since for $n > 0$ it holds that T^n, I^n, V^n are positive.

Theorem 6. *Let (T^n, I^n, V^n) be a solution of system (34). Then is uniformly bounded for all $n > 0$.*

Proof. From the first equation of scheme (33), one gets that

$$\frac{T^{n+1} - T^n}{\varphi(h)} = \Lambda - \beta T^{n+1} V^n - \mu_0 T^{n+1} \le \Lambda - \mu_0 T^{n+1}.$$

When $n \to \infty$ and since $\varphi(h) = h + O(h^2)$, then $\varphi(h)$ coincide with 0 in the limit as $h \to 0$. This implies that $\limsup_{n \to \infty} T^n \le \frac{\Lambda}{\mu_0}$.

Next, let $\mathcal{W}^n = T^{n-m_1} + e^{\delta_{I_E}\Delta} I^n$. From first and second equation of system (33), one obtains

$$\frac{\mathcal{W}^{n+1} - \mathcal{W}^n}{\varphi} = \frac{T^{n-m_1+1} - T^{n-m_1}}{\varphi(h)} + e^{\delta_{I_E}\Delta} \frac{I^{n+1} - I^n}{\varphi(h)} = \Lambda - \mu_0 T^{n-m_1+1} - \delta_I e^{\delta_{I_E}\Delta} I^{n+1}$$

$$\le \Lambda - d\mathcal{W}^{n+1},$$

where $d = \min\{\mu_0, \delta_I\}$. Thus, $\limsup_{n \to \infty} \mathcal{W}^n \le \frac{\Lambda}{d}$. Therefore, $\limsup_{n \to \infty} I^n \le \frac{\Lambda e^{-\delta_{I_E}\Delta}}{d}$. Then, give $\epsilon > 0$ we can choose a $M \in \mathbb{N}$ such that $I^n \le \frac{\Lambda e^{-\delta_{I_E}\Delta}}{d} + \epsilon$ for $n \ge M$. Using the last equation of (33) it is concluded that

$$\frac{V^{n+1} - V^n}{\varphi(h)} \le N \delta_I I^{n+1} - C V^{n+1} \le N \delta_I \left(\frac{\Lambda e^{-\delta_{I_E}\Delta}}{d} + \epsilon \right) - C V^{n+1}$$

for $n \geq M+1$. Then $\limsup\limits_{n\to\infty} V^n \leq \dfrac{N\delta_I}{C}\left(\dfrac{\Lambda e^{-\delta_I_E \Delta}}{d}+\epsilon\right)$, and as is for all $\epsilon > 0$ it follows that $\limsup\limits_{n\to\infty} V^n \leq \dfrac{N\delta_I}{C}\left(\dfrac{\Lambda e^{-\delta_I_E \Delta}}{d}\right)$. This completes the proof. Moreover, from Equation (27) it follows that I_E^n is bounded. □

4.1. Equilibrium Points of the NSFD Numerical Scheme

The equilibrium points of the scheme (34) are given by analyzing the behavior of system when n approaches to infinity. Thus, after a few calculations we find that

$$
\begin{aligned}
I^* &= \dfrac{\varphi(h)\beta T^* V^* e^{-\delta_{I_E}\Delta} + I^*}{1+\varphi(h)\delta_I} \\
V^* &= \dfrac{\varphi(h)N\delta_I I^* + V^*}{1+\varphi(h)(C+\beta T^*)} \\
T^* &= \dfrac{\varphi(h)\Lambda + T^*}{1+\varphi(h)(\beta V^* + \mu_0)}.
\end{aligned}
\qquad (35)
$$

Note that the equations of the scheme (35) correspond to Equations (14)–(16). Thus, the critical points of the discrete scheme will coincide in the limit $h \to 0$, with those of the continuous model.

4.2. Local Stability of the NSFD Numerical Scheme

For the study of the local stability of the critical points of the numerical scheme (34) it is necessary to use the following lemma:

Lemma 1. *The roots of the quadratic polynomial $\lambda^2 - a_1\lambda + a_2 = 0$, satisfy $|\lambda_i| < 1$ for $i = 1, 2$ if and only if the following conditions hold:*
i. $1 - a_1 + a_2 > 0$,
ii. $1 + a_1 + a_2 > 0$,
iii. $a_2 < 1$.

Proof. See [7]. □

Theorem 7. *The disease-free equilibrium point $P_f = \left(0, 0, \dfrac{\Lambda}{\mu_0}\right)$ of the scheme (34) is locally asymptotically stable if $\mathcal{R}_0 < 1$.*

Proof. Calculating the eigenvalues of the Jacobian matrix of the system (34) at the disease-free point, we obtain the following characteristic polynomial

$$\left(\dfrac{1}{1+\varphi(h)\mu_0} - \lambda\right) \begin{vmatrix} \dfrac{1}{1+\varphi(h)\delta_I} - \lambda & \dfrac{\varphi(h)\beta e^{-\delta_{I_E}\Delta}\Lambda}{\mu_0(1+\varphi(h)\delta_I)} \\ \dfrac{\mu_0 h N\delta_I}{\mu_0 + \mu_0 hC + h\beta\Lambda} & \dfrac{\mu_0}{\mu_0 + \mu_0 hC + h\beta\Lambda} - \lambda \end{vmatrix} = 0.$$

Thus, the first eigenvalue is

$$\lambda_1 = \dfrac{1}{1+\varphi(h)\mu_0} < 1.$$

The other two eigenvalues, are the roots of quadratic polynomial

$$\lambda^2 - \left(\dfrac{1}{p} + \dfrac{\mu_0}{q}\right)\lambda + \dfrac{\mu_0 - \varphi(h)^2 \delta_I N\beta\Lambda e^{-\delta_{I_E}\Delta}}{\mu_0 pq} = 0, \qquad (36)$$

where $p = 1 + \varphi(h)\delta_I > 0$ and $q = \mu_0 + \mu_0\varphi(h)C + \varphi(h)\beta\Lambda$. Next, let $a_1 = \dfrac{1}{p} + \dfrac{\mu_0}{q}$ and $a_2 = \dfrac{\mu_0 - \varphi(h)^2\delta_I N\beta\Lambda e^{-\delta_I_E\Delta}}{\mu_0 pq}$. We have the following affirmations.

1. If $1 > \mathcal{R}_0$, it follows that $\varphi(h)^2\delta_I(C\mu_0 + \beta\Lambda) > \varphi(h)^2\delta_I N\beta\Lambda e^{-\delta_{I_E}\Delta}$. Thus,

$$\delta_I(\mu_0 + \mu_0 hC + h\beta\Lambda) > \mu_0\delta_I + \varphi(h)\delta_I N\beta\Lambda e^{-\delta_{I_E}\Delta} \iff$$
$$(1 + \varphi(h)\delta_I)q > q + \varphi(h)\mu_0\delta_I + \varphi(h)^2\delta_I N\beta\Lambda e^{-\delta_{I_E}\Delta} \iff$$
$$pq + \mu_0 > q + \varphi(h)\mu_0\delta_I + \varphi(h)^2\delta_I N\beta\Lambda e^{-\delta_{I_E}\Delta} + \mu_0 \iff$$
$$pq + \mu_0 > q + \varphi(h)^2\delta_I N\beta\Lambda e^{-\delta_{I_E}\Delta} + p\mu_0 \iff$$
$$1 - \left(\dfrac{1}{p} + \dfrac{\mu_0}{q}\right) + \dfrac{\mu_0 - h^2\delta_I N\beta\Lambda e^{-\delta_{I_E}\Delta}}{pq} > 0$$

    Therefore, one gets that $1 + a_2 > a_1$.

2. Since $a_1 > 0$, it is sufficient to prove that $1 + a_2 > 0$. By hypothesis $1 > \mathcal{R}_0$, then

$$\mu_0 + q + \varphi(h)\delta_I\mu_0 + \varphi(h)^2\delta_I(\mu_0 C + \beta\Lambda) > \varphi(h)^2\delta_I(\mu_0 C + \beta\Lambda) > \varphi(h)^2\delta_I N\beta\Lambda e^{-\delta_{I_E}\Delta}.$$

    Accordingly

$$\mu_0 + q + \varphi(h)\delta_I(\mu_0 + \mu_0\varphi(h)C + \varphi(h)\beta\Lambda) > \varphi(h)^2\delta_I N\beta\Lambda e^{-\delta_{I_E}\Delta} \iff$$
$$\mu_0 + q + \varphi(h)\delta_I q > \varphi(h)^2\delta_I N\beta\Lambda e^{-\delta_{I_E}\Delta} \iff$$
$$\mu_0 + (1 + \varphi(h)\delta_I)q > \varphi(h)^2\delta_I N\beta\Lambda e^{-\delta_{I_E}\Delta} \iff$$
$$1 + \dfrac{\mu_0 - \varphi(h)^2\delta_I N\beta\Lambda e^{-\delta_{I_E}\Delta}}{pq} > 0$$

3. Given that

$$\mu_0 - \varphi(h)^2\delta_I N\beta\Lambda e^{-\delta_{I_E}\Delta} < \mu_0 + \mu_0\varphi(h)C + \varphi(h)\beta\Lambda + \varphi(h)\delta_I q,$$

    then

$$\dfrac{\mu_0 - \varphi(h)^2\beta e^{-\delta_{I_E}\Delta}\Lambda N\delta_I}{pq} < 1,$$

    that is $a_2 < 1$.

Next, by virtue of Lemma 1 we have that the eigenvalues of the Jacobian matrix of the system (35) evaluated at P_f satisfy $|\lambda_i| < 1$. for $i = 1, 2$. Then P_f is locally asymptotically stable if $\mathcal{R}_0 < 1$. □

4.3. Global Stability of the NSFD Numerical Scheme

Several authors have used discrete Lyapunov functions to study the global behavior of numerical solutions generated by non-standard finite difference schemes (*NSFD*), see [19,106–110]. For the study of the global stability of the critical points of the numerical scheme (34) it is necessary to use the Lyapunov functions and Equation (28).

Theorem 8. *The disease-free equilibrium point* $P_f = \left(0, 0, \dfrac{\Lambda}{\mu_0}\right)$ *of the scheme* (34) *is globally asymptotically stable if* $\mathcal{R}_0 \leq 1$.

Proof. Let the following Lyapunov function be

$$\mathcal{L}^n = \frac{T^0 g\left(\frac{T^n}{T^0}\right) + e^{\delta_{I_E}\Delta} I^n + \beta T^0 \left(\frac{1}{C} + \varphi(h)\right) V^n}{\varphi(h)} + \sum_{i=n-m_1}^{n-1} \beta T^{i+1} V^i. \tag{37}$$

Using the inequality $\ln z \leq z - 1$, the difference of $\mathcal{L}^n$ in (37) satisfies

$$\mathcal{L}^{n+1} - \mathcal{L}^n \leq \left(\frac{T^{n+1} - T^0}{T^{n+1}}\right)\left(\Lambda - \beta T^{n+1} V^n - \mu_0 T^{n+1}\right)$$
$$+ e^{\delta_{I_E}\Delta}\left(\beta T^{n-m_1+1} V^{n-m_1} e^{-\delta_{I_E}\Delta} - \delta_I I^{n+1}\right)$$
$$+ \frac{\beta T^0}{C}\left(N\delta_I I^{n+1} - CV^{n+1} - \beta T^n V^{n+1}\right) + \beta T^0\left(V^{n+1} - V^n\right)$$
$$+ \beta T^{n+1} V^n - \beta T^{n-m_1+1} V^{n-m_1}$$
$$\leq -\mu_0 \frac{T^{n+1} - T^0}{T^{n+1}} - e^{\delta_{I_E}\Delta} \delta_I I^{n+1}(1 - \mathcal{R}_0).$$

It follows that $\mathcal{L}^{n+1} - \mathcal{L}^n \leq 0$ for all $n \leq 0$ if $\mathcal{R}_0 \leq 1$. This means that $\mathcal{L}^n$ is monotone decreasing sequence and since $\mathcal{L}^n \geq 0$ for $n \geq 0$ then there exists a limit, i.e., $\lim_{n\to\infty} \mathcal{L}^n \geq 0$. Hence $\lim_{n\to\infty}(\mathcal{L}^{n+1} - \mathcal{L}^n) = 0$, which implies that $\lim_{n\to\infty} T^n = T^0$, $\lim_{n\to\infty} I^n = 0$, and from (34) $\lim_{n\to\infty} V^n = 0$. By the previous analysis, we conclude that P_f is globally asymptotically stable for scheme (34). □

5. Numerical Simulations Using the NSFD Scheme

In this section, we present some numerical solutions of the mathematical model of HIV. To carry out the simulations we use the constructed NSFD numerical scheme (34). We choose the parameter values based on existing experimental data and previous model studies, see [54,106,111,112]. The values of these parameters are given in Tables 1 and 2. For the first case we choose values such $\mathcal{R}_0 < 1$ and for the second we have the case $\mathcal{R}_0 > 1$.

Table 1. Parameter values for the numerical simulations when $\mathcal{R}_0 < 1$.

Parameters	Value
β	4.8×10^{-6} mm^{3}day^{-1}
C	3 mm^3 day^{-1}
Λ	2.3 day^{-1}
δ_I	0.24 day^{-1}
δ_{I_E}	0.05 day^{-1}
μ_0	0.0046 day^{-1}
Δ	0.4 day
N	500
V_0	1 mm^{-3}
T_0	100 mm^{-3}
I_{E_0}	0
I_0	0

Table 2. Parameter values for the numerical simulations when $\mathcal{R}_0 > 1$.

Parameters	Value
β	4.8×10^{-6} mm^{3}day^{-1}
C	2.4 mm^3 day^{-1}
Λ	23 day^{-1}
δ_I	0.2 day^{-1}
δ_{I_E}	0.05 day^{-1}
μ_0	0.0046 day^{-1}
Δ	0.4 day
N	500
V_0	1 mm^{-3}
T_0	100 mm^{-3}
I_{E_0}	1
I_0	1

For the numerical simulations we use a small step size, $h = 0.001$. For the discrete derivatives given in system (33), we have many options for the denominator function φ. We have chosen $\varphi(h) = (1 - exp(-hp))/p$, where $p = \max\{\Lambda, \Delta, \mu_0, \delta_I, \delta_{I_E}\}$ are parameters of the model and included in the numerical scheme (33). This particular φ usually provides better numerical results based on previous articles related to $NSFD$ schemes [113,114]. In addition, this option satisfies the asymptotic relation $\varphi(h) = h + O(h^2)$, and Rule 1. We performed numerical simulations to show that the solutions obtained by the proposed $NSFD$ scheme and the well-known MATLAB routine dde23 agree. A great advantage of the proposed $NSFD$ numerical scheme (34) is that the computation time is smaller. For the first case, the numerical solution using the $NSFD$ scheme is obtained in approximately 0.083992 s, while the dde23 routine spent 2.573003 s. For the second case, the numerical solution using the $NSFD$ scheme is obtained in approximately 0.176461 s, while the routine dde23 spent 11.181812 s. The results obtained are shown in Figures 2 and 3 respectively. It can be seen that the numerical solution of the proposed $NSFD$ numerical scheme (34) and the one obtained by means of the routine dde23 agree.

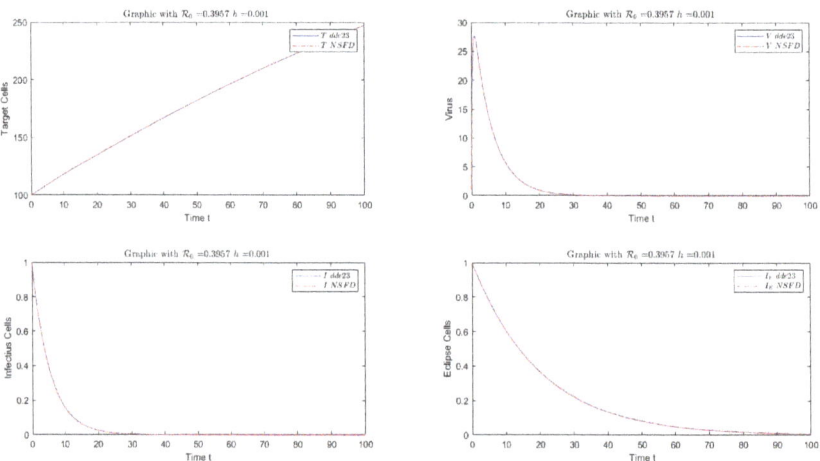

Figure 2. Simulation $NSFD$ versus dde23 when $\mathcal{R}_0 < 1$.

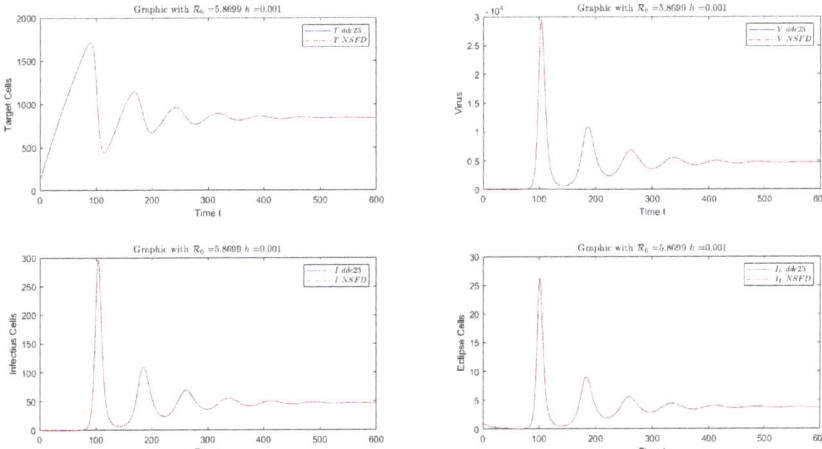

Figure 3. Simulation $NSFD$ versus dde23 when $\mathcal{R}_0 > 1$.

6. Conclusions

We proposed a mathematical model based on a set of delay differential equations that describe intracellular HIV infection. The model considers three different subpopulations of cells and the HIV virus. The mathematical model is formulated in such a way that takes into account the time between viral entry into a target cell and the production of new virions. Moreover, this time is included using a discrete time delay. We analyzed the local stability of the infection-free and endemic equilibrium states. By using a suitable Lyapunov functional and the LaSalle invariant principle, it is proved that if the basic reproduction ratio is less than unity, the infection-free equilibrium is globally asymptotically stable. In addition, we designed a non-standard difference scheme that preserves some properties of the continuous model. We prove that the constructed $NSFD$ scheme has the same equilibrium points of the continuous model, and the disease-free equilibrium holds the same stability properties. As required by the constraints of the real phenomena, the solutions given by the numerical scheme satisfy positivity and boundedness. The numerical simulations corroborate that the developed $NSFD$ numerical scheme preserves the properties of the continuous model and presents a robust behavior when working with a variety of parameter values.

Author Contributions: Conceptualization, G.G.-P.; Formal analysis, G.G.-P., A.J.A. and J.J.N.; Investigation, G.G.-P. and A.J.A.; Methodology, G.G.-P., A.J.A., M.C. and N.D.L.E.; Software, G.G.-P.; Supervision, G.G.-P. and A.J.A.; Validation, A.J.A.; Visualization, A.J.A., J.J.N., M.C. and N.D.L.E.; Writing—original draft, G.G.-P.; Writing—review and editing, G.G.-P., M.C. and N.D.L.E. All authors have read and agreed to the published version of the manuscript.

Funding: This research received no external funding.

Acknowledgments: The authors are grateful to the reviewers for their careful reading of this manuscript and their useful comments to improve the content of this paper.

Conflicts of Interest: The authors declare no conflict of interest.

References

1. Nelson, K.E.; Williams, C.M. *Early History of Infectious Disease: Epidemiology and Control of Infectious Diseases*; Jones & Bartlett Publishers: Burlington, MA, USA, 2007; Volume 2, pp. 3–23.
2. Dabis, F.; Bekker, L.G. We still need to beat HIV. *Science* **2017**, *357*, 335. [CrossRef]

3. Ayele, T.A.; Worku, A.; Kebede, Y.; Alemu, K.; Kasim, A.; Shkedy, Z. Choice of initial antiretroviral drugs and treatment outcomes among HIV-infected patients in sub-Saharan Africa: Systematic review and meta-analysis of observational studies. *System. Rev.* **2017**, *6*, 173. [CrossRef]
4. Duvergé, A.; Negroni, M. Pseudotyping Lentiviral Vectors: When the Clothes Make the Virus. *Viruses* **2020**, *12*, 1311. [CrossRef]
5. Dubrow, R.; Silverberg, M.J.; Park, L.S.; Crothers, K.; Justice, A.C. HIV infection, aging, and immune function: Implications for cancer risk and prevention. *Curr. Opin. Oncol.* **2012**, *24*, 506. [CrossRef]
6. Verma, M.; Erwin, S.; Abedi, V.; Hontecillas, R.; Hoops, S.; Leber, A.; Bassaganya-Riera, J.; Ciupe, S.M. Modeling the mechanisms by which HIV-associated immunosuppression influences HPV persistence at the oral mucosa. *PLoS ONE* **2017**, *12*, e0168133. [CrossRef]
7. Brauer, F.; Castillo-Chavez, C. *Mathematical Models in Population Biology and Epidemiology*; Springer: Berlin, Germany, 2001; Volume 40.
8. González-Parra, G.; Dobrovolny, H.M. Assessing uncertainty in A2 respiratory syncytial virus viral dynamics. *Comput. Math. Methods Med.* **2015**, *2015*. [CrossRef]
9. Beauchemin, C.A.; Handel, A. A review of mathematical models of influenza A infections within a host or cell culture: Lessons learned and challenges ahead. *BMC Public Health* **2011**, *11*, S7. [CrossRef] [PubMed]
10. Alade, T.O. On the generalized Chikungunya virus dynamics model with distributed time delays. *Int. J. Dyn. Control* **2020**, *2020*, 1–11. [CrossRef]
11. Dobrovolny, H.M.; Reddy, M.B.; Kamal, M.A.; Rayner, C.R.; Beauchemin, C.A. Assessing mathematical models of influenza infections using features of the immune response. *PLoS ONE* **2013**, *8*, e57088. [CrossRef]
12. Doekes, H.M.; Fraser, C.; Lythgoe, K.A. Effect of the Latent Reservoir on the Evolution of HIV at the Within-and Between-Host Levels. *PLoS Comput. Biol.* **2017**, *13*, e1005228. [CrossRef] [PubMed]
13. González-Parra, G.; Dobrovolny, H.M.; Aranda, D.F.; Chen-Charpentier, B.; Rojas, R.A.G. Quantifying rotavirus kinetics in the REH tumor cell line using in vitro data. *Virus Res.* **2018**, *244*, 53–63. [CrossRef] [PubMed]
14. González-Parra, G.; Dobrovolny, H.M. The rate of viral transfer between upper and lower respiratory tracts determines RSV illness duration. *J. Math. Biol.* **2019**, *79*, 467–483. [CrossRef] [PubMed]
15. Maheswari, M.; Krishnapriya, P.; Krishnan, K.; Pitchaimani, M. A mathematical model of HIV-1 infection within host cell to cell viral transmissions with RTI and discrete delays. *J. Appl. Math. Comput.* **2018**, *56*, 151–178. [CrossRef]
16. Pinky, L.; Gonzalez-Parra, G.; Dobrovolny, H.M. Effect of stochasticity on coinfection dynamics of respiratory viruses. *BMC Bioinf.* **2019**, *20*, 1–12. [CrossRef]
17. Song, H.; Jiang, W.; Liu, S. Virus dynamics model with intracellular delays and immune response. *Math. Biosci. Eng.* **2015**, *12*, 185. [CrossRef]
18. Wang, Y.; Liu, X. Stability and Hopf bifurcation of a within-host chikungunya virus infection model with two delays. *Math. Comput. Simul.* **2017**, *138*, 31–48. [CrossRef]
19. Zhou, J.; Yang, Y. Global dynamics of a discrete viral infection model with time delay, virus-to-cell and cell-to-cell transmissions. *J. Diff. Equ. Appl.* **2017**, *23*, 1853–1868. [CrossRef]
20. Nelson, P.W.; Perelson, A.S. Mathematical analysis of delay differential equation models of HIV-1 infection. *Math. Biosci.* **2002**, *179*, 73–94. [CrossRef]
21. Pawelek, K.A.; Liu, S.; Pahlevani, F.; Rong, L. A model of HIV-1 infection with two time delays: Mathematical analysis and comparison with patient data. *Math. Biosci.* **2012**, *235*, 98–109. [CrossRef]
22. Cooper, A.; García, M.; Petrovas, C.; Yamamoto, T.; Koup, R.A.; Nabel, G.J. HIV-1 causes CD4 cell death through DNA-dependent protein kinase during viral integration. *Nature* **2013**, *498*, 376. [CrossRef]
23. Doitsh, G.; Galloway, N.L.; Geng, X.; Yang, Z.; Monroe, K.M.; Zepeda, O.; Hunt, P.W.; Hatano, H.; Sowinski, S.; Muñoz-Arias, I.; et al. Cell death by pyroptosis drives CD4 T-cell depletion in HIV-1 infection. *Nature* **2014**, *505*, 509. [CrossRef] [PubMed]
24. Mickens, R.E. *Nonstandard Finite Difference Models of Differential Equations*; World Scientific: Singapore, 1994.
25. Mickens, R.E. *Applications of Nonstandard Finite Difference Schemes*; World Scientific: Singapore, 2000.
26. Mickens, R.E. Nonstandard finite difference schemes for differential equations. *J. Diff. Equ. Appl.* **2002**, *8*, 823–847. [CrossRef]
27. Mickens, R.E. Dynamic consistency: A fundamental principle for constructing nonstandard finite difference schemes for differential equations. *J. Diff. Equ. Appl.* **2005**, *11*, 645–653. [CrossRef]
28. Martin-Vaquero, J.; Queiruga-Dios, A.; del Rey, A.M.; Encinas, A.H.; Guillen, J.H.; Sanchez, G.R. Variable step length algorithms with high-order extrapolated non-standard finite difference schemes for a SEIR model. *J. Comput. Appl. Math.* **2018**, *330*, 848–854. [CrossRef]
29. Farooqi, A.; Ahmad, R.; Farooqi, R.; Alharbi, S.O.; Baleanu, D.; Rafiq, M.; Khan, I.; Ahmad, M. An Accurate Predictor-Corrector-Type Nonstandard Finite Difference Scheme for an SEIR Epidemic Model. *J. Math.* **2020**, *2020*. [CrossRef]
30. Khalsaraei, M.M.; Shokri, A.; Ramos, H.; Heydari, S. A positive and elementary stable nonstandard explicit scheme for a mathematical model of the influenza disease. *Math. Comput. Simul.* **2021**, *182*, 397–410. [CrossRef]
31. Dang, Q.A.; Hoang, M.T. Positive and elementary stable nonstandard Runge-Kutta methods for a class of autonomous dynamical systems. *Int. J. Comput. Math.* **2020**, *97*, 2036–2054. [CrossRef]
32. Sweilam, N.; Al-Mekhlafi, S.; Shatta, S.; Baleanu, D. Numerical Study for Two Types Variable-Order Burgers' Equations with Proportional Delay. *Appl. Numer. Math.* **2020**, *156*, 364–376. [CrossRef]

33. Hoang, M.T.; Egbelowo, O.F. Dynamics of a fractional-order hepatitis b epidemic model and its solutions by nonstandard numerical schemes. In *Mathematical Modelling and Analysis of Infectious Diseases*; Springer: Berlin, Germany, 2020; pp. 127–153.
34. Egbelowo, O.F.; Hoang, M.T. Global dynamics of target-mediated drug disposition models and their solutions by nonstandard finite difference method. *J. Appl. Math. Comput.* **2020**, 1–23. [CrossRef]
35. Alexander, M.E.; Summers, A.R.; Moghadas, S.M. Neimark–Sacker bifurcations in a non-standard numerical scheme for a class of positivity-preserving ODEs. *Proc. R. Soc. A Math. Phys. Eng. Sci.* **2006**, *462*, 3167–3184. [CrossRef]
36. Dumont, Y.; Lubuma, J.M.S. Non-standard finite-difference methods for vibro-impact problems. *Proc. R. Soc. A Math. Phys. Eng. Sci.* **2005**, *461*, 1927–1950. [CrossRef]
37. Bruggeman, J.; Burchard, H.; Kooi, B.W.; Sommeijer, B. A second-order, unconditionally positive, mass-conserving integration scheme for biochemical systems. *Appl. Numer. Math.* **2007**, *57*, 36–58. [CrossRef]
38. Arenas, A.J.; Moraño, J.A.; Cortés, J.C. Non-standard numerical method for a mathematical model of RSV epidemiological transmission. *Comput. Math. Appl.* **2008**, *56*, 670–678. [CrossRef]
39. Dimitrov, D.T.; Kojouharov, H.V. Stability-preserving finite-difference methods for general multi-dimensional autonomous dynamical systems. *Int. J. Numer. Anal. Model* **2007**, *4*, 282–292.
40. Dimitrov, D.T.; Kojouharov, H.V. Nonstandard finite-difference methods for predator–prey models with general functional response. *Math. Comput. Simul.* **2008**, *78*, 1–11. [CrossRef]
41. Dimitrov, D.T.; Kojouharov, H.V. Nonstandard finite-difference schemes for general two-dimensional autonomous dynamical systems. *Appl. Math. Lett.* **2005**, *18*, 769–774. [CrossRef]
42. Gumel, A.B. A competitive numerical method for a chemotherapy model of two HIV subtypes. *Appl. Math. Comput.* **2002**, *131*, 329–337. [CrossRef]
43. Jansen, H.; Twizell, E.H. An unconditionally convergent discretization of the SEIR model. *Math. Comput. Simul.* **2002**, *58*, 147–158. [CrossRef]
44. Obaid, H.A.; Ouifki, R.; Patidar, K.C. An unconditionally stable nonstandard finite difference method applied to a mathematical model of HIV infection. *Int. J. Appl. Math. Comput. Sci.* **2013**, *23*, 357–372. [CrossRef]
45. González-Parra, G.; Arenas, A.J.; Chen-Charpentier, B.M. Combination of nonstandard schemes and Richardson's extrapolation to improve the numerical solution of population models. *Math. Comput. Model.* **2010**, *52*, 1030–1036. [CrossRef]
46. Ahmad, A.; Farman, M.; Akgül, A.; Bukhari, N.; Imtiaz, S. Mathematical analysis and numerical simulation of co-infection of TB-HIV. *Arab J. Basic Appl. Sci.* **2020**, *27*, 431–441. [CrossRef]
47. Asai, Y.; Herrmann, E.; Kloeden, P.E. Stable integration of stiff random ordinary differential equations. *Stochast. Anal. Appl.* **2013**, *31*, 293–313. [CrossRef]
48. Baker, C.T.; Bocharov, G.; Ford, J.M.; Lumb, P.M.; Norton, S.J.; Paul, C.; Junt, T.; Krebs, P.; Ludewig, B. Computational approaches to parameter estimation and model selection in immunology. *J. Comput. Appl. Math.* **2005**, *184*, 50–76. [CrossRef]
49. Reinharz, V.; Churkin, A.; Dahari, H.; Barash, D. A robust and efficient numerical method for RNA-mediated viral dynamics. *Front. Appl. Math. Statist.* **2017**, *3*, 20. [CrossRef] [PubMed]
50. Bocharov, G.; Marchuk, G.; Romanyukha, A. Numerical solution by LMMs of stiff delay differential systems modelling an immune response. *Numer. Math.* **1996**, *73*, 131–148. [CrossRef]
51. Rihan, F.A.; Rahman, D.A.; Lakshmanan, S.; Alkhajeh, A. A time delay model of tumour–immune system interactions: Global dynamics, parameter estimation, sensitivity analysis. *Appl. Math. Comput.* **2014**, *232*, 606–623. [CrossRef]
52. Beauchemin, C.A.; McSharry, J.J.; Drusano, G.L.; Nguyen, J.T.; Went, G.T.; Ribeiro, R.M.; Perelson, A.S. Modeling amantadine treatment of influenza A virus in vitro. *J. Theor. Biol.* **2008**, *254*, 439–451. [CrossRef]
53. Hill, A.L.; Rosenbloom, D.I.; Nowak, M.A.; Siliciano, R.F. Insight into treatment of HIV infection from viral dynamics models. *Immunol. Rev.* **2018**, *285*, 9–25. [CrossRef]
54. Noecker, C.; Schaefer, K.; Zaccheo, K.; Yang, Y.; Day, J.; Ganusov, V. Simple mathematical models do not accurately predict early SIV dynamics. *Viruses* **2015**, *7*, 1189–1217. [CrossRef]
55. Perelson, A.S. Modelling viral and immune system dynamics. *Nat. Rev. Immunol.* **2002**, *2*, 28. [CrossRef]
56. Banerjee, S.; Guedj, J.; Ribeiro, R.M.; Moses, M.; Perelson, A.S. Estimating biologically relevant parameters under uncertainty for experimental within-host murine West Nile virus infection. *J. R. Soc. Interface* **2016**, *13*, 20160130. [CrossRef] [PubMed]
57. Miao, H.; Xia, X.; Perelson, A.S.; Wu, H. On identifiability of nonlinear ODE models and applications in viral dynamics. *SIAM Rev.* **2011**, *53*, 3–39. [CrossRef] [PubMed]
58. Kirschner, D.; Pienaar, E.; Marino, S.; Linderman, J.J. A review of computational and mathematical modeling contributions to our understanding of Mycobacterium tuberculosis within-host infection and treatment. *Curr. Opin. Syst. Biol.* **2017**, *3*, 170–185. [CrossRef]
59. Hethcote, H.W. The mathematics of infectious diseases. *SIAM Rev.* **2000**, *42*, 599–653. [CrossRef]
60. Hethcote, H.W.; Van den Driessche, P. An SIS epidemic model with variable population size and a delay. *J. Math. Biol.* **1995**, *34*, 177–194. [CrossRef]
61. Baltes, A.; Akpinar, F.; Inankur, B.; Yin, J. Inhibition of infection spread by co-transmitted defective interfering particles. *PLoS ONE* **2017**, *12*, e0184029. [CrossRef]
62. Liao, L.E.; Iwami, S.; Beauchemin, C.A. (In)validating experimentally derived knowledge about influenza A defective interfering particles. *J. R. Soc. Interface* **2016**, *13*, 20160412. [CrossRef]

63. Ho, D.D.; Neumann, A.U.; Perelson, A.S.; Chen, W.; Leonard, J.M.; Markowitz, M. Rapid turnover of plasma virions and CD4 lymphocytes in HIV-1 infection. *Nature* **1995**, *373*, 123. [CrossRef]
64. Goto, T.; Harada, S.; Yamamoto, N.; Nakai, M. Entry of human immunodeficiency virus (HIV) into MT-2, human T cell leukemia virus carrier cell line. *Arch. Virol.* **1988**, *102*, 29–38. [CrossRef]
65. Platt, E.J.; Kozak, S.L.; Durnin, J.P.; Hope, T.J.; Kabat, D. Rapid dissociation of HIV-1 from cultured cells severely limits infectivity assays, causes the inactivation ascribed to entry inhibitors, and masks the inherently high level of infectivity of virions. *J. Virol.* **2010**, *84*, 3106–3110. [CrossRef]
66. Bai, F.; Huff, K.E.; Allen, L.J. The effect of delay in viral production in within-host models during early infection. *J. Biol. Dyn.* **2019**, *13*, 47–73. [CrossRef] [PubMed]
67. Cao, P.; McCaw, J.M. The mechanisms for within-host influenza virus control affect model-based assessment and prediction of antiviral treatment. *Viruses* **2017**, *9*, 197. [CrossRef] [PubMed]
68. Holder, B.P.; Beauchemin, C.A. Exploring the effect of biological delays in kinetic models of influenza within a host or cell culture. *BMC Public Health* **2011**, *11*, S10. [CrossRef] [PubMed]
69. Dixit, N.M.; Markowitz, M.; Ho, D.D.; Perelson, A.S. Estimates of intracellular delay and average drug efficacy from viral load data of HIV-infected individuals under antiretroviral therapy. *Antivir. Ther.* **2004**, *9*, 237–246.
70. Kakizoe, Y.; Nakaoka, S.; Beauchemin, C.A.; Morita, S.; Mori, H.; Igarashi, T.; Aihara, K.; Miura, T.; Iwami, S. A method to determine the duration of the eclipse phase for in vitro infection with a highly pathogenic SHIV strain. *Scient. Rep.* **2015**, *5*, 10371. [CrossRef] [PubMed]
71. Keyfitz, B.L.; Keyfitz, N. The McKendrick partial differential equation and its uses in epidemiology and population study. *Math. Comput. Model.* **1997**, *26*, 1–9. [CrossRef]
72. Kuang, Y. *Delay Differential Equations: With Applications in Population Dynamics*, 1st ed.; Mathematics in Science and Engineering 191; Elsevier: Amsterdam, The Netherlands; Academic Press: Cambridge, MA, USA, 1993.
73. Driver, R.D. *Ordinary and Delay Differential Equations*, 1st ed.; Applied Mathematical Sciences 20; Springer: New York, NY, USA, 1977.
74. Anderson, R. Transmission dynamics and control of infectious disease agents. In *Population Biology of Infectious Diseases*; Springer: Berlin, Germany, 1982; pp. 149–176.
75. Diekmann, O.; Heesterbeek, J.; Roberts, M. The construction of next-generation matrices for compartmental epidemic models. *J. R. Soc. Interface* **2009**, *2009*, rsif20090386. [CrossRef]
76. González-Parra, G.; Arenas, A.; Diego, F.; Aranda, L.S. Modeling the epidemic waves of AH1N1/09 influenza around the world. *Spat. Spatio-Temp. Epidemiol.* **2011**, *2*, 219–226. [CrossRef]
77. Gonzalez-Parra, G.; Díaz-Rodríguez, M.; Arenas, A.J. Mathematical modeling to design public health policies for Chikungunya epidemic using optimal control. *Opt. Control Appl. Methods* **2020**, *41*, 1584–1603. [CrossRef]
78. Jang, S.R.J. On a discrete West Nile epidemic model. *Comput. Appl. Math.* **2007**, *26*, 397–414. [CrossRef]
79. Sekiguchi, M. Permanence of some discrete epidemic models. *Int. J. Biomath.* **2009**, *2*, 443–461. [CrossRef]
80. Chinviriyasit, S.; Chinviriyasit, W. Numerical modelling of an SIR epidemic model with diffusion. *Appl. Math. Comput.* **2010**, *216*, 395–409. [CrossRef]
81. Enatsu, Y.; Nakata, Y.; Muroya, Y. Global stability for a class of discrete SIR epidemic models. *Math. Biosci. Eng.* **2010**, *7*, 347–361.
82. Muroya, Y.; Nakata, Y.; Izzo, G.; Vecchio, A. Permanence and global stability of a class of discrete epidemic models. *Nonlinear Anal. Real World Appl.* **2011**, *12*, 2105–2117. [CrossRef]
83. Jódar, L.; Villanueva, R.J.; Arenas, A.J.; González, G.C. Nonstandard numerical methods for a mathematical model for influenza disease. *Math. Comput. Simul.* **2008**, *79*, 622–633. [CrossRef]
84. Lambert, J. *Computational Methods in Ordinary Differential Equations*; John Wiley & Sons: Hoboken, NJ, USA, 1973.
85. Shishkin, G.I. The Richardson scheme for the singularly perturbed parabolic reaction-diffusion equation in the case of a discontinuous initial condition. *Computat. Math. Math. Phys.* **2009**, *49*, 1348–1368. [CrossRef]
86. Al'shin, A.; Al'shina, E. Numerical diagnosis of blow-up of solutions of pseudoparabolic equations. *J. Math. Sci.* **2008**, *148*, 143–162. [CrossRef]
87. Munyakazi, J.B.; Patidar, K.C. On Richardson extrapolation for fitted operator finite difference methods. *Appl. Math. Comput.* **2008**, *201*, 465–480. [CrossRef]
88. Burg, C.; Erwin, T. Application of Richardson extrapolation to the numerical solution of partial differential equations. *Numer. Methods Part. Diff. Equ.* **2009**, *25*, 810–832. [CrossRef]
89. Gurski, K. A simple construction of nonstandard finite-difference schemes for small nonlinear systems applied to SIR models. *Comput. Math. Appl.* **2013**, *66*, 2165–2177. [CrossRef]
90. Munyakazi, J.B.; Patidar, K.C.; Sayi, M.T. A robust fitted operator finite difference method for singularly perturbed problems whose solution has an interior layer. *Math. Comput. Simul.* **2019**, *160*, 155–167. [CrossRef]
91. Clermont, G.; Zenker, S. The inverse problem in mathematical biology. *Math. Biosc.* **2015**, *260*, 11–15. [CrossRef] [PubMed]
92. Pollicott, M.; Wang, H.; Weiss, H. Extracting the time-dependent transmission rate from infection data via solution of an inverse ODE problem. *J. Biol. Dyn.* **2012**, *6*, 509–523. [CrossRef] [PubMed]
93. González-Parra, G.; Benincasa, T. Mathematical modeling and numerical simulations of Zika in Colombia considering mutation. *Math. Comput. Simul.* **2019**, *163*, 1–18.

94. Arthur, J.G.; Tran, H.T.; Aston, P. Feasibility of parameter estimation in hepatitis C viral dynamics models. *J. Inver. Ill-Posed Probl.* **2017**, *25*, 69–80. [CrossRef]
95. Hale, J.K.; Lunel, S.M.V. *Introduction to Functional Differential Equations*; Springer Science & Business Media: Berlin, Germany, 2013; Volume 99.
96. Smith, H.L. *An Introduction to Delay Differential Equations with Applications to the Life Sciences*; Springer: New York, NY, USA, 2011; Volume 57.
97. Mukandavire, Z.; Chiyaka, C.; Garira, W.; Musuka, G. Mathematical analysis of a sex-structured HIV/AIDS model with a discrete time delay. *Nonlinear Anal. Theory Methods Appl.* **2009**, *71*, 1082–1093. [CrossRef]
98. Asai, Y.; Kloeden, P.E. Numerical schemes for ordinary delay differential equations with random noise. *Appl. Math. Comput.* **2019**, *347*, 306–318. [CrossRef]
99. Engelborghs, K.; Luzyanina, T.; Roose, D. Numerical bifurcation analysis of delay differential equations using DDE-BIFTOOL. *ACM Trans. Math. Softw. (TOMS)* **2002**, *28*, 1–21. [CrossRef]
100. Jamilla, C.; Mendoza, R.; Mező, I. Solutions of neutral delay differential equations using a generalized Lambert W function. *Appl. Math. Comput.* **2020**, *382*, 125334. [CrossRef]
101. Singh, H. Numerical simulation for fractional delay differential equations. *Int. J. Dyn. Control* **2020**, 1–12. [CrossRef]
102. Baker, C.T.; Paul, C.A.; Willé, D.R. Issues in the numerical solution of evolutionary delay differential equations. *Adv. Comput. Math.* **1995**, *3*, 171–196. [CrossRef]
103. García, M.; Castro, M.; Martín, J.; Rodríguez, F. Exact and nonstandard numerical schemes for linear delay differential models. *Appl. Math. Comput.* **2018**, *338*, 337–345. [CrossRef]
104. Manna, K. A non-standard finite difference scheme for a diffusive HBV infection model with capsids and time delay. *J. Diff. Equ. Appl.* **2017**, *23*, 1901–1911. [CrossRef]
105. Patidar, K.C.; Sharma, K.K. ε-uniformly convergent non-standard finite difference methods for singularly perturbed differential difference equations with small delay. *Appl. Math. Comput.* **2006**, *175*, 864–890. [CrossRef]
106. Xu, J.; Geng, Y. Dynamic Consistent NSFD Scheme for a Delayed Viral Infection Model with Immune Response and Nonlinear Incidence. *Discrete Dyn. Nat. Soc.* **2017**, *2017*, 1–13. [CrossRef]
107. Xu, J.; Geng, Y. Stability preserving NSFD scheme for a delayed viral infection model with cell-to-cell transmission and general nonlinear incidence. *J. Diff. Equ. Appl.* **2017**, *23*, 893–916. [CrossRef]
108. Ding, D.; Ma, Q.; Ding, X. An unconditionally positive and global stability preserving NSFD scheme for an epidemic model with vaccination. *Int. J. Appl. Math. Comput. Sci.* **2014**, *24*, 635–646. [CrossRef]
109. Liu, J.; Peng, B.; Zhang, T. Effect of discretization on dynamical behavior of SEIR and SIR models with nonlinear incidence. *Appl. Math. Lett.* **2015**, *39*, 60–66. [CrossRef]
110. Hattaf, K.; Lashari, A.A.; Boukari, B.E.; Yousfi, N. Effect of Discretization on Dynamical Behavior in an Epidemiological Model. *Diff. Equ. Dyn. Syst.* **2015**, *23*, 403–413. [CrossRef]
111. Culshaw, R.; Ruan, S. A delay-differential equation model of HIV infection of CD4 T-cells. *Math. Biosci.* **2000**, *165*, 27–39. [CrossRef]
112. Toro-Zapata, H.; Caicedo-Casso, A.; Lee, S. The Role of Immune Response in Optimal HIV Treatment Interventions. *Processes* **2018**, *6*, 102. [CrossRef]
113. Mickens, R.E. Calculation of denominator functions for nonstandard finite difference schemes for differential equations satisfying a positivity condition. *Numer. Methods Part. Diff. Equ. An Int. J.* **2007**, *23*, 672–691. [CrossRef]
114. Mickens, R.E.; Smith, A. Finite-difference models of ordinary differential equations: Influence of denominator functions. *J. Franklin Inst.* **1990**, *327*, 143–149. [CrossRef]

Article

Dependence of Dynamics of a System of Two Coupled Generators with Delayed Feedback on the Sign of Coupling

Alexandra Kashchenko

Mathematical Department, P.G. Demidov Yaroslavl State University, Yaroslavl 150003, Russia; sa-ahr@yandex.ru

Received: 31 August 2020; Accepted: 12 October 2020; Published: 15 October 2020

Abstract: In this paper, we study the nonlocal dynamics of a system of delay differential equations with large parameters. This system simulates coupled generators with delayed feedback. Using the method of steps, we construct asymptotics of solutions. By these asymptotics, we construct a special finite-dimensional map. This map helps us to determine the structure of solutions. We study the dependence of solutions on the coupling parameter and show that the dynamics of the system is significantly different in the case of positive coupling and in the case of negative coupling.

Keywords: relaxation mode; delay differential equation; large parameter; asymptotics

MSC: 34K13; 34K25

1. Introduction

Consider equation

$$\dot{u}(t) = -\nu u(t) + \lambda F(u(t-T)), \tag{1}$$

where u is a scalar function, parameters ν, T, and λ are positive, $F(u)$ is some nonlinear compactly supported function. This equation is a mathematical model in problems of radiophysics and biology. It simulates a generator with nonlinear delayed feedback with a first-order RC low-pass filter (see, for example, [1–3]). Such generators are used in the manufacture of sonars, noise radars, and D-amplifiers [2]. Equation (1) models a biological process where the single state variable u decays with a rate ν proportional to u in the present and is produced with a rate dependent on the value of u some time in the past [4]. Such processes arise in a variety of problems in various areas in biology (see Table 1 and references in [4]). In addition, the dynamics of Equation (1) is of general scientific interest [5–13]. The authors find complicated periodic solutions [5–7] and chaos [8] in this model in the case of "step-like" nonlinearity. In Ref. [9], authors study properties of solutions and find a global attractor of model (1) with delayed positive feedback and in the paper [10] existence and stability of relaxation cycle of the multidimensional system (1) in the case of large λ is studied. In Refs. [11–13], the authors study properties of solutions of normalized Equation (1) (parameters $\nu = \lambda = 1$) in the case of sufficiently large T ($T \gg 1$). They deal with equation

$$\varepsilon \dot{u}(t) = -u(t) + f(u(t-1)), \tag{2}$$

where $\varepsilon = 1/T$ and study how the dynamics of this equation when ε is small (when T is large in (1)) is related with dynamics of this equation in the case $\varepsilon = 0$.

In this paper, we deal with a system of two coupled normalized ($\nu = 1$) equations of the form (1)

$$\begin{cases} \dot{u}_1 + u_1 = \lambda F(u_1(t-T)) + \gamma(u_2 - u_1), \\ \dot{u}_2 + u_2 = \lambda F(u_2(t-T)) + \gamma(u_1 - u_2). \end{cases} \tag{3}$$

Here, delay time T is a positive constant, a nonlinear sufficiently smooth function $F(u)$ is compactly supported:

$$F(u) = \begin{cases} f(u), & |u| \leq p, \\ 0, & |u| > p, \end{cases}$$

where p is some positive constant.

We assume that function $f(u)$ on the segment $u \in [-p, p]$ satisfies the conditions:

$$\begin{aligned} & f(p) = f(-p) = 0; \\ & f(u) \neq 0 \text{ except for a finite number of points;} \\ & \text{if } f(u^*) = 0, \text{ then } f'(u^*) \neq 0 \text{ or } f''(u^*) \neq 0. \end{aligned} \quad (4)$$

and that coefficient λ is large enough: $\lambda \gg 1$.

This model simulates two coupled D-amplifiers or two noise-radars with a large amount of feedback. If coupling parameter γ is asymptotically small at $\lambda \to +\infty$, then exponentially orbitally stable relaxation cycles coexist in model (3) (see [14,15]). Now, we are interested in nonlocal dynamics of this model in the case γ is some nonzero constant and we study how the dynamic properties of the system differ in the cases of positive and negative coupling.

The paper is organized as follows. In Section 2, we introduce some set of initial conditions and integrating by steps system (3) under some non-degeneracy conditions we construct solutions with initial conditions from the chosen set. By formulas of solution, we obtain the operator of translation along the trajectories Π and map describing dynamics of this operator. Using this map, we clarify asymptotics of solutions of system (3) in the case $\gamma > 0$ in Section 3 and in the case $\gamma < 0$ in Section 4. In Section 5, as an example, we consider a narrower class of functions f and prove that asymptotic formulas of solution given in Sections 2–4 are valid for a wide set of initial conditions (for all initial conditions from this set, non-degeneracy conditions hold) and prove the existence of relaxation cycles in system (3). We show that the dynamics of system (3) is significantly different in the case of positive and negative coupling in Section 6 and, in Section 7, we draw conclusions.

2. Constructing the Asymptotics of Solutions

Let's find relaxation solutions of (3) and study the dynamics of this system. For this purpose, we consider initial conditions $(u_1(s), u_2(s))^T \in C_{[-T,0]}(\mathbb{R}^2)$ outside of the strip $|u_j(s)| < p$ ($s \in [-T, 0]$, $j = 1, 2$) and construct asymptotics of all solutions of system (3) for this set of initial conditions.

Due to the choice of initial conditions on the segment $t \in [0, T]$, system (3) has the form

$$\begin{cases} \dot{u}_1 + u_1 = \gamma(u_2 - u_1), \\ \dot{u}_2 + u_2 = \gamma(u_1 - u_2). \end{cases} \quad (5)$$

Moreover, system (3) has form (5) until at least one of the components of the solution comes into the strip $|u_j| < p$. Thus, for $t \geq 0$, until at least one of the components of the solution of system (3) for the first time comes into the strip $|u_j| < p$, a solution of system (3) has form

$$\begin{aligned} u_1(t) &= \tfrac{1}{2}(u_1(0) + u_2(0))e^{-t} + \tfrac{1}{2}(u_1(0) - u_2(0))e^{-(1+2\gamma)t}, \\ u_2(t) &= \tfrac{1}{2}(u_1(0) + u_2(0))e^{-t} - \tfrac{1}{2}(u_1(0) - u_2(0))e^{-(1+2\gamma)t}. \end{aligned} \quad (6)$$

It follows from (6) that, in the case $\gamma < -\tfrac{1}{2}$, there exist solutions of system (3) tending to infinity, and, in the case $\gamma = -\tfrac{1}{2}$, there exist solutions of system (3) tending to a constant at $t \to +\infty$. We are interested in relaxation solutions, which is why we assume further that $\gamma > -\tfrac{1}{2}$.

If $\gamma > -\frac{1}{2}$, then at least one component of a solution eventually comes into the strip $|u_j| < p$ ($j = 1$ or 2). Let $t_1 \geq 0$ be the first time moment such that some component of the solution (we denote it as u_i) gets inside the strip $|u_i(t)| \leq p$:

$$|u_1(s+t_1)| \geq p, \quad |u_2(s+t_1)| \geq p \text{ for } s \in [-T, 0), \tag{7}$$

$|u_i(t_1)| = p$ and $|u_i(t)| < p$ if $t_1 < t < t_1 + \delta$ (where $\delta > 0$ is some constant and i equals 1 or 2). Then,

$$u_i(t_1) = kp, \quad u_{3-i}(t_1) = xp, \tag{8}$$

where k denotes the sign of $u_i(t_1)$ (parameter k takes values -1 or 1) and x is some value such that $|x| \geq 1$. We denote the set of pairs of initial functions $(u_1(s), u_2(s))^T \in C_{[-T,0]}(\mathbb{R}^2)$ satisfying conditions (7) and (8) as $IC(i, k, x)$.

We will integrate system (3) using a method of steps. It follows from (7) that, on the first step (time segment $t \in [t_1, t_1 + T]$), system (3) has form (5) and the solution has a form

$$\begin{array}{rcl} u_i(t) &=& \frac{(k+x)p}{2}e^{-(t-t_1)} + \frac{(k-x)p}{2}e^{-(1+2\gamma)(t-t_1)}, \\ u_{3-i}(t) &=& \frac{(k+x)p}{2}e^{-(t-t_1)} + \frac{(x-k)p}{2}e^{-(1+2\gamma)(t-t_1)}. \end{array} \tag{9}$$

Since function u_i is inside the strip $|u_i| < p$ for $t \in [t_1, t_1 + \delta]$, then, for $t \in [t_1 + T, t_1 + 2T]$, we have that $F(u_i(t-T))$ is not identically equal to 0. In addition, $F(u_{3-i}(t-T))$ may be identically equal to 0 or not (it depends on value of x). Then, on the second step ($t \in [t_1 + T, t_1 + 2T]$), we consider system (3) as an inhomogeneous system of ordinary differential equations (here functions $F(u_i(t-T))$ and $F(u_{3-i}(t-T))$ are known from the previous step and we consider them as inhomogeneity). Thus, the following formula for solution of system (3) holds:

$$\begin{array}{rcl} u_i(t) &=& \frac{(k+x)p}{2}e^{-(t-t_1)} + \frac{(k-x)p}{2}e^{-(1+2\gamma)(t-t_1)} + \frac{\lambda}{2}A(k, x, t, t_1), \\ u_{3-i}(t) &=& \frac{(k+x)p}{2}e^{-(t-t_1)} + \frac{(x-k)p}{2}e^{-(1+2\gamma)(t-t_1)} + \frac{\lambda}{2}B(k, x, t, t_1), \end{array} \tag{10}$$

where

$$A(k, x, t, t_1) = \int_{T+t_1}^{t} \left(e^{s-t} + e^{(1+2\gamma)(s-t)}\right) F\left(\frac{(k+x)p}{2}e^{t_1+T-s} + \frac{(k-x)p}{2}e^{(1+2\gamma)(t_1+T-s)}\right) ds$$
$$+ \int_{T+t_1}^{t} \left(e^{s-t} - e^{(1+2\gamma)(s-t)}\right) F\left(\frac{(k+x)p}{2}e^{t_1+T-s} + \frac{(x-k)p}{2}e^{(1+2\gamma)(t_1+T-s)}\right) ds,$$

$$B(k, x, t, t_1) = \int_{T+t_1}^{t} \left(e^{s-t} - e^{(1+2\gamma)(s-t)}\right) F\left(\frac{(k+x)p}{2}e^{t_1+T-s} + \frac{(k-x)p}{2}e^{(1+2\gamma)(t_1+T-s)}\right) ds$$
$$+ \int_{T+t_1}^{t} \left(e^{s-t} + e^{(1+2\gamma)(s-t)}\right) F\left(\frac{(k+x)p}{2}e^{t_1+T-s} + \frac{(x-k)p}{2}e^{(1+2\gamma)(t_1+T-s)}\right) ds.$$

Let's introduce the following conditions on the functions A and B:

Assumption 1. *Number of points $t^* \in [t_1 + T, t_1 + 2T]$ for which $A(k, x, t^*, t_1) = 0$ ($B(k, x, t^*, t_1) = 0$) is finite. If $A(k, x, t^*, t_1) = 0$ ($B(k, x, t^*, t_1) = 0$), then there exists $j \in \mathbb{N}$ such that $\frac{\partial^j A(k, x, t, t_1)}{\partial t^j}\Big|_{t=t^*} \neq 0$ ($\frac{\partial^j B(k, x, t, t_1)}{\partial t^j}\Big|_{t=t^*} \neq 0$, respectively).*

Assumption 2. *Inequality $A(k, x, t_1 + 2T, t_1) B(k, x, t_1 + 2T, t_1) \neq 0$ holds.*

Under Assumption 2, we obtain that

$$u_i(t_1 + 2T) = \frac{\lambda}{2}\Big(A(k,x,t_1+2T,t_1) + o(1)\Big),$$
$$u_{3-i}(t_1 + 2T) = \frac{\lambda}{2}\Big(B(k,x,t_1+2T,t_1) + o(1)\Big) \quad (11)$$

at $\lambda \to +\infty$ and that both functions $u_i(t)$ and $u_{3-i}(t)$ at the point $t = t_1 + 2T$ are outside of the strip $|u_j| < p$.

Lemma 1. *If Assumptions 1 and 2 hold, then on the segment $t \in [t_1 + 2T, t_1 + 3T]$ functions $u_i(t)$ and $u_{3-i}(t)$ have the form*

$$\begin{aligned}
u_i(t) &= \tfrac{\lambda}{4}(A(k,x,2T+t_1,t_1) + B(k,x,2T+t_1,t_1) + o(1))e^{-(t-t_1-2T)} \\
&+ \tfrac{\lambda}{4}(A(k,x,2T+t_1,t_1) - B(k,x,2T+t_1,t_1) + o(1))e^{-(1+2\gamma)(t-t_1-2T)}, \\
u_{3-i}(t) &= \tfrac{\lambda}{4}(A(k,x,2T+t_1,t_1) + B(k,x,2T+t_1,t_1) + o(1))e^{-(t-t_1-2T)} \\
&- \tfrac{\lambda}{4}(A(k,x,2T+t_1,t_1) - B(k,x,2T+t_1,t_1) + o(1))e^{-(1+2\gamma)(t-t_1-2T)}.
\end{aligned} \quad (12)$$

Proof. Let $t \in [t_1 + 2T, t_1 + 3T]$. On this segment, we consider system (3) as a system of inhomogeneous linear ordinary differential equations (on this time segment we consider known functions $\lambda F(u_i(t-T))$ and $\lambda F(u_{3-i}(t-T))$ as inhomogeneity). Therefore, a solution of this system on the time segment $t \in [t_1 + 2T, t_1 + 3T]$ has the form of a sum of particular integral (PI) and complementary function (CF, solution of linear part of system (3)–system (5)) with constants determined from the initial conditions (11):

$$u_i(t) = u_{i_{CF}}(t) + u_{i_{PI}}(t),$$
$$u_{3-i}(t) = u_{(3-i)_{CF}}(t) + u_{(3-i)_{PI}}(t).$$

Let's find asymptotics of particular integral of this system at $\lambda \to +\infty$. A particular integral of the system (3) on the time segment $t \in [t_1 + 2T, t_1 + 3T]$ has the form

$$u_{i_{PI}}(t) = \tfrac{\lambda}{2}\int_{t_1+2T}^{t}(e^{s-t} + e^{(1+2\gamma)(s-t)})F(u_i(s-T)) + (e^{s-t} - e^{(1+2\gamma)(s-t)})F(u_{3-i}(s-T))ds,$$
$$u_{(3-i)_{PI}}(t) = \tfrac{\lambda}{2}\int_{t_1+2T}^{t}(e^{s-t} - e^{(1+2\gamma)(s-t)})F(u_i(s-T)) + (e^{s-t} + e^{(1+2\gamma)(s-t)})F(u_{3-i}(s-T))ds. \quad (13)$$

Suppose a particular integral (13) is non-zero. This integral on some segment is non-zero only if functions $F(u_i(s-T))$ or $F(u_{3-i}(s-T))$ are non-zero on this segment. Function $F(u_i(t-T))$ ($F(u_{3-i}(t-T))$) is non-zero only if $|u_i(t-T)| < p$ ($|u_{3-i}(t-T)| < p$). For sufficiently large values of λ this condition holds only if $A(k,x,t-T,t_1)$ ($B(k,x,t-T,t_1)$ respectively) is in the neighborhood of zero. Function $A(k,x,\cdot,t_1)$ ($B(k,x,\cdot,t_1)$) is continuous; consequently, there exists point $t^* \in [t_1+T, t_1+2T]$ such that $A(k,x,t^*,t_1) = 0$ ($B(k,x,t^*,t_1) = 0$, respectively).

Consider the point $t^* \in [t_1+T, t_1+2T]$ such that $A(k,x,t^*,t_1) = 0$. It follows from Assumption 1 that there exist $j \in \mathbb{N}$ such that $\left.\dfrac{\partial^j A(k,x,t,t_1)}{\partial t^j}\right|_{t=t^*} \neq 0$. Let q be the minimum from these numbers j. Consequently, it follows from (10) that, in the neighborhood of t^*, we have

$$u_i(t-T) = \frac{(k+x)p}{2}e^{-(t-T-t_1)} + \frac{(k-x)p}{2}e^{-(1+2\gamma)(t-T-t_1)}$$
$$+ \frac{\lambda}{2}\Big(\frac{\partial^q A(k,x,t^*,t_1)}{\partial t^q} + o(1)\Big)\frac{(t-T-t^*)^q}{q!}. \quad (14)$$

Let's estimate "time of living" Δt^* of function $u_i(t-T)$ in the strip $|u_i| < p$ in the neighborhood of the point $t - T = t^*$ ("time of living" means here length of the maximal interval of values t such that t^* belongs to this segment and inequality $|u_i(t)| < p$ is true for all points t from this segment). From (14), under the condition that λ is sufficiently large, we get that $\Delta t^* \leq M_1 \lambda^{-\frac{1}{q}}$, where $M_1 = M_1(k, x, \gamma)$ is some positive value. From Assumption 1, we know that number of points t^* such that $A(k, x, t^*, t_1) = 0$ is finite, which is why there exists $Q = q_{max}$—maximum from values q for all points t^*. Then, on the whole segment $t - T \in [t_1 + T, t_1 + 2T]$ "time of living" Δt_{total} of function $u_i(t - T)$ in the strip $|u_i| < p$ has estimate $\Delta t_{total} \leq M_2 \lambda^{-\frac{1}{Q}}$, where $M_2 = M_2(k, x, \gamma)$ is some positive value. Similarly, for function $u_{3-i}(t-T)$, we have estimate $\Delta t_{total} \leq M_3 \lambda^{-\frac{1}{P}}$, where M_3 and P are some positive values. Function F is bounded, which is why, for a particular integral (13), we have the following estimate:

$$|u_{i_{PI}}(t)| \leq M\lambda^{\frac{\max\{P,Q\}-1}{\max\{P,Q\}}}, \quad |u_{(3-i)_{PI}}(t)| \leq M\lambda^{\frac{\max\{P,Q\}-1}{\max\{P,Q\}}},$$

where M is some positive value, $t \in [t_1 + 2T, t_1 + 3T]$.

A solution of linear part of system (3) satisfying initial conditions (11) on this segment has form

$$u_{i_{CF}}(t) = \frac{\lambda}{4}(A(k, x, 2T + t_1, t_1) + B(k, x, 2T + t_1, t_1) + o(1))e^{-(t-t_1-2T)}$$
$$+ \frac{\lambda}{4}(A(k, x, 2T + t_1, t_1) - B(k, x, 2T + t_1, t_1) + o(1))e^{-(1+2\gamma)(t-t_1-2T)},$$
$$u_{(3-i)_{CF}}(t) = \frac{\lambda}{4}(A(k, x, 2T + t_1, t_1) + B(k, x, 2T + t_1, t_1) + o(1))e^{-(t-t_1-2T)}$$
$$- \frac{\lambda}{4}(A(k, x, 2T + t_1, t_1) - B(k, x, 2T + t_1, t_1) + o(1))e^{-(1+2\gamma)(t-t_1-2T)}.$$

Thus, a complementary function gives us the leading term of asymptotics of solution of system (3) on the segment $t \in [t_1 + 2T, t_1 + 3T]$ and thus a solution on this segment has form (12). □

Corollary 1. *The leading term of asymptotics of solution of system (3) coincides with solution of system (5) with initial conditions (11) on the segment $t \in [t_1 + 2T, t_1 + 3T]$.*

Let's study asymptotics of solutions of system (3) for values $t > t_1 + 3T$. While both components of solution are outside of the strip $|u_j| < p$ ($j = 1, 2$), system (3) has form (5) and solution has form (12). If some component of solution comes to the strip $|u_j| < p$ at the point $t = t_0 > t_1 + 2T$, then on the next step $t \in [t_0 + T, t_0 + 2T]$ nonlinearity F is non-zero and the leading term of asymptotics of solution may change. Whether it changes or not is determined by the values of the functions

$$G_{\pm}(t) = (A(k, x, 2T + t_1, t_1) + B(k, x, 2T + t_1, t_1))$$
$$\pm (A(k, x, 2T + t_1, t_1) - B(k, x, 2T + t_1, t_1))e^{-2\gamma(t-t_1-2T)}$$

in the neighborhood of the point t_0.

Note that, in terms of functions G_+ and G_- on the segment $t \in [t_1 + 2T, t_0]$, we have the following representation of functions u_i and u_{3-i}:

$$u_i(t) = \frac{\lambda}{4}\Big(G_+(t) + o(1)\Big)e^{-(t-t_1-2T)}, \tag{15}$$

$$u_{3-i}(t) = \frac{\lambda}{4}\Big(G_-(t) + o(1)\Big)e^{-(t-t_1-2T)}. \tag{16}$$

There exists two principally different cases when function $u_i(t)$ (or $u_{3-i}(t)$) comes into the strip $|u_j| < p$ at the point $t = t_0 > t_1 + 2T$:

1. The second multiplier in Formula (15) or Formula (16) at some point from an asymptotically small at $\lambda \to +\infty$ neighborhood of the point $t = t_0$ is equal to zero.
2. The second multiplier in Formulas (15) and (16) in some (independent from λ) neighborhood of the point $t = t_0$ is non-zero and the third multiplier is asymptotically small on λ at $\lambda \to +\infty$ in the neighborhood of the point $t = t_0$.

Note that, for some functions F and values of parameters k, x, and γ, Case 1 does not take place. Suppose we have function F and values of parameters k, x, and γ such that this Case occurs. Then, we have the following Lemma.

Lemma 2. *Suppose some component of solution comes into the strip $|u_j| < p$ at the point $t = t_0 > t_1 + 2T$ and Formula (12) is valid for the leading term of asymptotics of solution on the segment $t \in [t_1 + 2T, t_0]$. If there exists a point from an asymptotically small at $\lambda \to +\infty$ neighborhood of the point $t = t_0$ such that the second multiplier in (15) or (16) is equal to zero, then asymptotics of solution on the segment $t \in [t_0 + T, t_0 + 2T]$ has form (12).*

Proof. First, note that, if the second multiplier in (15) or (16) is equal to zero at some point from the small neighborhood of the point $t = t_0$, then there exists value t_* such that $|t_* - t_0| = o(1)$ at $\lambda \to +\infty$ and $G_+(t_*)G_-(t_*) = 0$.

Each equation $G_+(t) = 0$ and $G_-(t) = 0$ has at most one root and, if one equation has a root, then another equation has no roots. This root does not depend on λ, and it follows from Assumption 2 that if $G_+(t_*) = 0$ ($G_-(t_*) = 0$), then $G'_+(t_*) \neq 0$ ($G'_-(t_*) \neq 0$, respectively).

Assume without loss of generality that function u_i comes into the strip $|u_i| < p$ at the point $t = t_0$ and $G_+(t_*) = 0$. Acting like in the proof of Lemma 1, we obtain that "time of living" Δt_* of function $u_i(t)$ in the strip $|u_i| < p$ in the neighborhood of the point $t = t_*$ has estimate $\Delta t_* \leq const \lambda^{-1}$. This is why a particular integral of the system (3) on the segment $t \in [t_0 + T, t_0 + 2T]$ has estimate

$$|u_{i_{PI}}(t)| \leq const_1, \quad |u_{(3-i)_{PI}}(t)| \leq const_2,$$

and a complementary function has estimate

$$|u_{i_{CF}}(t)| \geq const_3 \lambda, \quad |u_{(3-i)_{CF}}(t)| \geq const_4 \lambda,$$

where $const_3 > 0$ and $const_4 > 0$.

Thus, on the segment $t \in [t_0 + T, t_0 + 2T]$, Formula (12) is valid. □

For the further reasoning, we need a notation of the time moment of leaving the strip $|u_j| < p$ in Case 1 (if this Case occurs). We denote it as t_{leave}. It follows from Lemma 2 that $t_{leave} < t_* + T$. If Case 1 does not take place, then we define $t_{leave} = t_1 + 2T$. Thus, there exists a constant $M_{t.l.} > 0$ independent on λ such that $t_{leave} < M_{t.l.}$.

Lemma 2 implies the following statement.

Corollary 2. *For all $t > t_{leave}$, both functions $u_i(t)$ and $u_{3-i}(t)$ are outside of the strip $|u_j| < p$ until Case 2 occurs.*

Let's study Case 2 in more detail.

First, consider the case $\gamma > 0$. If non-degeneracy condition

$$A(k, x, 2T + t_1, t_1) + B(k, x, 2T + t_1, t_1) \neq 0 \tag{17}$$

holds, then there exist positive constants c_{min}, c_{Max}, such that

$$0 < c_{min} < |G_\pm(t) + o(1)| < c_{Max}$$

in some independent on λ neighborhood of the point $t = t_0$. Therefore, $|\lambda e^{-(t_0-t_1-2T)}| < M_4$ at $\lambda \to +\infty$, where M_4 is some positive constant. This is why

$$t_0 - t_1 = (1 + o(1))\ln \lambda \tag{18}$$

at $\lambda \to +\infty$. In addition, in the neighborhood of the point $t = t_0$, solution of system (3) has form

$$\begin{aligned}u_i(t) &= \tfrac{\lambda}{4}(A(k,x,2T+t_1,t_1) + B(k,x,2T+t_1,t_1) + o(1))e^{-(t-t_1-2T)},\\ u_{3-i}(t) &= \tfrac{\lambda}{4}(A(k,x,2T+t_1,t_1) + B(k,x,2T+t_1,t_1) + o(1))e^{-(t-t_1-2T)}.\end{aligned} \tag{19}$$

Consider the case $-\tfrac{1}{2} < \gamma < 0$. If non-degeneracy condition

$$A(k,x,2T+t_1,t_1) - B(k,x,2T+t_1,t_1) \neq 0 \tag{20}$$

holds, then, for some positive constants d_{min} and d_{Max} in some independent on λ neighborhood of the point $t = t_0$, we have

$$0 < d_{min} < |(G_{\pm}(t) + o(1))e^{2\gamma(t-t_1-2T)}| < d_{Max}.$$

Therefore, we obtain that $|\lambda e^{-(1+2\gamma)(t_0-t_1-2T)}| < M_5$ at $\lambda \to +\infty$, where M_5 is some positive constant. Consequently,

$$t_0 - t_1 = ((1+2\gamma)^{-1} + o(1))\ln \lambda \tag{21}$$

at $\lambda \to +\infty$ and in the neighborhood of the point $t = t_0$ solution of system (3) has form

$$\begin{aligned}u_i(t) &= \tfrac{\lambda}{4}(A(k,x,2T+t_1,t_1) - B(k,x,2T+t_1,t_1) + o(1))e^{-(1+2\gamma)(t-t_1-2T)},\\ u_{3-i}(t) &= -\tfrac{\lambda}{4}(A(k,x,2T+t_1,t_1) - B(k,x,2T+t_1,t_1) + o(1))e^{-(1+2\gamma)(t-t_1-2T)}.\end{aligned} \tag{22}$$

From Formulas (18) and (21), we get that $t_0 - t_{leave} > T$. In addition, it follows from Formulas (19) and (22) that if $|u_j(t_0)| = p$, then there exists $\delta > 0$ such that $|u_j(t)| < p$ for all $t \in (t_0, t_0 + \delta)$. Thus, there exists t_2 (it is equal to t_0 from the Case 2), such that

$$t_2 - t_1 = \begin{cases}(1+o(1))\ln \lambda, & \gamma > 0,\\ ((1+2\gamma)^{-1} + o(1))\ln \lambda, & -\tfrac{1}{2} < \gamma < 0,\end{cases} \tag{23}$$

$$|u_1(s+t_2)| > p, \quad |u_2(s+t_2)| > p \text{ for all } s \in [-T,0), \tag{24}$$

and

$$u_{\bar{i}}(t_2) = \bar{k}p, \quad u_{3-\bar{i}}(t_2) = \bar{x}p \tag{25}$$

at $\lambda \to +\infty$.

It follows from Lemmas 1 and 2, Corollaries 1 and 2 and from the reasoning given above that the next statement is true.

Corollary 3. *On the time segment $t \in [t_1 + 2T, t_2]$, a solution of system (3) has form (12).*

It follows from Formulas (24) and (25) that we obtain an operator of translation along the trajectories that map our set of initial conditions $IC(i, k, x)$ to a set $IC(\bar{i}, \bar{k}, \bar{x})$. Thus, at the point t_2, we return to the initial situation with replacement k, x, i, and t_1 by $\bar{k}$, $\bar{x}$, $\bar{i}$, and t_2. If we do the same steps as in this section and in all the next iterations, Assumptions 1 and 2 and non-degeneracy condition (17) in the case $\gamma > 0$ (non-degeneracy condition (20) in the case $-\tfrac{1}{2} < \gamma < 0$, respectively) hold (with new values $k = k_n$, $x = x_n$, $i = i_n$ and replacing t_1 with t_n ($n = 2, 3, \ldots$)), then, from an operator of translation along the trajectories, we obtain a map on i_n, k_n, and x_n. This map determines dynamics of the system (3) because on the segments $t \in [t_n, t_{n+1}]$ solution satisfies Formulas (9), (10) amd (12) with $i = i_n, k = k_n, x = x_n, t_1 = t_n$.

In the next two sections, we construct an exact form of maps on $i = i_n, k = k_n,$ and $x = x_n$ in the case $\gamma > 0$ (see Section 3) and in the case $-\frac{1}{2} < \gamma < 0$ (see Section 4) and using dynamical properties of these maps clarify asymptotics of solution on the intervals $t \in [t_n, t_{n+1}]$ ($n = 2, 3, \ldots$).

3. Dynamics in the Case of the Positive Coupling

In this section, we construct a map on k_n, x_n, and i_n and make conclusions on dynamics of system (3) in the case of positive coupling ($\gamma > 0$).

Define $C(n)$ and $D(n)$ as

$$C(n) = A(k_n, x_n, 2T + t_n, t_n) + B(k_n, x_n, 2T + t_n, t_n),$$
$$D(n) = A(k_n, x_n, 2T + t_n, t_n) - B(k_n, x_n, 2T + t_n, t_n),$$

where $n \in \mathbb{N}$. Suppose that

$$C(n) \neq 0, \tag{26}$$

((26) is condition (17) with $k = k_n, x = x_n, t_1 = t_n$) and Assumptions 1 and 2 hold for values k_n, x_n and t_n for all $n \in \mathbb{N}$. Then, acting like in Section 2, we get that in the case of positive coupling values $u_i(t_{n+1})$ and $u_{3-i}(t_{n+1})$ have form

$$u_i(t_{n+1}) = \tfrac{\lambda}{4}(C(n) + o(1))e^{-(t_{n+1} - t_n - 2T)},$$
$$u_{3-i}(t_{n+1}) = \tfrac{\lambda}{4}(C(n) + o(1))e^{-(t_{n+1} - t_n - 2T)}.$$

Thus, we obtain that, in the case $\gamma > 0$, values t_n ($n = 1, 2, \ldots$) satisfy

$$t_{n+1} - t_n = (1 + o(1))\ln \lambda \tag{27}$$

at $\lambda \to +\infty$.

From (12) and (27), we get that the mapping on k_n, x_n, and i_n has form

$$k_{n+1} = \text{sign}(C(n)),$$
$$i_{n+1} = \begin{cases} i_n, & \text{sign}(C(n)D(n)) = -1, \\ 3 - i_n, & \text{sign}(C(n)D(n)) = 1, \end{cases} \tag{28}$$
$$x_{n+1} = k_{n+1} + O(\lambda^{-2\gamma})$$

at $\lambda \to +\infty$.

It follows from (28) that we have $k_n - x_n = o(1)$ for all $n = 2, 3, \ldots$ under the condition that Assumptions 1 and 2 and inequality (26) are fulfilled. Thus, starting from the second iteration Assumption 1 should be satisfied for parameters $k = k_n, x = k_n + o(1)$, and $t_1 = t_n$. Let's formulate this assumption for these values of parameters k, x, and t_1. Functions $A(k_n, k_n + o(1), t, t_n)$ and $B(k_n, k_n + o(1), t, t_n)$ have form

$$A(k_n, k_n + o(1), t, t_n) = B(k_n, k_n + o(1), t, t_n) + o(1) = 2 \int_{T+t_n}^{t} e^{s-t} F\left(k_n p e^{t_n + T - s}\right) ds + o(1).$$

In Assumption 1 value $t \in [t_n + T, t_n + 2T]$, so, for each n value, $\tilde{t} = t - t_n$ is in the segment $[T, 2T]$. Since

$$\int_{T+t_n}^{t} e^{s-t} F\left(k_n p e^{t_n + T - s}\right) ds = \int_{T}^{\tilde{t}} e^{s-\tilde{t}} F\left(k_n p e^{T-s}\right) ds,$$

then Assumption 1 for any $n = 2, 3, \ldots$ is the same (only k_n may change, but it takes two values only). Thus, if the following assumption holds, then Assumption 1 holds for all $n = 2, 3, \ldots$

Assumption 3. *Number of points $t^* \in [T, 2T]$ such that $h(k, t^*) = 0$ is finite. If $h(k, t^*) = 0$, then there exists $j \in \mathbb{N}$ such that $\dfrac{\partial^j h(k, \tilde{t})}{\partial \tilde{t}^j}\bigg|_{\tilde{t}=t^*}$ is non-zero. Here, $k = 1$ or -1 and*

$$h(k, \tilde{t}) = \int_T^{\tilde{t}} e^{s-\tilde{t}} F\left(kpe^{T-s}\right) ds.$$

Under Assumption 3, the asymptotics of the solution has form

$$\begin{aligned} u_{i_n}(t) &= k_n p e^{-(t-t_n)} + o(1), \\ u_{3-i_n}(t) &= k_n p e^{-(t-t_n)} + o(1) \end{aligned} \qquad (29)$$

on the time segments $t \in [t_n, t_n + T]$, where $n = 2, 3, \ldots$ ((29) is Formula (9) with $i = i_n$, $k = k_n$, $x = x_n = k_n + o(1)$, and $t_1 = t_n$). On the segments $t \in [t_n + T, t_n + 2T]$, the main terms of asymptotics of solution is given by the formula

$$\begin{aligned} u_{i_n}(t) &= \lambda\big(h(k_n, t - t_n) + o(1)\big), \\ u_{3-i_n}(t) &= \lambda\big(h(k_n, t - t_n) + o(1)\big) \end{aligned} \qquad (30)$$

((30) is Formula (10) with $i = i_n$, $k = k_n$, $x = x_n = k_n + o(1)$, and $t_1 = t_n$, where functions A and B are rewritten in terms of function h).

We assume that the following non-degeneracy condition holds:

$$h(1, 2T)h(-1, 2T) \neq 0 \qquad (31)$$

(the fulfillment of this inequality guarantees that the Assumption 2 and (26) are satisfied for all $n = 2, 3, \ldots$).

Then, on the segments, a $t \in [t_n + 2T, t_{n+1}]$ solution satisfies equalities

$$\begin{aligned} u_{i_n}(t) &= \lambda\Big(h(k_n, 2T) + o(1)\Big) e^{-(t-t_n-2T)}, \\ u_{3-i_n}(t) &= \lambda\Big(h(k_n, 2T) + o(1)\Big) e^{-(t-t_n-2T)}. \end{aligned} \qquad (32)$$

at $\lambda \to +\infty$ ((32) is Formula (12) with $i = i_n$, $k = k_n$, $x = x_n = k_n + o(1)$, and $t_1 = t_n$, where functions A and B are rewritten in terms of function h).

Thus, we have the following theorem:

Theorem 1. *Suppose $\gamma > 0$ and for values of k_1 and x_1 Assumptions 1, 2, and inequality (17) hold. Suppose Assumption 3 and inequality (31) hold. Then, for any sufficiently large $\lambda > 0$, there exists $t_2 = t_2(k_1, x_1) > 0$ such that for all $t > t_2$ solution of system (3) satisfies Formulas (29), (30), and (32).*

In Figure 1, an example of a solution of system (3) in the case of $\gamma > 0$ is shown.

Since F is smooth and $x_{n+1} - k_{n+1} = O(\lambda^{-2\gamma})$ at $\lambda \to +\infty$, then, in the case $\gamma > \frac{1}{2}$, we have the following statement.

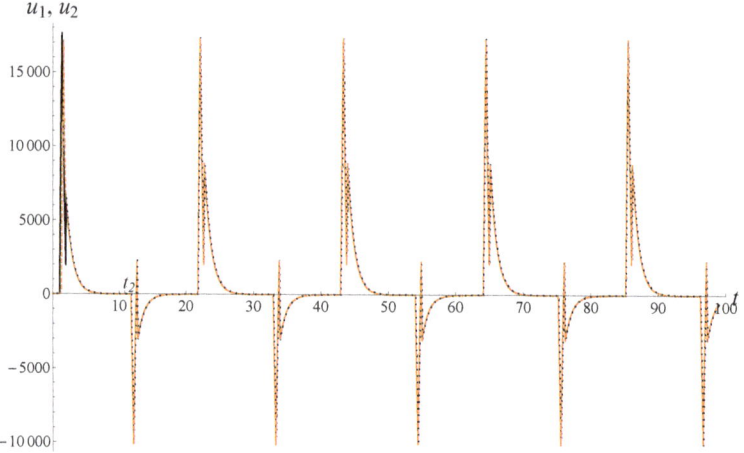

Figure 1. Example of solution. Values of parameters: $T = 1$, $\gamma = 0.1$, $p = 1$, $\lambda = 10{,}000$. Black line—$u_1(t)$, orange dashed line—$u_2(t)$.

Corollary 4. *Suppose $\gamma > \frac{1}{2}$ and for values k_1 and x_1 Assumptions 1 and 2 hold and inequality (17) is true. Suppose Assumption 3 and inequality (31) are true. Then, for any sufficiently large $\lambda > 0$, there exists $t_2(k_1, x_1) > 0$ such that for all $t > t_2$ inequality $|u_1(t) - u_2(t)| = o(1)$ is true.*

4. Dynamics in the Case of Negative Coupling

In this section, we assume that $-\frac{1}{2} < \gamma < 0$. We construct map on k_n, x_n, and i_n for these values of γ and make conclusions about dynamics of system (3).

Suppose inequality

$$D(n) \neq 0 \qquad (33)$$

and Assumptions 1 and 2 for values k_n, x_n, and t_n hold for all $n \in \mathbb{N}$. Then, like in Section 2, we obtain that, in the case $-\frac{1}{2} < \gamma < 0$, values $u_i(t_{n+1})$ and $u_{3-i}(t_{n+1})$ have the form

$$u_i(t_{n+1}) = \tfrac{\lambda}{4}(D(n) + o(1))e^{-(1+2\gamma)(t_{n+1} - t_n - 2T)},$$
$$u_{3-i}(t_{n+1}) = \tfrac{\lambda}{4}(-D(n) + o(1))e^{-(1+2\gamma)(t_{n+1} - t_n - 2T)}.$$

Thus, we obtain that, in the case of negative coupling,

$$t_{n+1} - t_n = \left(\frac{1}{1+2\gamma} + o(1)\right)\ln\lambda \qquad (34)$$

at $\lambda \to +\infty$. It follows from (12) and (34) that the mapping on k_n, x_n, and i_n has form

$$k_{n+1} = \begin{cases} \operatorname{sign}(D(n)), & \operatorname{sign}(C(n)D(n)) = -1, \\ -\operatorname{sign}(D(n)), & \operatorname{sign}(C(n)D(n)) = 1, \end{cases}$$
$$i_{n+1} = \begin{cases} i_n, & \operatorname{sign}(C(n)D(n)) = -1, \\ 3 - i_n, & \operatorname{sign}(C(n)D(n)) = 1, \end{cases} \qquad (35)$$
$$x_{n+1} = -k_{n+1} + O\left(\lambda^{\frac{2\gamma}{1+2\gamma}}\right),$$

at $\lambda \to +\infty$.

Thus, under Assumptions 1, 2 and (33) on the n-th (where $n \geq 2$) iteration of mapping, we have $k_n + x_n = o(1)$ at $\lambda \to +\infty$. Thus, starting from the second iteration, Assumption 1 should be satisfied for $k = k_n$, $x = -k_n + o(1)$, and $t_1 = t_n$. Let's formulate this assumption for these values of parameters. Functions $A(k_n, -k_n + o(1), t, t_n)$ and $B(k_n, -k_n + o(1), t, t_n)$ have the form

$$A(k_n, -k_n + o(1), t, t_n) = \int_{T+t_n}^{t} \left(e^{s-t} + e^{(1+2\gamma)(s-t)}\right) F\left(k_n p e^{(1+2\gamma)(t_n+T-s)}\right) ds$$

$$+ \int_{T+t_n}^{t} \left(e^{s-t} - e^{(1+2\gamma)(s-t)}\right) F\left(-k_n p e^{(1+2\gamma)(t_n+T-s)}\right) ds + o(1),$$

$$B(k_n, -k_n + o(1), t, t_n) = \int_{T+t_n}^{t} \left(e^{s-t} - e^{(1+2\gamma)(s-t)}\right) F\left(k_n p e^{(1+2\gamma)(t_n+T-s)}\right) ds$$

$$+ \int_{T+t_n}^{t} \left(e^{s-t} + e^{(1+2\gamma)(s-t)}\right) F\left(-k_n p e^{(1+2\gamma)(t_n+T-s)}\right) ds + o(1).$$

Value t in Assumption 1 on the n-th iteration of steps described in Section 2 is in the segment $[t_n + T, t_n + 2T]$; therefore, for each step value, $\tilde{t} = t - t_n$ is in the segment $[T, 2T]$. Note that

$$\int_{T+t_n}^{t} \left(e^{s-t} + e^{(1+2\gamma)(s-t)}\right) F\left(k_n p e^{(1+2\gamma)(t_n+T-s)}\right) ds$$

$$+ \int_{T+t_n}^{t} \left(e^{s-t} - e^{(1+2\gamma)(s-t)}\right) F\left(-k_n p e^{(1+2\gamma)(t_n+T-s)}\right) ds$$

$$= \int_{T}^{\tilde{t}} \left(e^{s-\tilde{t}} + e^{(1+2\gamma)(s-\tilde{t})}\right) F\left(k_n p e^{(1+2\gamma)(T-s)}\right) ds + \int_{T}^{\tilde{t}} \left(e^{s-\tilde{t}} - e^{(1+2\gamma)(s-\tilde{t})}\right) F\left(-k_n p e^{(1+2\gamma)(T-s)}\right) ds$$

and

$$\int_{T+t_n}^{t} \left(e^{s-t} - e^{(1+2\gamma)(s-t)}\right) F\left(k_n p e^{(1+2\gamma)(t_n+T-s)}\right) ds$$

$$+ \int_{T+t_n}^{t} \left(e^{s-t} + e^{(1+2\gamma)(s-t)}\right) F\left(-k_n p e^{(1+2\gamma)(t_n+T-s)}\right) ds$$

$$= \int_{T}^{\tilde{t}} \left(e^{s-\tilde{t}} - e^{(1+2\gamma)(s-\tilde{t})}\right) F\left(k_n p e^{(1+2\gamma)(T-s)}\right) ds + \int_{T}^{\tilde{t}} \left(e^{s-\tilde{t}} + e^{(1+2\gamma)(s-\tilde{t})}\right) F\left(-k_n p e^{(1+2\gamma)(T-s)}\right) ds.$$

Thus, for each $n = 2, 3, \ldots$, Assumption 1 is the same (only k_n may change). Thus, if the following assumption holds, then Assumption 1 holds for all $n = 2, 3, \ldots$.

Assumption 4. *Number of points* $t^* \in [T, 2T]$ *such that* $g_1(k, t^*) = 0$ ($g_2(k, t^*) = 0$) *is finite. If* $g_1(k, t^*) = 0$ ($g_2(k, t^*) = 0$), *then there exists* $j \in \mathbb{N}$ *such that* $\dfrac{\partial^j g_1(k, \tilde{t})}{\partial \tilde{t}^j}\Big|_{\tilde{t}=t^*}$ ($\dfrac{\partial^j g_2(k, \tilde{t})}{\partial \tilde{t}^j}\Big|_{\tilde{t}=t^*}$, *respectively) is non-zero. Here,* $k = 1$ *or* $k = -1$ *and*

$$g_1(k, \tilde{t}) = \int_{T}^{\tilde{t}} \left(e^{s-\tilde{t}} + e^{(1+2\gamma)(s-\tilde{t})}\right) F\left(k p e^{(1+2\gamma)(T-s)}\right) ds$$

$$+ \int_{T}^{\tilde{t}} \left(e^{s-\tilde{t}} - e^{(1+2\gamma)(s-\tilde{t})}\right) F\left(-k p e^{(1+2\gamma)(T-s)}\right) ds,$$

$$g_2(k, \tilde{t}) = \int_{T}^{\tilde{t}} \left(e^{s-\tilde{t}} - e^{(1+2\gamma)(s-\tilde{t})}\right) F\left(k p e^{(1+2\gamma)(T-s)}\right) ds$$

$$+ \int_{T}^{\tilde{t}} \left(e^{s-\tilde{t}} + e^{(1+2\gamma)(s-\tilde{t})}\right) F\left(-k p e^{(1+2\gamma)(T-s)}\right) ds.$$

Thus, under Assumption 4, the asymptotics of the solution has form

$$u_{i_n}(t) = k_n p e^{-(1+2\gamma)(t-t_n)} + o(1),$$
$$u_{3-i_n}(t) = -k_n p e^{-(1+2\gamma)(t-t_n)} + o(1) \qquad (36)$$

on the segments $t \in [t_n, t_n + T]$ ((36) is Formula (9) with $i = i_n$, $k = k_n$, $x = x_n = -k_n + o(1)$, and $t_1 = t_n$). On the segments, the $t \in [t_n + T, t_n + 2T]$ solution satisfies equalities

$$u_{i_n}(t) = \tfrac{\lambda}{2}\Big(g_1(k_n, t - t_n) + o(1)\Big),$$
$$u_{3-i_n}(t) = \tfrac{\lambda}{2}\Big(g_2(k_n, t - t_n) + o(1)\Big) \qquad (37)$$

((37) is Formula (10) with $i = i_n$, $k = k_n$, $x = x_n = -k_n + o(1)$, and $t_1 = t_n$, where functions A and B are rewritten in terms of functions g_1 and g_2).

Suppose that the following non-degeneracy condition holds:

$$g_1(1, 2T) g_1(-1, 2T) g_2(1, 2T) g_2(-1, 2T) \neq 0,$$
$$g_1(1, 2T) \neq g_2(1, 2T), \qquad (38)$$
$$g_1(-1, 2T) \neq g_2(-1, 2T),$$

(the fulfillment of these inequalities leads to fulfillment of Assumption 2 and inequality (33) for all $n = 2, 3, \ldots$). Thus, under condition (38) on the segments $t \in [t_n + 2T, t_{n+1}]$, we have the following asymptotics of solution:

$$u_{i_n}(t) = \tfrac{\lambda}{2}\left(\int_T^{2T} e^s \left(F\left(k_n p e^{(1+2\gamma)(T-s)}\right) + F\left(-k_n p e^{(1+2\gamma)(T-s)}\right)\right) ds + o(1)\right) e^{t_n - t}$$
$$+ \tfrac{\lambda}{2}\left(\int_T^{2T} e^{(1+2\gamma)s} \left(F\left(k_n p e^{(1+2\gamma)(T-s)}\right) - F\left(-k_n p e^{(1+2\gamma)(T-s)}\right)\right) ds + o(1)\right) e^{(1+2\gamma)(t_n - t)},$$
$$u_{3-i_n}(t) = \tfrac{\lambda}{2}\left(\int_T^{2T} e^s \left(F\left(k_n p e^{(1+2\gamma)(T-s)}\right) + F\left(-k_n p e^{(1+2\gamma)(T-s)}\right)\right) ds + o(1)\right) e^{t_n - t}$$
$$- \tfrac{\lambda}{2}\left(\int_T^{2T} e^{(1+2\gamma)s} \left(F\left(k_n p e^{(1+2\gamma)(T-s)}\right) - F\left(-k_n p e^{(1+2\gamma)(T-s)}\right)\right) ds + o(1)\right) e^{(1+2\gamma)(t_n - t)} \qquad (39)$$

((39) is Formula (12) with $i = i_n$, $k = k_n$, $x = x_n = -k_n + o(1)$, and $t_1 = t_n$, where functions A and B are rewritten in terms of function F).

We obtain the following result on dynamics of system (3).

Theorem 2. *Suppose $-\tfrac{1}{2} < \gamma < 0$ and for values of k_1 and x_1 Assumptions 1, 2, and inequality (20) hold. Suppose Assumption 4 and inequalities (38) hold. Then, for any sufficiently large $\lambda > 0$, there exists $t_2 = t_2(k_1, x_1) > 0$ such that for all $t > t_2$ solution of system (3) satisfies Formulas (36), (37), and (39).*

In Figure 2, an example of the solution in the case of $-\tfrac{1}{2} < \gamma < 0$ is shown.

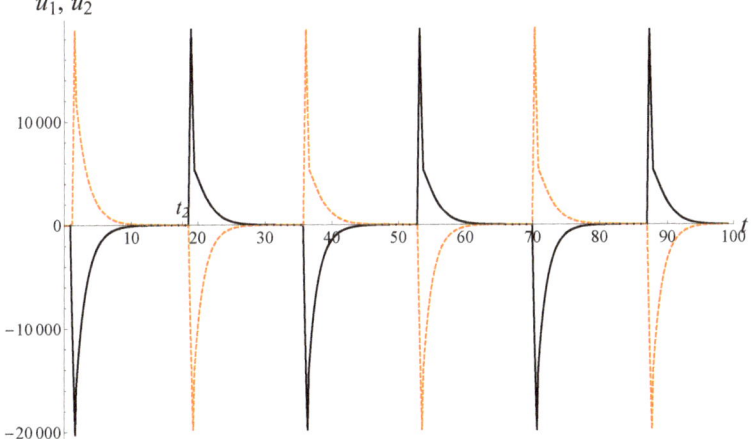

Figure 2. Example of solution. Values of parameters: $T = 0.9$, $\gamma = -0.2$, $p = 1$, $\lambda = 10{,}000$. Black line—$u_1(t)$, orange dashed line—$u_2(t)$.

5. Example

In this section, we show how method described in Sections 2–4 works in the case when function f satisfies conditions (4) and inequality

$$uf(u) > 0 \text{ if } 0 < |u| < p \tag{40}$$

and initial conditions satisfy inequalities

$$\begin{aligned} kx > 0 & \text{ if } \gamma > 0, \\ kx < 0 & \text{ if } -\tfrac{1}{2} < \gamma < 0 \end{aligned} \tag{41}$$

(here k and x are defined as in Section 2).

As in Section 2, we construct asymptotics of all solutions of system (3) with initial conditions outside of the strip $|u_j| < p$ ($j = 1, 2$) and satisfying inequality (41). Let t_1 and i be defined as in Section 2. Then, the following lemmas hold.

Lemma 3. *If initial conditions fulfill (41), then functions $u_i(t)$ and $u_{3-i}(t)$ do not change their signs on the segment $t \in [t_1, t_1 + T]$ and for all $t \in [t_1, t_1 + T]$ inequalities*

$$\begin{aligned} u_i(t)u_{3-i}(t) > 0 & \text{ if } \gamma > 0, \\ u_i(t)u_{3-i}(t) < 0 & \text{ if } -\tfrac{1}{2} < \gamma < 0 \end{aligned} \tag{42}$$

hold.

Proof. Consider the case $k = 1$. If $\gamma > 0$, then $x \geq 1 > 0$. For these values of k, x, and γ system of inequalities,

$$\begin{cases} |k + x| > |x - k|, \\ e^{-(t - t_1)} \geq e^{-(1 + 2\gamma)(t - t_1)} \end{cases} \tag{43}$$

holds. Since $u_i(t)$ and $u_{3-i}(t)$ have form (9), $k + x > 0$ and (43) holds, then we get that

$$u_i(t) > 0, \ u_{3-i}(t) > 0 \tag{44}$$

on the interval $t \in [t_1, t_1 + T]$. If $-\frac{1}{2} < \gamma < 0$, then $x \leq -1 < 0$. This is why we obtain that

$$\begin{cases} |k+x| < |x-k|, \\ e^{-(t-t_1)} \leq e^{-(1+2\gamma)(t-t_1)}. \end{cases} \quad (45)$$

It follows from (9), $k - x > 0$, and (45) that

$$u_i(t) > 0, \ u_{3-i}(t) < 0 \quad (46)$$

on the interval $t \in [t_1, t_1 + T]$.

Consider the case $k = -1$. If $\gamma > 0$, then $x \leq -1 < 0$. Then, from (9), $k + x < 0$, and (43), we obtain that

$$u_i(t) < 0, \ u_{3-i}(t) < 0 \quad (47)$$

on the interval $t \in [t_1, t_1 + T]$. In addition, in the case $-\frac{1}{2} < \gamma < 0$, we get that $x \geq 1 > 0$ and from (9), $k - x > 0$, and (45), we get

$$u_i(t) < 0, \ u_{3-i}(t) > 0 \quad (48)$$

on the interval $t \in [t_1, t_1 + T]$.

It follows from (44), (46)–(48) that inequalities (42) hold. □

Lemma 4. *If function $u_i(t)$ comes into the strip $|u_i(t)| < p$ at the point $t = t_1$, then (1) x satisfies inequality*

$$|x| \leq |1 + 1/\gamma|; \quad (49)$$

(2) function u_i is in the strip $|u_i(t)| < p$ for all $t \in (t_1, t_1 + T]$.

Proof. It follows from (9) that

$$u_i'(t) = -\frac{(k+x)p}{2}e^{-(t-t_1)} - (1+2\gamma)\frac{(k-x)p}{2}e^{-(1+2\gamma)(t-t_1)}, \quad (50)$$

therefore

$$u_i'(t_1) = -\left(\frac{k+x}{2}p + (1+2\gamma)\frac{k-x}{2}p\right).$$

Consider the case $k = 1$. For $k = 1$ value, $u_i(t_1)$ is equal to p. If this function comes into the strip $|u_i(t)| < p$ at the point $t = t_1$, then derivative $u_i'(t_1)$ is non-positive. For $k = 1$ inequality, $u_i'(t_1) \leq 0$ is equivalent to $1 + \gamma \geq \gamma x$. It follows from condition (41) that $\gamma x > 0$ in the case $k = 1$. Thus, in the case $k = 1$, inequality (49) holds.

Consider the case $k = -1$. For $k = -1$ value $u_i(t_1) = -p$ and if this function comes into the strip $|u_i(t)| < p$ at the point $t = t_1$, then derivative $u_i'(t_1)$ is non-negative. For $k = -1$ condition, $u_i'(t_1) \geq 0$ is equivalent to inequality $-1 - \gamma \leq \gamma x$. From (41), we get that $\gamma x < 0$, so inequality (49) is true in this case, too.

It follows from (41) and (49) that in the case $\gamma > 0$ system of inequalities

$$\begin{cases} |k+x| \geq |(1+2\gamma)(x-k)|, \\ e^{-(t-t_1)} > e^{-(1+2\gamma)(t-t_1)} \end{cases} \quad (51)$$

holds and in the case $-\frac{1}{2} < \gamma < 0$ system of inequalities

$$\begin{cases} |k+x| \leq |(1+2\gamma)(x-k)|, \\ e^{-(t-t_1)} < e^{-(1+2\gamma)(t-t_1)} \end{cases} \quad (52)$$

is true on the interval $t \in (t_1, t_1 + T]$

Using (50)–(52), and (41), we obtain that

$$\begin{aligned} u_i'(t) < 0 \text{ if } k = 1, \\ u_i'(t) > 0 \text{ if } k = -1 \end{aligned} \tag{53}$$

on the interval $t \in (t_1, t_1 + T]$. Combining (44), (46)–(48) with (53), we get that function $u_i(t)$ is in the strip $|u_i(t)| < p$ for all $t \in (t_1, t_1 + T]$. □

Lemma 5. *If function f satisfies (4) and (40), initial conditions satisfy (41) and*

$$|x| < |1 + 1/\gamma|, \tag{54}$$

then Assumptions 1–4 hold.

Proof. Consider some function $f(u)$, satisfying conditions (4) and (40).

Let us prove that for this function Assumption 1 holds. From Lemmas 3 and 4, we obtain that $u_i(t)$ is in the strip $|u_i(t)| < p$ and it does not change sign on the interval $t \in (t_1, t_1 + T]$. This is why from condition (40) we get that the first summands in $A(k, x, t, t_1)$ and $B(k, x, t, t_1)$ are non-zero. Thus, from formulas (44), (46)–(48), and assumption (40), we obtain that the following inequalities hold

$$\begin{aligned} A(k,x,t,t_1) > 0, \ B(k,x,t,t_1) > 0 & \quad \text{if} \quad k = 1, \ \gamma > 0 \\ A(k,x,t,t_1) > 0, \ B(k,x,t,t_1) < 0 & \quad \text{if} \quad k = 1, \ -\tfrac{1}{2} < \gamma < 0 \\ A(k,x,t,t_1) < 0, \ B(k,x,t,t_1) < 0 & \quad \text{if} \quad k = -1, \ \gamma > 0 \\ A(k,x,t,t_1) < 0, \ B(k,x,t,t_1) > 0 & \quad \text{if} \quad k = -1, \ -\tfrac{1}{2} < \gamma < 0 \end{aligned} \tag{55}$$

on the interval $t \in (t_1 + T, t_1 + 2T]$. Thus, we have proved that under condition (40) functions $A(k, x, t, t_1)$ and $B(k, x, t, t_1)$ are non-zero on the interval $t \in (t_1 + T, t_1 + 2T]$. If $t^* = t_1 + T$, then $A(k, x, t^*, t_1) = B(k, x, t^*, t_1) = 0$. Derivatives $\left. \dfrac{\partial^j A(k,x,t,t_1)}{\partial t^j} \right|_{t=t_1+T} = 0$ for $j = 1, 2$ and derivatives $\left. \dfrac{\partial^j B(k,x,t,t_1)}{\partial t^j} \right|_{t=t_1+T} = 0$ for $j = 1, 2, 3$. Expressions

$$\left. \frac{\partial^3 A(k,x,t,t_1)}{\partial t^3} \right|_{t=t_1+T} = 2 f''(kp) \left(\frac{k+x}{2} p + (1 + 2\gamma) \frac{k-x}{2} p \right)^2$$

and

$$\left. \frac{\partial^4 B(k,x,t,t_1)}{\partial t^4} \right|_{t=t_1+T} = 2\gamma f''(kp) \left(\frac{k+x}{2} p + (1 + 2\gamma) \frac{k-x}{2} p \right)^2,$$

are non-zero: under condition (54) last factor in these derivatives is non-zero and $f''(kp) \neq 0$ because of (4) (if $x = \pm(1 + 1/\gamma)$, then for all $j \in \mathbb{N}$ expressions $\left. \dfrac{\partial^j A(k,x,t,t_1)}{\partial t^j} \right|_{t=t_1+T}$ and $\left. \dfrac{\partial^j B(k,x,t,t_1)}{\partial t^j} \right|_{t=t_1+T}$ equal zero). Consequently, Assumption 1 holds under condition (54). This assumption holds for $x = \pm k + o(1)$ at $\lambda \to +\infty$, so Assumptions 3 and 4 hold.

Since the system of inequalities (55) is true for $t = t_1 + 2T$, then Assumption 2 holds. □

Note that if function $u_i(t)$ comes to the strip $|u_i(t)| < p$, then x satisfies inequality (49), and for all x such that (54) hold, Assumption 1 is true. Thus, only for two values of parameter $x : x_{1,2} = \pm(1 + 1/\gamma)$ is Assumption 1 false.

Lemma 6. *If function f satisfies (4) and (40), then inequalities (26) and (31) are true in the case $\gamma > 0$ and inequalities (33) and (38) hold in the case $-\tfrac{1}{2} < \gamma < 0$.*

Proof. It follows from Lemma 5 that $A(k, x, t_1 + 2T, t_1)$ and $B(k, x, t_1 + 2T, t_1)$ have the same sign in the case $\gamma > 0$ and the opposite signs in the case $-\tfrac{1}{2} < \gamma < 0$. Therefore, in the case $\gamma > 0$

($-\frac{1}{2} < \gamma < 0$) inequality (26) (inequality (33) respectively) holds for all $n = 1, 2, 3, \ldots$. Thus, inequalities (31) and (38) are fulfilled because they are equivalent to Assumption 2 and conditions (26) and (33) for $n = 2, 3, \ldots$. □

Thus, we have proved that all assumptions in Theorems 1 and 2 are true if function f satisfies (4) and (40) and for x_1 conditions (41) and (54) hold. Therefore, for class of functions f considered in this section, the following theorems are true.

Theorem 3. *Suppose $\gamma > 0$ and inequalities (41) and $x_1 \neq \pm(1 + 1/\gamma)$ hold. Then, for any sufficiently large $\lambda > 0$ there exists $t_2 = t_2(k_1, x_1) > 0$ such that for all $t > t_2$ solution of system (3) satisfies Formulas (29), (30) and (32).*

Theorem 4. *Suppose $-\frac{1}{2} < \gamma < 0$ and inequalities (41) and $x_1 \neq \pm(1 + 1/\gamma)$ hold. Then, for any sufficiently large $\lambda > 0$ there exists $t_2 = t_2(k_1, x_1) > 0$ such that for all $t > t_2$ solution of system (3) satisfies Formulas (36), (37) and (39).*

Remark 1. *If $x_1 = \pm(1 + 1/\gamma)$, then Assumption 1 is not true, so Theorems 3 and 4 are not proven. However, probably, they are true because for all initial conditions in the neighborhood of these values they are true.*

Consider the map (28). If we take set $\{1\} \times [1, 1 + 1/\gamma - \delta]$ (where δ is a small positive constant ($0 < \delta < 1/\gamma$)) of pairs (k, x), then it follows from Lemmas 3–6 that the image of this set under the map (28) is set $\{1\} \times [1, 1 + a]$, where $a = o(1)$ at $\lambda \to +\infty$. Therefore, there exists at least one fixed point of the operator of translation along the trajectories and positive relaxation cycle of system (3) corresponds to this fixed point (if k_1 and x_1 fulfill (41) and function f satisfies (40), then in the case of positive coupling solution of system (3) does not change its sign). Similarly, there exists at least one negative relaxation cycle of system (3) in the case of positive coupling.

In Figure 3, there are examples of two coexisting relaxation cycles of system (3).

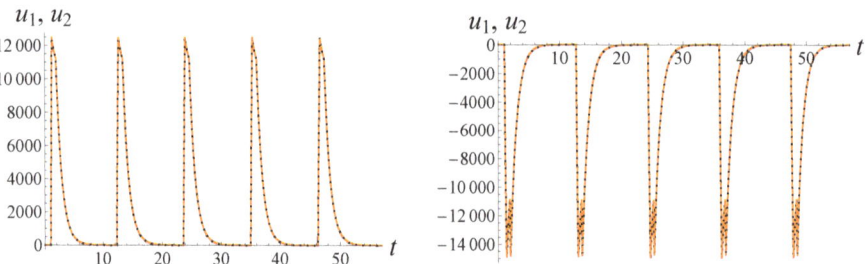

Figure 3. Two coexisting relaxation cycles of the system (3). Values of parameters: $T = 1$, $\gamma = 0.4$, $p = 1$, $\lambda = 10{,}000$. Black line—$u_1(t)$, orange dashed line—$u_2(t)$.

If $-\frac{1}{2} < \gamma < 0$, then it follows from (35) that $x_{n+1} = -k_{n+1} + o(1)$ at $\lambda \to +\infty$. It follows from Lemmas 3–6 that for all $(k_n, x_n) \in \{-1\} \times [1, 1 + 1/\gamma - \delta]$ and $(k_n, x_n) \in \{1\} \times [-1 - 1/\gamma + \delta, -1]$ Theorem 4 is true. Therefore, there exists at least one $q \in \mathbb{N}$, such that image of the set $\{-1\} \times [1, 1 + 1/\gamma - \delta]$ (or $\{1\} \times [-1 - 1/\gamma + \delta, -1]$) under the q-th iteration of map (35) belongs to the set $\{-1\} \times [1, 1 + 1/\gamma - \delta]$ (or $\{1\} \times [-1 - 1/\gamma + \delta, -1]$ respectively). Thus, in the case of $-\frac{1}{2} < \gamma < 0$, there exists at least one relaxation cycle.

Thus, the following statement holds.

Corollary 5. *Suppose conditions (4) and (40) are true. Then, in the case $\gamma > 0$, there exists at least two relaxation cycles of system (3) and in the case of $-\frac{1}{2} < \gamma < 0$ there exists at least one relaxation cycle of system (3).*

6. Dependence of Dynamics of System (3) on the Sign of Coupling

In this section, we show how asymptotics and difference $t_{n+1} - t_n$ (analog of period) of solutions of system (3) depends on the value γ in the case $\gamma > 0$ and in the case $-\frac{1}{2} < \gamma < 0$ (in this section below, we discuss only such solutions of system (3) for those assumptions of Theorem 1 or 2 fulfill).

First, consider the case $\gamma > 0$. From Formulas (29), (30), and (32), we obtain that components $u_1(t)$ and $u_2(t)$ have the same leading terms of asymptotics on the interval $t \in [t_2, +\infty)$ and that these leading terms of asymptotics do not depend on γ. Thus, from Formulas (9), (10), (12), (29), (30) and (32), we obtain that the leading term of asymptotics of solution of system (3) depends on γ only for $t \in [0, t_2]$ (see Figure 4). From Corollary 4, we get that in the case $\gamma > \frac{1}{2}$ difference $u_1(t) - u_2(t)$ has order $o(1)$ at $\lambda \to +\infty$ for all $t \geq t_2$, so we may say that in the case $\gamma > \frac{1}{2}$ oscillators $u_1(t)$ and $u_2(t)$ "synchronize" (for smaller values of γ oscillators $u_1(t)$ and $u_2(t)$ may "synchronize", too, but in the case $\gamma > \frac{1}{2}$ they must "synchronize").

The leading term of asymptotics of the difference $t_{n+1} - t_n$ does not depend on γ, too.

Figure 4 illustrates dependence of solutions of system (3) on γ in the case $\gamma > 0$. There are solutions of system (3) with identical function F, parameters λ and T, and initial conditions for different parameters γ in Figure 4.

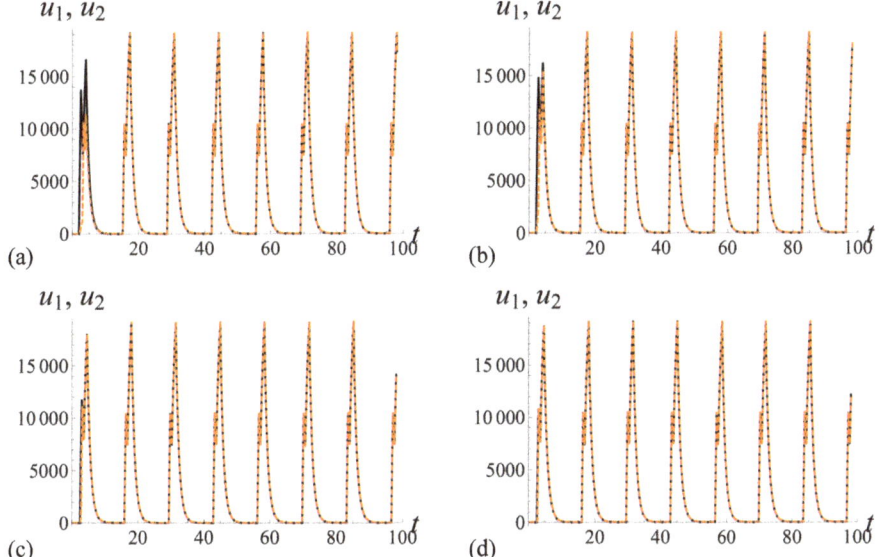

Figure 4. Solutions of system (3) for different values of parameter γ. Values of parameters: $T = 2$, $p = 1.5$, $\lambda = 1000$, $k = 1$, $x = 3$, (**a**) $\gamma = 0.2$; (**b**) $\gamma = 0.6$; (**c**) $\gamma = 1$; (**d**) $\gamma = 1.5$. Black line—$u_1(t)$, orange dashed line—$u_2(t)$.

Now, consider the case $-\frac{1}{2} < \gamma < 0$.

From (9), (10), (12), (36), (37), and (39), we get that asymptotics of solutions of system (3) depends crucially on the value of parameter γ for all $t \geq 0$ in the case $-\frac{1}{2} < \gamma < 0$ and that oscillators $u_1(t)$ and $u_2(t)$ are not close to each other (the leading terms of their asymptotics are different for all $t \geq t_2$).

It follows from (34) that difference $t_{n+1} - t_n$ increases with the decreasing of parameter γ (see Figure 5).

Thus, asymptotics and shape of solution and difference $t_{n+1} - t_n$ depend crucially on the value of γ in the case $-\frac{1}{2} < \gamma < 0$ (see Figure 5).

Figure 5 illustrates the dependence of solutions of system (3) on γ in the case $-\frac{1}{2} < \gamma < 0$. Solutions of system (3) with identical function F, parameters λ and T, and initial conditions for different parameters γ are presented in Figure 5.

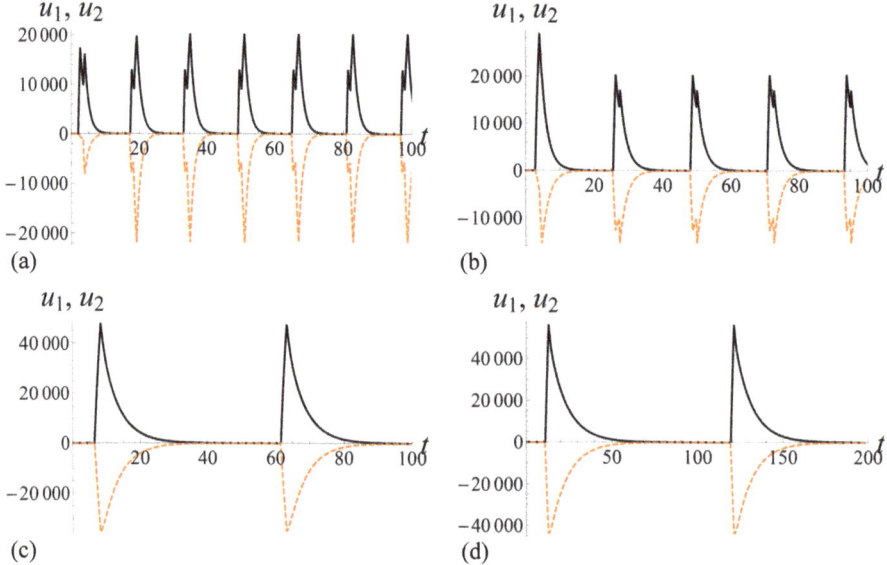

Figure 5. Solutions of system (3) for different values of parameter γ. Values of parameters: $T = 2$, $p = 1.5$, $\lambda = 1000$, $k = 1$, $x = -4$, (**a**) $\gamma = -0.1$; (**b**) $\gamma = -0.25$; (**c**) $\gamma = -0.4$; (**d**) $\gamma = -0.45$. Black line—$u_1(t)$, orange dashed line—$u_2(t)$.

7. Conclusions

In this paper, we have studied the nonlocal dynamics of a system of two coupled generators with delayed feedback and dependence of solutions on the value of coupling.

For a wide set of initial conditions from the phase space of system (3) using method of steps and special constructed finite dimensional map, we get asymptotics of relaxation solutions. We obtain relaxation cycles of system (3).

We prove that the dynamics of system (3) are qualitatively different in case $\gamma > 0$ and case $-\frac{1}{2} < \gamma < 0$: in the case $\gamma > 0$, there exists a moment of time t_2 after that both components of solution have the same leading term of asymptotics and this leading term does not depend on γ if $t > t_2$, generators $u_1(t)$ and $u_2(t)$ "synchronize" if $\gamma > \frac{1}{2}$; in the case of $-\frac{1}{2} < \gamma < 0$, the leading term of asymptotics and shape of solution depend on γ, oscillators $u_1(t)$ and $u_2(t)$ are not close to each other; the leading term of asymptotics of the value $t_{n+1} - t_n$ (this value serves us an analog of period) increase with decreasing of the value γ in the case $-\frac{1}{2} < \gamma < 0$ and remains unchanged with changing γ in the case $\gamma > 0$.

The method of research used in this paper is applicable for systems of higher dimensions (case of n identically diffusion coupled oscillators, where $n > 2$) and for systems of n ($n \geq 2$) coupled oscillators with other types of coupling.

Funding: Research funded by the Council on grants of the President of the Russian Federation (MK-1028.2020.1).

Conflicts of Interest: The author declares no conflict of interest. The funders had no role in the design of the study; in the collection, analyses, or interpretation of data; in the writing of the manuscript, or in the decision to publish the results.

References

1. Kislov, V.Y.; Dmitriev, A.S. *Problems of Modern Radio Engineering and Electronics*; Chapter Nonlinear Stochastization of Oscillations in Radio Engineering and Electronic Systems; Nauka Publishers: Nauka, Moscow, 1987; pp. 154–169.
2. Kilias, T.; Kelber, K.; Mogel, A.; Schwarz, W. Electronic chaos generators—Design and applications. *Int. J. Electron.* **1995**, *79*, 737–753. [CrossRef]
3. Kilias, T.; Mogel, A.; Schwarz, W. *Nonlinear Dynamics: New Theoretical and Applied Results*; Chapter Generation and Application of Broadband Signals Using Chaotic Electronic Systems; Akademie Verlag: Berlin, Germany, 1995; pp. 92–111.
4. An der Heiden, U.; Mackey, M.C. The dynamics of production and destruction: analytic insight into complex behavior. *J. Math. Biol.* **1982**, *16*, 75–101. [CrossRef]
5. An der Heiden, U.; Walther, H.O. Existence of chaos in control systems with delayed feedback. *J. Differ. Equ.* **1983**, *47*, 273–295. [CrossRef]
6. Stoffer, D. Delay equations with rapidly oscillating stable periodic solutions. *J. Dyn. Differ. Equ.* **2008**, *20*, 201–238. [CrossRef]
7. Krisztin, T.; Vas, G. Large-amplitude periodic solutions for differential equations with delayed monotone positive feedback. *J. Dyn. Differ. Equ.* **2011**, *23*, 727–790. [CrossRef]
8. Lakshmanan, M.; Senthilkumar, D.V. *Dynamics of Nonlinear Time-Delay Systems*; Springer Science & Business Media: Berlin/Heidelberg, Germany, 2011. [CrossRef]
9. Krisztin, T.; Walther, H.O. Unique periodic orbits for delayed positive feedback and the global attractor. *J. Dyn. Differ. Equ.* **2001**, *13*, 1–57. [CrossRef]
10. Kaschenko, S.A. Asymptotics of Relaxational Oscillations for Systems of Difference-Differential Equations with Compactly Supported Nonlinearity. I. *Differ. Equ.* **1995**, *31*, 1275–1285.
11. Mallet-Paret, J.; Nussbaum, R.D. Global continuation and asymptotic behaviour for periodic solutions of a differential-delay equation. *Ann. Mat. Pura Appl.* **1986**, *145*, 33–128. [CrossRef]
12. Ivanov, A.F.; Sharkovsky, A.N. *Dynamics Reported*; Chapter Oscillations in Singularly Perturbed Delay Equations; Springer: Berlin/Heidelberg, Germany, 1992; pp. 164–224. [CrossRef]
13. Kashchenko, I.; Kaschenko, S. Normal and quasinormal forms for systems of difference and differential-difference equations. *Commun. Nonlinear Sci. Numer. Simul.* **2016**, *38*, 243–256. [CrossRef]
14. Kashchenko, A.A. Relaxation Cycles in a Model of Two Weakly Coupled Oscillators with Sign-Changing Delayed Feedback. *Theor. Math. Phys.* **2020**, *202*, 381–389. [CrossRef]
15. Kashchenko, A.A. Relaxation modes of a system of diffusion coupled oscillators with delay. *Commun. Nonlinear Sci. Numer. Simul.* **2020**, *93*, 105488. [CrossRef]

Publisher's Note: MDPI stays neutral with regard to jurisdictional claims in published maps and institutional affiliations.

© 2020 by the author. Licensee MDPI, Basel, Switzerland. This article is an open access article distributed under the terms and conditions of the Creative Commons Attribution (CC BY) license (http://creativecommons.org/licenses/by/4.0/).

Article

A Class of Fractional Degenerate Evolution Equations with Delay

Amar Debbouche [1,*] **and Vladimir E. Fedorov** [2,3,4]

1. Department of Mathematics, Guelma University, Guelma 24000, Algeria
2. Department of Mathematical Analysis, Chelyabinsk State University, 129 Kashirin Brothers St., Chelyabinsk 454001, Russia; kar@csu.ru
3. Laboratory of Functional Materials, South Ural State University (National Research University), Lenin Av. 76, Chelyabinsk 454080, Russia
4. Department of Differential Equations, N.N. Krasovskii Institute of Mathematics and Mechanics of the Ural Branch of the Russian Academy of Sciences, 16 S. Kovalevskaya St., Yekaterinburg 620108, Russia
* Correspondence: amar_debbouche@yahoo.fr

Received: 31 August 2020; Accepted: 28 September 2020; Published: 3 October 2020

Abstract: We establish a class of degenerate fractional differential equations involving delay arguments in Banach spaces. The system endowed by a given background and the generalized Showalter–Sidorov conditions which are natural for degenerate type equations. We prove the results of local unique solvability by using, mainly, the method of contraction mappings. The obtained theory via its abstract results is applied to the research of initial-boundary value problems for both Scott–Blair and modified Sobolev systems of equations with delays.

Keywords: Gerasimov–Caputo fractional derivative; differential equation with delay; degenerate evolution equation; fixed point theorem

---

1. Introduction

During the last decades, fractional differential equations and their potential applications have gained a lot of importance, mainly because fractional calculus has become a powerful tool with more accurate and successful results when modeling several complex phenomena in numerous seemingly diverse and widespread fields of science and engineering [1]. It was found that various, especially interdisciplinary applications, can be elegantly modeled with the help of fractional derivatives which provide an excellent instrument for the description of memory and hereditary properties of various materials and processes [2,3]. Advanced analysis and numerical simulations of several fractional-order systems have been shown to be very interesting, producing more useful results in applied sciences [4,5].

Delay differential equations are a type of equations in which the derivative of the unknown function at a certain time is given in terms of the values of the function at previous times. It arises in many biological and physical applications, and it often forces us to consider variable or state-dependent delays [6–8]. Integer or fractional-order degenerate differential equations, i.e., evolution equations not solved with respect to the highest order derivative, are often used to describe various processes in science and engineering: in [9,10] certain classes of the time-fractional order partial differential equations with polynomials differential with respect to the spatial variables elliptic self-adjoint operator, which contain some equations from hydrodynamics and the filtration theory, are studied. In [11] approximate controllability issues for such models are investigated; the unique solvability of similar equations with distributed order time derivatives are researched in [12].

In applications, fractional-order degenerate evolution equations with a delay are often successful. Such kinds of equations with a degenerate operator at the highest-order fractional derivative describe the dynamics of some fractional models of viscoelastic fluids (see the application in the last section of this work). There are very few papers dealing with essentially degenerate fractional-order equations with delay. Motivated by this fact, the purpose of this work is a step towards eliminating this gap.

We are concerned with the following fractional differential equations with delay

$$LD_t^\alpha x(t) = Mx(t) + \int_{-r}^{0} \mathcal{K}(s)x(t+s)ds + g(t), \quad t \in [0,T], \qquad (1)$$

where $\mathcal{X}, \mathcal{Y}$ are Banach spaces, $L, M : \mathcal{X} \to \mathcal{Y}$ are linear operators, L is continuous, $\ker L \neq \{0\}$ (for this reason such equations are called Sobolev type equations [13,14], or degenerate [15]), operator M is closed and densely defined in $\mathcal{X}$, D_t^α is the Gerasimov-Caputo derivative of the order $\alpha \in (m-1, m]$, $m \in \mathbb{N}$. Equation (1) is endowed by a given background

$$Px(t) = h(t), \quad t \in [-r, 0], \qquad (2)$$

and by the generalized Showalter–Sidorov conditions

$$(Px)^{(k)}(0) = x_k, \; k = 0, 1, \ldots, m-1, \qquad (3)$$

which are natural for degenerate evolution equations. Here, P is a projector along the degeneration space of the homogeneous equation $LD_t^\alpha x(t) = Mx(t)$, it will be defined below. By the contraction mappings method, the local unique solvability of problems (1)–(3) is established.

Degenerate first-order evolution equations in Banach spaces were studied in [16,17] under various conditions on the operators L, M and on the delay term. The unique solvability results for problems (1) and (2) with a strongly (L,p)-radial operator M, $g \equiv 0$ at $\alpha = 1$ were obtained in [18]. Here we use a similar approach, which is adapted to the case of a fractional derivative. The second section contains the preliminary results which are needed for supporting our results, in particular, the theorem on unique solvability of the Cauchy problem to the inhomogeneous linear Equation (1) with $\mathcal{K} \equiv 0$. In the third section, we obtain the proof of the main result by means of the Banach fixed point theorem. The fourth and fifth sections demonstrate the applications of the obtained abstract results to the study of the unique solvability of initial-boundary value problems for time-fractional systems of partial differential equations with delay.

2. Solvability of Degenerate Inhomogeneous Equation

Let for $\delta > 0$, $t > 0$ $g_\delta(t) := \Gamma(\delta)^{-1} t^{\delta - 1}$, $J_t^\delta h(t) := \int_0^t g_\delta(t-s)h(s)ds$, $m - 1 < \alpha \leq m \in \mathbb{N}$, D_t^m is the usual derivative of the order $m \in \mathbb{N}$, J_t^0 be the identical operator. The Gerasimov-Caputo derivative of a function h is defined as

$$D_t^\alpha h(t) = D_t^m J_t^{m-\alpha} \left(h(t) - \sum_{k=0}^{m-1} h^{(k)}(0) g_{k+1}(t) \right).$$

Lemma 1. *Ref. [19]. Let $\mathcal{Z}$ be a Banach space, $l - 1 < \beta \leq l \in \mathbb{N}$, $t > 0$. Then*

$$\exists C_\beta > 0 \quad \forall h \in C^l([0,t];\mathcal{Z}) \quad \|D_t^\beta h\|_{C([0,t];\mathcal{Z})} \leq C_\beta \|h\|_{C^l([0,t];\mathcal{Z})}.$$

For Banach spaces, $\mathcal{X}$ and $\mathcal{Y}$ denote as $\mathcal{L}(\mathcal{X};\mathcal{Y})$ the Banach space of all linear continuous operators, acting from $\mathcal{X}$ to $\mathcal{Y}$. Let $\mathcal{C}l(\mathcal{X};\mathcal{Y})$ be the set of all linear closed operators, densely defined in $\mathcal{X}$, with the image in $\mathcal{Y}$.

In further consideration, we will assume that $L \in \mathcal{L}(\mathcal{X};\mathcal{Y})$, $\ker L \neq \{0\}$, $M \in \mathcal{C}l(\mathcal{X};\mathcal{Y})$, D_M is the domain of M with the graph norm $\|\cdot\|_{D_M} := \|\cdot\|_{\mathcal{X}} + \|M\cdot\|_{\mathcal{Y}}$. Denote $\rho^L(M) := \{\mu \in \mathbb{C} :$

$(\mu L - M)^{-1} \in \mathcal{L}(\mathcal{Y}; \mathcal{X})\}$. An operator M is called (L, σ)-bounded, if a ball $B_a(0) := \{\mu \in \mathbb{C} : |\mu| < a\}$ with some $a > 0$ contains the set $\rho^L(M)$. If M is (L, σ)-bounded, we have the projections

$$P := \frac{1}{2\pi i} \int_{|\mu|=a} (\mu L - M)^{-1} L \, d\mu \in \mathcal{L}(\mathcal{X}), \quad Q := \frac{1}{2\pi i} \int_{|\mu|=a} L(\mu L - M)^{-1} \, d\mu \in \mathcal{L}(\mathcal{Y})$$

(see [14] (pp. 89–90)). Set $\mathcal{X}^0 := \ker P$, $\mathcal{X}^1 := \operatorname{im} P$, $\mathcal{Y}^0 := \ker Q$, $\mathcal{Y}^1 := \operatorname{im} Q$. Denote by L_k (or M_k) the restriction of the operator L (or M) on $\mathcal{X}^k$ (or $D_{M_k} := D_M \cap \mathcal{X}^k$ respectively), $k = 0, 1$.

Theorem 1. *Ref. [14] (pp. 90–91). Let an operator M be (L, σ)-bounded. Then*
(i) $M_1 \in \mathcal{L}(\mathcal{X}^1; \mathcal{Y}^1)$, $M_0 \in \mathcal{C}l(\mathcal{X}^0; \mathcal{Y}^0)$, $L_k \in \mathcal{L}(\mathcal{X}^k; \mathcal{Y}^k)$, $k = 0, 1$;
(ii) there exist operators $M_0^{-1} \in \mathcal{L}(\mathcal{Y}^0; \mathcal{X}^0)$, $L_1^{-1} \in \mathcal{L}(\mathcal{Y}^1; \mathcal{X}^1)$.

Denote $\mathbb{N}_0 := \{0\} \cup \mathbb{N}$, $G := M_0^{-1} L_0$. For $p \in \mathbb{N}_0$ operator M is called (L, p)-bounded, if it is (L, σ)-bounded, $G^p \neq 0$, $G^{p+1} = 0$.

Consider the degenerate inhomogeneous equation

$$LD_t^\alpha x(t) = Mx(t) + f(t), \quad t \in [0, T]. \tag{4}$$

A solution of this equation is a function $x \in C([0, T]; D_M)$, such that $D_t^\alpha x \in C([0, T]; \mathcal{X})$ and equality (4) holds. A solution of the generalized Showalter–Sidorov problem

$$(Px)^{(k)}(0) = x_k, \quad k = 0, 1, \ldots, m - 1, \tag{5}$$

to Equation (4) is a solution of the equation, such that conditions (5) are true.

Denote by $E_{\alpha,\beta}(z) = \sum_{n=0}^{\infty} \frac{z^n}{\Gamma(\alpha n + \beta)}$ the Mittag-Leffler function.

Theorem 2. *Refs. [20,21]. Let $p \in \mathbb{N}_0$, an operator M be (L, p)-bounded, $Qf \in C([0, T]; \mathcal{Y})$, for all $l = 0, 1, \ldots, p$ there exist $(GD_t^\alpha)^l M_0^{-1}(I - Q)f$, $D_t^\alpha (GD_t^\alpha)^l M_0^{-1}(I - Q)f \in C([0, T]; \mathcal{X})$, $x_0, x_1, \ldots, x_{m-1} \in \mathcal{X}^1$. Then, problems (4) and (5) have a unique solution*

$$x_f(t) = \sum_{k=0}^{m-1} t^k E_{\alpha, k+1}(L_1^{-1} M_1 t^\alpha) P x_k + \int_0^t E_{\alpha,\alpha}(L_1^{-1} M_1 (t-s)^\alpha) L_1^{-1} Qf(s) ds - \sum_{l=0}^{p} (GD_t^\alpha)^l M_0^{-1}(I - Q)f(t). \tag{6}$$

Remark 1. *Due to Lemma 1 a function $f \in C^{m(p+1)}([0, T]; \mathcal{Y})$ satisfies the conditions of Theorem 2.*

3. Main Result

Consider the problem

$$Px(t) = h(t), \quad t \in [-r, 0], \quad (Px)^{(k)}(0) = x_k, \quad k = 0, 1, \ldots, m - 1, \tag{7}$$

for the degenerate fractional evolution equation with delay

$$LD_t^\alpha x(t) = Mx(t) + \int_{-r}^0 \mathcal{K}(s) x(t+s) ds + g(t), \quad t \in [0, T], \tag{8}$$

where $h(0) = x_0$, $\mathcal{K} : [-r, 0] \to \mathcal{L}(\mathcal{X}; \mathcal{Y})$, $g : [0, T] \to \mathcal{Y}$.

A function $x \in C([0, T]; D_M) \cap C([-r, T]; \mathcal{X})$ is called a solution of problems (7) and (8), if $D_t^\alpha x \in C([0, T]; \mathcal{X})$, it satisfies Equalities (7) and (8).

Theorem 3. Let $p \in \mathbb{N}_0$, an operator M be (L, p)-bounded, $h \in C([-r, 0]; \mathcal{X}^1)$, $x_k \in \mathcal{X}^1$, $k = 0, 1, \ldots, m-1$, $h(0) = x_0$, $T_0 > 0$, $g \in C^{m(p+1)}([0, T_0]; \mathcal{Y})$, $\mathcal{K} \in C^{m(p+1)}([-r, 0]; \mathcal{L}(\mathcal{X}; \mathcal{Y}))$, $\mathcal{K}^{(n)}(-r) = \mathcal{K}^{(n)}(0) = 0$ at $n = 0, 1, \ldots, m(p+1) - 1$. Then there exists $T \in (0, T_0)$, such that problems (7) and (8) have a unique solution.

Proof. Fix $T > 0$ and consider on the segment $[0, T]$ Equation (4) with some $f \in C^{m(p+1)}([0, T]; \mathcal{Y})$. Due to Theorem 2 and Remark 1 we have the solution x_f of problems (4) and (5) with the given x_k, $k = 0, 1, \ldots, m-1$. For brevity denote $X_\beta(t) := E_{\alpha, \beta}(L_1^{-1} M_1 t^\alpha) P$, $\beta > 0$, put at $t \in [-r, 0)$ $x_f(t) = h(t) + h_0(t)$ with some $h_0 \in C([-r, 0]; \mathcal{X}^0)$ and define the operator

$$[\Phi f](t) := \int_{-r}^{0} \mathcal{K}(s) x_f(t+s) ds + g(t) = \int_{-t}^{0} \mathcal{K}(s) \sum_{k=0}^{m-1}(t+s)^k X_{k+1}(t+s) x_k ds +$$

$$+ \int_{-t}^{0} \mathcal{K}(s) \int_{0}^{t+s} X_\alpha(t+s-\tau) L_1^{-1} Q f(\tau) d\tau ds - \int_{-t}^{0} \mathcal{K}(s) \sum_{l=0}^{p} (G D_t^\alpha)^l M_0^{-1} (I - Q) f(t+s) ds +$$

$$+ \int_{t-r}^{0} \mathcal{K}(s-t)(h(s) + h_0(s)) ds + g(t), \quad t \in [0, r),$$

$$[\Phi f](t) := \int_{-r}^{0} \mathcal{K}(s) x_f(t+s) ds + g(t) = \int_{-r}^{0} \mathcal{K}(s) \sum_{k=0}^{m-1}(t+s)^k X_{k+1}(t+s) x_k ds +$$

$$+ \int_{-r}^{0} \mathcal{K}(s) \int_{0}^{t+s} X_\alpha(t+s-\tau) L_1^{-1} Q f(\tau) d\tau ds - \int_{-r}^{0} \mathcal{K}(s) \sum_{l=0}^{p} (D_t^\alpha G)^l M_0^{-1} (I - Q) f(t+s) ds + g(t), \quad t \in [r, T],$$

By induction, we can prove that at $t \in [0, T]$, $n = 0, 1, \ldots, m(p+1)$

$$[\Phi f]^{(n)}(t) = \frac{d^n}{dt^n} \int_{t-r}^{t} \mathcal{K}(\tau - t) x_f(\tau) d\tau + g^{(n)}(t) = (-1)^n \int_{-r}^{0} \mathcal{K}^{(n)}(s) x_f(t+s) ds + g^{(n)}(t), \quad (9)$$

since $\mathcal{K}^{(n)}(-r) = \mathcal{K}^{(n)}(0) = 0$ at $n = 0, 1, \ldots, m(p+1) - 1$. Therefore, for every f from the Banach space $C^{m(p+1)}([0, T]; \mathcal{Y})$ with the standard norm $\|\cdot\|_{m(p+1)}$ we have $\Phi f \in C^{m(p+1)}([0, T]; \mathcal{Y})$.

Let $t_r := \min\{t, r\}$. For $f_1, f_2 \in C^{m(p+1)}([0, T]; \mathcal{Y})$, $t \in [0, T]$, $n = 0, 1, \ldots, m(p+1)$, due to (9)

$$\frac{d^n}{dt^n}([\Phi f_1](t) - [\Phi f_2](t)) = (-1)^n \int_{-t_r}^{0} \mathcal{K}^{(n)}(s) \int_{0}^{t+s} X_\alpha(t+s-\tau) L_1^{-1} Q(f_1(\tau) - f_2(\tau)) d\tau ds -$$

$$- (-1)^n \int_{-t_r}^{0} \mathcal{K}^{(n)}(s) \sum_{l=0}^{p} (G D_t^\alpha)^l M_0^{-1} (I - Q)(f_1(t+s) - f_2(t+s)) ds,$$

therefore, using Theorem 1 and Lemma 1, we obtain

$$\|\Phi f_1 - \Phi f_1\|_{m(p+1)} \leq C_1 \int_{-t_r}^{0} (t+s) \sum_{n=0}^{m(p+1)} \|\mathcal{K}^{(n)}(s)\|_{\mathcal{L}(\mathcal{X}; \mathcal{Y})} ds \|f_1 - f_2\|_0 +$$

$$+ C_2 \int_{-t_r}^{0} \sum_{n=0}^{m(p+1)} \|\mathcal{K}^{(n)}(s)\|_{\mathcal{L}(\mathcal{X}; \mathcal{Y})} ds \|f_1 - f_2\|_{mp} \leq C F(T) \|f_1 - f_2\|_{m(p+1)},$$

where, for the monotonously non-decreasing non-negative function

$$F(t) := \int_{-t_r}^{0} \sum_{n=0}^{m(p+1)} \|\mathcal{K}^{(n)}(s)\|_{\mathcal{L}(\mathcal{X};\mathcal{Y})} ds$$

we have $F(t) \to 0$ as $t \to 0+$. So, the inequality $\|\Phi f_1 - \Phi f_1\|_{m(p+1)} \le q\|f_1 - f_2\|_{m(p+1)}$ with some $q \in (0,1)$ is valid for sufficiently small $T > 0$ and there exists a unique fixed point f_0 of the operator Φ in $C^{m(p+1)}([0,T];\mathcal{Y})$. Therefore,

$$LD_t^\alpha x_{f_0}(t) - M x_{f_0}(t) = f_0(t) = [\Phi f_0](t) = \int_{-r}^{0} \mathcal{K}(s) x_{f_0}(t+s) ds + g(t),$$

and the function x_{f_0}, which is defined as in the beginning of this proof, is a solution of problems (7) and (8).

Note that the choice of function h_0 does not affect the proof, hence, we can choose $h_0(t) \equiv (I - P) x_{f_0}(0)$, then the obtained x_{f_0} is continuous on $[-r, T]$.

Let there exist two solutions x_1, x_2 of the problem, denoted as $f_i(t) = \int_{-r}^{0} \mathcal{K}(s) x_i(t+s) ds$, $i = 1,2$. As before, we have $f_i \in C^{m(p+1)}([0,T];\mathcal{Y})$ and $LD_t^\alpha x_i - M x_i = f_i$, hence, by the construction $\Phi f_i = f_i$, $i = 1,2$. Thus, Φ has two fixed points, it is a contradiction. Consequently, $f_1 \equiv f_2$, for $y := x_1 - x_2$ we have $LD_t^\alpha y - My = 0$, $y^{(k)}(0) = 0$, $k = 0, 1, \ldots, m-1$, therefore, $y \equiv 0$ due to Theorem 2. So, the solution of problems (7) and (8) is unique. □

4. A Scott–Blair Type System

Consider the problem

$$\frac{\partial^k v}{\partial t^k}(x, 0) = z_k(x), \quad x \in \Omega, \quad k = 0, 1, \ldots, m-1, \tag{10}$$

$$v(x, t) = h(x, t), \quad x \in \Omega, \quad t \in [-r, 0], \tag{11}$$

$$v(x, t) = 0, \quad (x, t) \in \partial\Omega \times [0, T], \tag{12}$$

$$(1 - \chi\Delta) D_t^\alpha v(x, t) = -(\tilde{v} \cdot \nabla) v(x, t) - (v \cdot \nabla) \tilde{v}(x, t) - r(x, t) +$$

$$+ \int_{-r}^{0} (K_1(s) v(t+s) + K_2(s) r(t+s)) ds, \quad (x, t) \in \Omega \times [0, T], \tag{13}$$

$$\nabla \cdot v(x, t) = 0, \quad (x, t) \in \Omega \times [0, T], \tag{14}$$

where $\Omega \subset \mathbb{R}^n$ is a bounded region with a smooth boundary $\partial\Omega$, $\chi \in \mathbb{R}$, $\tilde{v}$ is a given function. Function of the fluid velocity $v = (v_1, v_2, \ldots, v_n)$ and of the pressure gradient $r = (r_1, r_2, \ldots, r_n) = \nabla p$ are unknown.

This system without delay can be obtained, if the dynamics of a Scott–Blair medium [22] are described by using a fractional derivative of the same order as in the rheological relation for this medium, with subsequent linearization.

Let $\mathbb{L}_2 := (L_2(\Omega))^n$, $\mathbb{H}^1 := (W_2^1(\Omega))^n$, $\mathbb{H}^2 := (W_2^2(\Omega))^n$. The closure of $\{v \in (C_0^\infty(\Omega))^n : \nabla \cdot v = 0\}$ in the space $\mathbb{L}_2$ will be denoted by $\mathbb{H}_\sigma$, and in the space $\mathbb{H}^1$ it will be $\mathbb{H}_\sigma^1$. We have the decomposition $\mathbb{L}_2 = \mathbb{H}_\sigma \oplus \mathbb{H}_\pi$, where $\mathbb{H}_\pi$ is the orthogonal complement for $\mathbb{H}_\sigma$. Denote by $\Pi : \mathbb{L}_2 \to \mathbb{H}_\pi$ the corresponding to this decomposition orthoprojector, $\Sigma = I - \Pi$, $\mathbb{H}_\sigma^2 = \mathbb{H}_\sigma^1 \cap \mathbb{H}^2$.

The operator $A := \Sigma\Delta$ with the domain $\mathbb{H}_\sigma^2$ in the space $\mathbb{H}_\sigma$ has a real, negative, discrete spectrum with a finite multiplicity, condensing at $-\infty$ [23].

At $\tilde{v} \in \mathbb{H}^1$ by the formula $Dw = -(\tilde{v} \cdot \nabla)w - (w \cdot \nabla)\tilde{v}$ operator $D \in \mathcal{L}(\mathbb{H}^2_\sigma; \mathbb{L}_2)$ is defined.
Put

$$\mathcal{X} = \mathbb{H}^2_\sigma \times \mathbb{H}_\pi, \quad \mathcal{Y} = \mathbb{L}_2 = \mathbb{H}_\sigma \times \mathbb{H}_\pi, \tag{15}$$

$$L = \begin{pmatrix} I - \chi A & \mathbb{O} \\ -\chi \Pi \Delta & \mathbb{O} \end{pmatrix} \in \mathcal{L}(\mathcal{X}; \mathcal{Y}), \quad M = \begin{pmatrix} \Sigma D & \mathbb{O} \\ \Pi D & -I \end{pmatrix} \in \mathcal{L}(\mathcal{X}; \mathcal{Y}). \tag{16}$$

By the choice of the space $\mathcal{X}$ we take into account Equation (14) and condition (12). The function $r(\cdot, t)$ is a gradient, since it belongs to the space $\mathbb{H}_\pi$ at $t \geq 0$.

Lemma 2. *Ref. [24]. Let $\chi \neq 0$, $\chi^{-1} \notin \sigma(A)$, the spaces $\mathcal{X}$ and $\mathcal{Y}$ and the operators L and M be defined by (15) and (16) respectively. Then the operator M is $(L,0)$-bounded and the projectors have the form*

$$P = \begin{pmatrix} I & \mathbb{O} \\ \chi \Pi \Delta (I - \chi A)^{-1} \Sigma D + \Pi D & \mathbb{O} \end{pmatrix}, \quad Q = \begin{pmatrix} I & \mathbb{O} \\ -\chi \Pi \Delta (I - \chi A)^{-1} & \mathbb{O} \end{pmatrix}.$$

The form of the projectors P and Q implies that $\mathcal{X}^0 = \{0\} \times \mathbb{H}_\pi$, $\mathcal{X}^1 = \{(w_1, w_2) \in \mathbb{H}^2_\sigma \times \mathbb{H}_\pi : w_2 = (\chi \Pi \Delta (I - \chi A)^{-1} \Sigma D + \Pi D)w_1\}$, $\mathcal{Y}^0 = \{0\} \times \mathbb{H}_\pi$, $\mathcal{Y}^1 = \{(w_1, w_2) \in \mathbb{H}_\sigma \times \mathbb{H}_\pi : w_2 = -\chi \Pi \Delta (I - \chi A)^{-1} w_1\}$.

Theorem 4. *Let $h \in C([-r,0]; \mathbb{H}^2_\sigma)$, $z_k \in \mathbb{H}^2_\sigma$, $k = 0,1,\ldots,m-1$, $h(\cdot,0) = z_0(\cdot)$, $\mathcal{K}_i \in C^m([-r,0]; \mathbb{R})$, $\mathcal{K}_i^{(n)}(-r) = \mathcal{K}_i^{(n)}(0) = 0$ at $n = 0,1,\ldots,m-1$, $i = 1,2$. Then there exists $T > 0$, such that problems (10)–(14) have a unique solution.*

Proof. Due to Lemma 2 and Theorem 3 at $p = 0$, $g \equiv 0$ we obtain the required statement. □

5. A Modified Sobolev System

Consider another problem

$$\frac{\partial^k v}{\partial t^k}(x,0) = z_k(x), \quad x \in \Omega, \quad k = 0,1,\ldots,m-1, \tag{17}$$

$$v(x,t) = h(x,t), \quad x \in \Omega, \quad t \in [-r,0], \tag{18}$$

$$v_n(x,t) := \sum_{i=1}^3 v_i(x,t) n_i(x) = 0, \quad (x,t) \in \partial\Omega \times [0,T], \tag{19}$$

$$D_t^\alpha v(x,t) = [v(x,t), \overline{\omega}] - r(x,t) + \int_{-r}^0 (K_1(s) v(t+s) + K_2(s) r(t+s)) ds, \quad (x,t) \in \Omega \times [0,T], \tag{20}$$

$$\nabla \cdot v(x,t) = 0, \quad (x,t) \in \Omega \times [0,T], \tag{21}$$

where $\Omega \subset \mathbb{R}^3$, is a bounded region with a smooth boundary $\partial\Omega$, $\overline{\omega} \in \mathbb{R}^3$.

Such a system without delay and at $\alpha = 1$ describes the dynamics of small internal movements of a stratified fluid in an equilibrium state [25].

Following the approach of S.L. Sobolev [25], we use the generalized statement of the problem (17)–(21), replacing incompressibility Equation (21) and boundary condition (19) with the equation

$$\Pi v(\cdot, t) = 0, \quad t \in [0, T], \tag{22}$$

where Π is the same orthoprojector as in the previous section. Indeed, the set $\{\nabla\varphi : \varphi \in C^\infty(\Omega)\}$ is dense in the subspace $\mathbb{H}_\pi$ and the integral identity

$$\int_\Omega \langle v, \nabla\varphi\rangle_{\mathbb{R}^3} = \int_{\partial\Omega} v_n\varphi ds - \int_\Omega (\nabla\cdot v)\varphi dx$$

is true for all $\varphi \in C^\infty(\Omega)$, $v \in \mathbb{H}^1$, hence, for every $v \in \mathbb{H}^1$ the satisfaction of conditions (19), (21) is equivalent to the inclusion $v \in \mathbb{H}_\sigma$. Rejecting the restriction $\mathbb{H}^1$ we obtain condition (22).

Define by $Bw = [w, \overline{w}]$ at a fixed $\overline{w} \in \mathbb{R}^3$ the linear operator $B \in \mathcal{L}(\mathbb{L}_2; \mathbb{L}_2)$. Put $\mathcal{X} = \mathcal{Y} = \mathbb{L}_2 = \mathbb{H}_\sigma \times \mathbb{H}_\pi$,

$$L = \begin{pmatrix} I & \mathbb{O} \\ \mathbb{O} & \mathbb{O} \end{pmatrix} \in \mathcal{L}(\mathcal{X}; \mathcal{Y}), \quad M = \begin{pmatrix} \Sigma B & \mathbb{O} \\ \Pi B & -I \end{pmatrix} \in \mathcal{L}(\mathcal{X}; \mathcal{Y}).$$

Then it can be shown directly (see [26]), that the operator M is $(L, 0)$-bounded and the projectors have the form

$$P = \begin{pmatrix} I & \mathbb{O} \\ \Pi B & \mathbb{O} \end{pmatrix}, \quad Q = \begin{pmatrix} I & \mathbb{O} \\ \mathbb{O} & \mathbb{O} \end{pmatrix}.$$

Therefore, $\mathcal{X}^0 = \{0\} \times \mathbb{H}_\pi$, $\mathcal{X}^1 = \{(w_1, w_2) \in \mathbb{H}^2_\sigma \times \mathbb{H}_\pi : w_2 = \Pi B w_1\}$, $\mathcal{Y}^0 = \{0\} \times \mathbb{H}_\pi$, $\mathcal{Y}^1 = \mathbb{H}_\sigma \times \{0\}$. As in the previous section Theorem 3 at $p = 0$, $g \equiv 0$ implies the next result.

Theorem 5. *Let $h \in C([-r, 0]; \mathbb{H}_\sigma)$, $z_k \in \mathbb{H}_\sigma$, $k = 0, 1, \ldots, m-1$, $h(\cdot, 0) = z_0(\cdot)$, $\mathcal{K}_i \in C^m([-r, 0]; \mathbb{R})$, $\mathcal{K}_i^{(n)}(-r) = \mathcal{K}_i^{(n)}(0) = 0$ at $n = 0, 1, \ldots, m-1$, $i = 1, 2$. Then there exists $T > 0$, such that problems (17), (18), (20), (22) have a unique solution.*

6. Conclusions

We studied the local unique solvability of the problem with the generalized Showalter–Sidorov conditions, which is associated by a given background for degenerate fractional evolution equations in Banach spaces with delay, including the Gerasimov–Caputo derivative and a relatively bounded pair of linear operators. The complexity of the studied problem is the simultaneous presence of a fractional derivative, a degenerate operator at it, and a delay argument in the equation. The obtained result shows that by the methods of the theory of resolving families of operators for degenerate evolution equations, this complex problem can be solved. Abstract results can be used for investigating problems for partial differential equations, demonstrated on a problem for Scott–Blair and modified Sobolev systems of equations with delays.

Author Contributions: Conceptualization, A.D.; methodology, V.E.F.; validation, A.D.; formal analysis, V.E.F.; investigation, V.E.F.; writing—original draft preparation, A.D., V.E.F.; writing—review and editing, A.D., V.E.F.; supervision, A.D.; project administration, A.D. All authors have read and agreed to the published version of the manuscript.

Funding: The work of the second author is done in the framework of the Ural Mathematical Center, it is also supported by Russian Foundation of Basic Research, project 19-41-450001, by Act 211 of Government of the Russian Federation, contract 02.A03.21.0011.

Conflicts of Interest: The authors declare no conflict of interest. The funders had no role in the design of the study; in the collection, analyses, or interpretation of data; in the writing of the manuscript, or in the decision to publish the results.

References

1. Samko, S.G.; Kilbas, A.A.; Marichev O.I. *Fractional Integrals and Derivatives: Theory and Applications*; Gordon and Breach: Amsterdam, The Netherlands, 1993.
2. Kilbas, A.A.; Srivastava, H.M.; Trujillo, J.J. *Theory and Applications of Fractional Differential Equations*; Elsevier Sci. Publ.: Amsterdam, The Netherlands; Boston, MA, USA; Heidelberg, Germany, 2006.
3. Podlubny, I. *Fractional Differential Equations*; Academic Press: San Diego, CA, USA, 1999.

4. Diethelm, K. *The Analysis of Fractional Differential Equations. An Application-Oriented Exposition Using Differential Operators of Caputo Type*; Springer: Berlin/Heidelberg, Germany, 2010.
5. Zhou, Y. *Fractional Evolution Equations and Inclusions: Analysis and Control*; Academic Press: San Diego, CA, USA, 2016.
6. Cesarano, C.; Pinelas, S.; Al-Showaikh, F.; Bazighifan, O. Asymptotic properties of solutions of fourth-order delay differential equations. *Symmetry* **2019**, *11*, 628. [CrossRef]
7. Cesarano, C.; Bazighifan, O. Oscillation of fourth-order functional differential equations with distributed delay. *Axioms* **2019**, *8*, 61. [CrossRef]
8. Fang, J.; Liu, C.; Simos, T.E.; Famelis, I.T. Neural Network Solution of Single-Delay Differential Equations. *Mediterr. J. Math.* **2020**, *17*, 30. [CrossRef]
9. Fedorov, V.E.; Debbouche, A. A class of degenerate fractional evolution systems in Banach spaces. *Differ. Equ.* **2013**, *49*, 1569–1576. [CrossRef]
10. Fedorov, V.E.; Romanova, E.A.; Debbouche, A. Analytic in a sector resolving families of operators for degenerate evolution equations of a fractional order. *J. Math. Sci.* **2018**, *228*, 380–394. [CrossRef]
11. Baleanu, D.; Fedorov, V.E.; Gordievskikh, D.M.; Taş, K. Approximate controllability of infinite-dimensional degenerate fractional order systems in the sectorial case. *Mathematics* **2019**, *7*, 735. [CrossRef]
12. Fedorov V.E. Generators of analytic resolving families for distributed order equations and perturbations. *Mathematics* **2020**, *8*, 1306. [CrossRef]
13. Demidenko, G.V.; Uspenskii, S.V. *Partial Differential Equations and Systems not Solvable with Respect to the Highest-Order Derivative*; Marcel Dekker, Inc.: New York, NY, USA; Basel, Switzerland, 2003.
14. Sviridyuk, G.A.; Fedorov, V.E. *Linear Sobolev Type Equations and Degenerate Semigroups of Operators*; VSP: Utrecht, The Netherlands; Boston, MA, USA, 2003.
15. Favini, A.; Yagi, A. *Degenerate Differential Equations in Banach Spaces*; Marcel Dekker, Inc.: New York, NY, USA; Basel, Switzerland; Hong Kong, China, 1999.
16. Fedorov, V.E.; Omelchenko, E.A. On solvability of some classes of Sobolev type equations with delay. *Funct. Differ. Equ.* **2011**, *18*, 187–199.
17. Fedorov, V.E.; Omel'chenko, E.A. Inhomogeneous degenerate Sobolev type equations with delay. *Sib. Math. J.* **2012**, *53*, 335–344. [CrossRef]
18. Fedorov, V.E.; Omel'chenko, E.A. Linear equations of the Sobolev type with integral delay operator. *Russ. Math.* **2014**, *58*, 60–69. [CrossRef]
19. Plekhanova, M.V.; Baybulatova, G.D. Semilinear equations in Banach spaces with lower fractional derivatives. In *Springer Proceedings in Mathematics & Statistics, Proceedings of the Nonlinear Analysis and Boundary Value Problems, Santiago de Compostela, Spain, 4–7 September 2019*; Springer: Cham, Switzerland, 2019; Volume 292, pp. 81–93.
20. Fedorov, V.E.; Gordievskikh, D.M.; Plekhanova, M.V. Equations in Banach spaces with a degenerate operator under a fractional derivative. *Differ. Equ.* **2015**, *51*, 1367–1375. [CrossRef]
21. Plekhanova, M.V. Nonlinear equations with degenerate operator at fractional Caputo derivative. *Math. Methods Appl. Sci.* **2016**, *2*, 58–71. [CrossRef]
22. Scott-Blair, G.M. *Survey of General and Applied Rheology*; Pitman: London, UK, 1949.
23. Ladyzhenskaya, O.A. *The Mathematical Theory of Viscous Incompressible Flow*; Gordon and Breach: New York, NY, USA, 1969.
24. Fedorov, V.E.; Plekhanova, M.V.; Nazhimov, R.R. Degenerate linear evolution equations with the Riemann—Liouville fractional derivative. *Sib. Math. J.* **2018**, *59*, 136–146. [CrossRef]
25. Sobolev, S.L. On a new problem of the mathematical physics. *News USSR Acad. Sci. Ser. Math.* **1954**, *18*, 3–50.
26. Urazaeva, A.V.; Fedorov, V.E. Prediction-control problem for some systems of equations of fluid dynamics. *Differ. Equ.* **2008**, *44*, 1147–1156. [CrossRef]

© 2020 by the authors. Licensee MDPI, Basel, Switzerland. This article is an open access article distributed under the terms and conditions of the Creative Commons Attribution (CC BY) license (http://creativecommons.org/licenses/by/4.0/).

Article

Theoretical Analysis (Convergence and Stability) of a Difference Approximation for Multiterm Time Fractional Convection Diffusion-Wave Equations with Delay

A. S. Hendy [1,2] and R. H. De Staelen [3,4,*]

1. Department of Computational Mathematics and Computer Science, Institute of Natural Sciences and Mathematics, Ural Federal University, 19 Mira St., 620002 Yekaterinburg, Russia; ahmed.hendy@fsc.bu.edu.eg
2. Department of Mathematics, Faculty of Science Benha University, Benha 13511, Egypt
3. Department of Electronics and Information Systems, Ghent University, 9000 Gent, Belgium
4. Beheer en Algemene Directie, Ghent University Hospital, C. Heymanslaan 10, 9000 Gent, Belgium
* Correspondence: rob.destaelen@ugent.be

Received: 31 August 2020; Accepted: 23 September 2020; Published: 03 October 2020

Abstract: In this paper, we introduce a high order numerical approximation method for convection diffusion wave equations armed with a multiterm time fractional Caputo operator and a nonlinear fixed time delay. A temporal second-order scheme which is behaving linearly is derived and analyzed for the problem under consideration based on a combination of the formula of $L_2 - 1_\sigma$ and the order reduction technique. By means of the discrete energy method, convergence and stability of the proposed compact difference scheme are estimated unconditionally. A numerical example is provided to illustrate the theoretical results.

Keywords: fractional convection diffusion-wave equations; compact difference scheme; nonlinear delay; spatial variable coefficients; convergence and stability

MSC: 65M06; 35K15; 35K55; 35K57

1. Introduction

Fractional derivatives and integrals have recently gained high interest in many fields of science. The ability of classifying and capturing the memory and hereditary properties of various materials and processes is an advantage of fractional derivatives over their integer counterparts, e.g., the modeling of anomalous diffusion by fractional differential equations gives more informative and interesting models [1]. For time-fractional differential equations, the memory feature implies that all previous information is needed to evaluate the time fractional derivative at the current time level. Accordingly, designing a numerical differentiation formula of good accuracy is as ever paramount, but especially hard. The approximation formulas based on the interpolation approximation, such as $L1$ [2] and $L_2 - 1_\sigma$ [3], are of significance to design numerical algorithms to solve time-fractional differential equations. Demonstrated applications in numerous seemingly diverse and widespread fields of physics, such as in porous and glassy materials, in percolation clusters over fractals to semi-conductors, polymers, random media, and beyond, like geophysical and biological systems or processes (e.g., [4,5]), can be effectively modeled by time-fractional diffusion-wave equations of different types. Here, we are seeking to design a compact difference scheme that behaves linearly to numerically solve the non-linear delayed multiterm time fractional convection diffusion-wave equation (dmfCDWEs) with spatial variable coefficients. More specifically, we consider

$$\sum_{r=0}^{m} p_r \frac{\partial^{\alpha_r} u(x,t)}{\partial t^{\alpha_r}} = \frac{\partial}{\partial x}\left(q_1(x)\frac{\partial u}{\partial x}\right) + q_2(x)\frac{\partial u}{\partial x} + f(u(x,t), u(x,t-s), x, t), \ 0 < t \leq T, \ 0 \leq x \leq L, \quad (1a)$$

with the following initial and boundary conditions

$$u(x,t) = d(x,t), \quad 0 \leq x \leq L, \quad t \in [-s, 0), \quad \frac{\partial u(x,0)}{\partial t} = \psi(x) = \lim_{t \to -0} \frac{\partial d(x,t)}{\partial t}, \quad (1b)$$

$$u(0,t) = \phi_0(t), \quad u(L,t) = \phi_L(t), \quad 0 < t \leq T, \quad (1c)$$

where $s > 0$ is a fixed delay parameter, $q_1(x)$ and $q_2(x)$ are functions chosen, respectively, to be sufficiently and arbitrary differentiable functions. The fractional derivative is defined in Caputo sense and the fractional orders $\{\alpha_r \mid 1 \leq r \leq m\}$ are specified in the manner $\{1 < \alpha_m \leq \alpha_{m-1} < \cdots < \alpha_0 = 2\}$. The existence and uniqueness of the global mild solutions for the problem of nonlinear fractional reaction–diffusion equations with delay and Caputo's fractional derivatives are addressed in [6].

This work can be considered to be an extension of our previously published work [7], in which we discussed a single term time fractional wave equation with spatial constant coefficients. The scheme was of $2 - \alpha$ order in time and fourth in space. Here, we treat the multiterm time fractional order case with spatial variable coefficients and seek to have temporal second order of convergence. Accordingly, we hinged on the proposed numerical formula in [8] to approximate the multiterm Caputo fractional derivatives of order $\alpha_r (0 < \alpha_r \leq 1)$ at the super-convergent point. The formula $L_2 - 1_\sigma$ can achieve at least second-order accuracy at this point. We rely on the compact operator proposed in [9] in order to attain a fourth order accuracy with respect to space in case of having spatial variable coefficients.

There are some results and findings available regarding the theoretical analysis and numerical computation of single term time fractional sub or super diffusion equations with delay. In [10], the authors introduced a satisfactory numerical method for time fractional diffusion equations with delay. In [11], a novel discrete Grönwall inequality is used to simplify the analysis of difference schemes for time-fractional multi-delayed sub-diffusion equations. Convergence and stability of a compact finite difference method for nonlinear time fractional reaction-diffusion equations with a fixed delay are proposed in [12] by the aid of a new discrete form of the fractional Grönwall inequalities. A numerical solution for a class of time fractional diffusion equations with delay is proposed in [13] that is based on a smooth difference approximation of specific L1 type. Additionally, there are many difference and spectral approaches proposed for multiterm or distributed order time fractional differential equations. An efficient spectral method that is based on Jacobi–Gauss–Radau collocation is applied in order to solve a system of multi-dimensional distributed-order generalized Schrödinger equations in [14]. A combined difference and Galerkin–Legendre spectral method in [15] is used to solve time fractional diffusion equations with nonlinear source term. A Legendre spectral-collocation method for the numerical solution of distributed-order fractional initial value problems is designed in [16]. In [17], the authors proposed a spectral τ-scheme to discretize the fractional diffusion equation with distributed-order fractional derivative in time and Dirichlet boundary conditions. The model solution is expanded in multi-dimensions in terms of Legendre polynomials and the discrete equations are obtained with the τ-method. The two-dimensional distributed-order time fractional cable equation is numerically solved based on the finite difference/spectral method, as clarified in [18]. Two classes of finite difference methods that are based on backward differential formula discretization in the temporal direction are proposed in [19] to efficiently solve the semilinear space fractional reaction–diffusion equation with time-delay. The coefficients of the problem are constants, a fractional centered difference approximation is employed for the space fractional derivative [20], and it gains a fourth order approximation in space due to the use of a specific compact operator [21].

The main purpose of our work is the manufacturing of a difference scheme for problems of the kind of (1). Until now, few works hace paid close attention to the multiterm fractional wave equation with variable coefficients and delay argument simultaneously. It is well known that it may be a more challenging task to solve the fractional partial differential equation with delay effectively, since the evolution of a fractional partial differential equation with delay at time t not only depends on its value at $t-s$, but also depends on all previous solutions due to the non-locality of the fractional operator. Higher order numerical schemes are extremely scarce and difficult in regard to the analysis and implementation for the variable coefficient multiterm time fractional convection-diffusion wave equation with delay. For a single temporal term for that kind of problem, we refer to [22]. For the multiterm problem, we design the difference scheme at a super-convergent point to gain a high order of convergence [8] which is more challenging than the single term problem at the level of theoretical analysis.

In order to simplify the problem, an exponential transformation technique [23] can be applied to system (1) to avoid the difficulties resulting from the spatial variable when constructing high order compact difference methods. That technique is used to eliminate the convection term. Assume that $\frac{q_1(x)}{q_2(x)}$ is integrable in the spatial interval $[0, L]$ and let $u(x,t) = \exp\left(\int_0^x \tilde{r}(s)ds\right) w(x,t)$, this yields

$$\frac{\partial^{\alpha_r} u}{\partial t^{\alpha_r}} = \exp\left(\int_0^x \tilde{r}(s)ds\right) \frac{\partial^{\alpha_r} w}{\partial t^{\alpha_r}}, \tag{2}$$

$$\frac{\partial u}{\partial x} = \exp\left(\int_0^x \tilde{r}(s)ds\right) \left[\tilde{r}(x)w(x,t) + \frac{\partial w}{\partial x}\right], \tag{3}$$

$$\frac{\partial^2 u}{\partial x^2} = \exp\left(\int_0^x \tilde{r}(s)ds\right) \left[\tilde{r}^2(x)w(x,t) + 2\tilde{r}(x)\frac{\partial w}{\partial x} + \tilde{r}'(x)w(x,t) + \frac{\partial^2 w}{\partial x^2}\right]. \tag{4}$$

When substituting (2)–(4) into (1), one directly obtains

$$\sum_{r=0}^{m} p_r \frac{\partial^{\alpha_r} w(x,t)}{\partial t^{\alpha_r}} = q_1(x)\frac{\partial^2 w}{\partial x^2} + [q_1'(x) + 2q_1(x)\tilde{r}(x) + q_2(x)]\frac{\partial w}{\partial x}$$

$$+ \left[q_1(x)\tilde{r}^2(x) + q_1(x)\tilde{r}'(x) + q_1'(x)\tilde{r}(x) + q_2(x)\tilde{r}(x)\right] w(x,t)$$

$$+ \exp\left(-\int_0^x \tilde{r}(s)ds\right) g(w(x,t), w(x,t-s), x, t),$$

where, $g(w(x,t), w(x,t-s), x, t) = f(u(x,t), u(x,t-s), x, t)$. By selecting $\tilde{r}(x) = -\frac{1}{2}\frac{q_2(x)}{q_1(x)}$, the system (1) is transformed in

$$\sum_{r=0}^{m} p_r \frac{\partial^{\alpha_r} w(x,t)}{\partial t^{\alpha_r}} = \frac{\partial}{\partial x}\left(q_1(x)\frac{\partial w}{\partial x}\right) + \tilde{f}(w(x,t), w(x,t-s), x, t), \quad 0 < t \leq T,\ 0 \leq x \leq L, \tag{5a}$$

with the following initial and boundary conditions

$$w(x,t) = \exp\left(-\int_0^x \tilde{r}(s)ds\right) d(x,t) := \tilde{d}(x,t), \quad t \in [-s, 0), \quad \frac{\partial w(x,0)}{\partial t} = \lim_{t\to -0}\frac{\partial \tilde{d}(x,t)}{\partial t} := \tilde{\psi}(x), \tag{5b}$$

$$w(0,t) = \phi_0(t), \quad w(L,t) = \exp\left(-\int_0^x \tilde{r}(s)ds\right) \phi_L(t) := \tilde{\phi}_L(t), \quad 0 < t \leq T, \tag{5c}$$

where

$$\tilde{f}(w(x,t), w(x,t-s), x, t) = \left[q_1(x)\tilde{r}^2 + q_1(x)\tilde{r}'(x) + q_1'(x)\tilde{r}(x) + q_2(x)\tilde{r}(x)\right] w(x,t)$$

$$+ \exp\left(-\int_0^x \tilde{r}(s)ds\right) g(w(x,t), w(x,t-s), x, t).$$

In order to transform (5) to a system with zero Dirichlet boundary conditions, we define $h(x,t) := \phi_0(t) + \frac{x}{L}(\tilde{\phi}_L(t) - \phi_0(t))$ and introduce the new function $v(x,t) = w(x,t) - h(x,t)$. Hence, we have

$$\sum_{r=0}^{m} p_r \frac{\partial^{\alpha_r} v(x,t)}{\partial t^{\alpha_r}} = \frac{\partial}{\partial x}\left(q_1(x)\frac{\partial v}{\partial x}\right) + \tilde{f}(v(x,t), v(x,t-s), x, t), \; 0 < t \leq T, \; 0 \leq x \leq L, \quad (6a)$$

with the following initial and boundary conditions

$$v(x,t) = \hat{r}(x,t), \quad 0 \leq x \leq L, \quad t \in [-s, 0], \quad \frac{\partial v(x,0)}{\partial t} = \hat{\psi}(x) = \lim_{t \to -0}\frac{\partial r(x,t)}{\partial t}, \quad (6b)$$

$$v(0,t) = v(L,t) = 0, \quad t > 0. \quad (6c)$$

According to the new transformed system (6), we analytically overcame the first degree of complexity by the elimination of the convection term. There are still two degrees of complexity to be numerically overcome; the multiterm fractional order on the one hand and the nonlinear delay on the other.

Throughout this work, we assume that the function $f(\mu, v, x, t)$ and the solution $v(x,t)$ of (6) are sufficiently smooth in the following sense:

- Assume that $v(x,t) \in C^{(6,4)}([0,L] \times [0,T])$,
- The partial derivatives $\tilde{f}_\mu(\mu, v, x, t)$ and $\tilde{f}_v(\mu, v, x, t)$ are continuous in the ϵ_0-neighborhood of the solution. Define

$$c_1 = \sup_{\substack{0<x<L, 0<t\leq T \\ |\epsilon_1|\leq \epsilon_0, |\epsilon_2|\leq \epsilon_0}} |\tilde{f}_\mu(v(x,t) + \epsilon_1, v(x,t-s) + \epsilon_2, x, t)|, \quad (7a)$$

$$c_2 = \max_{\substack{0<x<L, 0<t\leq T \\ |\epsilon_1|\leq \epsilon_0, |\epsilon_2|\leq \epsilon_0}} |\tilde{f}_v(u(x,t) + \epsilon_1, v(x,t-s) + \epsilon_2, x, t)|. \quad (7b)$$

The structure of this paper is arranged, as follows. First, we introduce a derivation of the compact difference scheme. Next, in the third section, convergence and stability for the compact difference scheme are carried out. Finally, the paper ends with a numerical illustration and a conclusion.

2. A Compact Difference Scheme

A linearized numerical method that combines the super-convergence approximation $L_2 - 1_\sigma$ with the order reduction method is derived. Some further notations are fixed before we continue. Take two positive integers M and n_0, let $h = \frac{L}{M}$, $\tau = \frac{s}{n_0}$ and denote $x_i = ih$ for $i = 0, \ldots, M$; $t_k = k\tau$ and $t_{k+\sigma} = (k+\sigma)\tau$, for $k = -n_0, \ldots, N$, where $N = \left\lfloor \frac{T}{\tau} \right\rfloor$. Using the points x_i in space and t_k in time, we cover the space-time domain by $\Omega_{h\tau} = \Omega_h \times \Omega_\tau$, where $\Omega_h = \{x_i \mid 0 \leq i \leq M\}$ and $\Omega_\tau = \{t_k \mid -n_0 \leq k \leq N\}$.

2.1. $L_2 - 1_\sigma$ Super-Convergence Scheme

Here, we give a preliminary for the Alikhanov scheme. Denote $\gamma_r = \alpha_r - 1$ ($0 \leq r \leq m$) and

$$\mathcal{F}(\sigma) = \sum_{r=0}^{m} \frac{p_r}{\Gamma(3-\gamma_r)}\sigma^{1-\gamma_r}\left[\sigma - (1-\frac{\gamma_r}{2})\right]\tau^{2-\gamma_r},$$

such that $0 < \gamma_m < \gamma_{m-1} < \cdots < \gamma_0 \leq 1$. Additionally, now we invoke the following lemmas from Alikhanov work.

Lemma 2.1 ([8]). *A unique positive root $\sigma^* \in [1 - \gamma_0/2, 1 - \gamma_m/2]$ exists for $\mathcal{F}(\sigma) = 0$.*

Lemma 2.2 ([8]). For $m = 0$, the root of $\mathcal{F}(\sigma) = 0$ is $1 - \gamma_0/2$. However, if $m \geq 1$, the root σ^* can be obtained by the Newton iteration method. The Newton iteration sequence $\{\sigma_k\}_{k=0}^{\infty}$ generated by $\sigma_0 = 1 - \gamma_m/2$ and $\sigma_{k+1} = \sigma_k - \frac{\mathcal{F}(\sigma_k)}{\mathcal{F}'(\sigma_k)}$ for $k = 0, 1, 2, \ldots$ is monotonically decreasing and convergent to σ^*.

For the sake of simplicity, let $\sigma = \sigma^*$ here and later. Define [3]

$$a_0 = \sigma^{1-\gamma}, \quad a_l^{\gamma} = (\sigma + l)^{1-\gamma} - (\sigma + l - 1)^{1-\gamma},$$

$$b_l^{\gamma} = \frac{1}{2-\gamma}\left[(l+\sigma)^{2-\gamma} - (l-1+\sigma)^{2-\gamma}\right] - \frac{1}{2}\left[(l+\sigma)^{1-\gamma} + (l-1+\sigma)^{1-\gamma}\right],$$

for each $l \in \mathbb{N}^+$. Next, we define $\{C_n^{(k+1,\gamma)}\}$, as follows

$$C_n^{(k+1,\gamma)} = \begin{cases} a_0, & n = 0, k = 0, \\ a_0^{\gamma} + b_1^{\gamma}, & n = 0, k \geq 1, \\ a_n^{\gamma} + b_{n+1}^{\gamma} - b_n^{\gamma}, & 1 \leq n \leq k-1, k \geq 1, \\ a_k^{\gamma} - b_k^{\gamma}, & n = k, k \geq 1. \end{cases} \quad (8)$$

Denote

$$\hat{C}_n^{(k+1)} = \sum_{r=0}^{m} p_r \frac{\tau^{-\gamma_r}}{\Gamma(2-\gamma_r)} C_n^{(k+1,\gamma_r)}, \quad \hat{b}_n = \sum_{r=0}^{m} p_r \frac{\tau^{-\gamma_r}}{\Gamma(2-\gamma_r)} b_n^{(\gamma_r)}, \quad n = 0, 1, \ldots, k.$$

The next two lemmas are devoted to the properties of the coefficients $\hat{C}_n^{(k)}$ and $\hat{b}_n$.

Lemma 2.3 ([8]). Given any non-negative integer m and positive constants $p_0, p_1, \ldots, p_m$, for any $\{\gamma_r \in (0,1] \mid r = 0, 1, \cdots, m\}$ it holds

$$\hat{C}_1^{(k+1)} > \hat{C}_2^{(k+1)} > \cdots > \hat{C}_{k-2}^{(k+1)} > \hat{C}_{k-1}^{(k+1)} > \sum_{r=0}^{m} p_r \frac{\tau^{-\gamma_r}}{\Gamma(2-\gamma_r)} \frac{1-\gamma_r}{2}(k-1+\sigma)^{-\gamma_r}.$$

In addition, there exists a $\tau_0 > 0$, such that $(2\sigma - 1)\hat{C}_0^{(k+1)} - \sigma \hat{C}_1^{(k+1)} > 0$, when $\tau \leq \tau_0, n = 2, 3, \cdots$, and $\hat{C}_0^{(k+1)} > \hat{C}_1^{(k+1)}$.

Lemma 2.4 ([24]). The sequences $\hat{C}_n^{(k)}$ and $\hat{b}_n$ satisfy

$$\hat{C}_n^{(k+1)} = \begin{cases} \hat{C}_n^{(k)}, & 0 \leq n \leq k-2, \\ \hat{C}_n^{(k)} + \hat{b}_{n+1}, & n = k-1. \end{cases} \quad (9)$$

Additionally, the following estimates hold:

$$\sum_{n=1}^{k} \hat{C}_n^{(k+1)} \leq \sum_{r=0}^{m} p_r \frac{3\tau^{-\gamma_r}}{2\Gamma(2-\gamma_r)}(k+\sigma)^{1-\gamma_r}$$

and

$$\sum_{n=1}^{k} \hat{b}_n \leq \sum_{r=0}^{m} p_r \frac{\gamma_r \tau^{-\gamma_r}}{2\Gamma(3-\gamma_r)}(k+\sigma)^{1-\gamma_r}.$$

Let $\mathcal{W}_h = \{w : \Omega_{h\tau} \to \mathbb{R} \mid w(x_i, t_k) = w_i^k; i = 0, 1, \ldots, M; k = -n_0, -n_0+1, \ldots, N\}$ be a grid function space on $\Omega_{h\tau}$ and also define $\mathring{\mathcal{W}}_h = \{w \in \mathcal{W}_h, w(x_0, \cdot) = w(x_M, \cdot) = 0\}$. Introduce the following notations

$$w_{i+1/2}^k = \frac{1}{2}\left[w_{i+1}^k + w_i^k\right], \quad \delta_x w_{i+1/2}^k = \frac{1}{h}\left[w_{i+1}^k - w_i^k\right], \quad \delta_t w_i^{\frac{1}{2}} = \frac{1}{\tau}\left[w_i^1 - w_i^0\right], \tag{10}$$

$$\delta_{\hat{x}} w_i^k = \frac{1}{2h}\left[w_{i+1}^k - w_{i-1}^k\right], \quad \delta_x^2 w_i^k = \frac{1}{h}\left[\delta_x w_{i+1/2}^k - \delta_x w_{i-1/2}^k\right], \tag{11}$$

$$w_i^{k+\sigma} = \sigma w_i^{k+1} + (1-\sigma) w_i^k, \quad \partial_{\hat{t}} w_i^k = \frac{1}{2\tau}\left[(2\sigma+1)w_i^{k+1} - 4\sigma w_i^k + (2\sigma-1)w_i^{k-1}\right]. \tag{12}$$

Moreover, denote $\varkappa_1 = (q_1')^2/q_1 - \frac{1}{2}q_1''$ and $\varkappa_2 = q_1 - \frac{h^2}{12}\varkappa_1$. The compact operator acting on the spatial variable is defined as [9],

$$\mathcal{A} w_i = \begin{cases} w_i + \frac{h^2}{12}\left[\delta_x^2 w_i - \delta_{\hat{x}}\left(\frac{q_1'}{q_1}w\right)_i\right], & i = 1, \ldots, M-1, \\ w_i, & i = 0, M. \end{cases}$$

Lemma 2.5 ([9]). *Let $g(x) \in C^6[0, L]$, such that $f(x) = \partial_x(q_1(x)g(x))$, and then it holds*

$$\mathcal{A} F_i = \delta_x(\varkappa_2(x)\delta_x G)_i + O(h^4),$$

where $F_i = f(x_i)$ and $G_i = g(x_i)$.

For any $v, w \in \mathcal{W}_h$, we define the following inner products

$$\langle v, w \rangle = h\left[\frac{1}{2}v_0 w_0 + \sum_{i=1}^{M-1} v_i w_i + \frac{1}{2}v_M w_M\right], \quad \langle v, w \rangle_1 = h\sum_{i=1}^{M-1}(\delta_x w_{i+1/2})(\delta_x v_{i+1/2}),$$

$$\langle v, w \rangle_{\varkappa_2} = h\sum_{i=1}^{M-1}\varkappa_2(x_{i+1/2})(\delta_x w_{i+1/2})(\delta_x v_{i+1/2}), \quad \langle w, v \rangle_{\mathcal{A}} = \langle w, v \rangle - \frac{h^2}{12}\langle w, v \rangle.$$

and their corresponding norms and semi-norms

$$\|w\|^2 = \langle w, w \rangle, \quad |w|_1^2 = \langle w, w \rangle_1, \quad \|w\|_{\mathcal{A}}^2 = \langle w, w \rangle_{\mathcal{A}}, \quad |w|_{1,\varkappa_2}^2 = \langle w, w \rangle_{\varkappa_2}, \quad \|w\|_\infty = \max_{0 \le i \le M}|w_i|.$$

Furthermore, assume that the coefficients satisfy

$$b_0 \le q_1(x) \le b_1, \quad b_2 \le \varkappa_2(x) \le b_3, \quad \left|\frac{q_1'}{q_1}\right| \le b_3, \tag{13}$$

where all b_i are positive constants.

Lemma 2.6 ([25]). *For any grid function $w \in \hat{\mathcal{W}}_h$, it holds that*

$$\|w\| \le \frac{L}{\sqrt{6}}|w|_1, \quad \|w\|_\infty \le \frac{\sqrt{L}}{2}|w|_1, \quad \sqrt{\frac{2}{3}}\|w\| \le \|w\|_{\mathcal{A}} \le \|w\|.$$

Lemma 2.7 ([26]). *For any grid function $w_1, w_2 \in \hat{\mathcal{W}}_h$, it holds that*

$$|\langle \mathcal{A}w, w \rangle| \ge \|w\|_{\mathcal{A}}^2 - \frac{c_3 h}{12}\|w\|^2, \quad |\langle \mathcal{A}w_1, w_2 \rangle| \le \frac{1}{2}\left[\|w_1\|_{\mathcal{A}}^2 + \|w_2\|_{\mathcal{A}}^2\right] + \frac{c_3 h}{24}\left[\|w_1\|^2 + \|w_2\|^2\right].$$

Lemma 2.8 ([8]). *For any $h(t) \in C^3[0, T]$ and $\gamma_r \in (0, 1]$, $0 \le r \le m$, such that $\gamma_0 > \gamma_1 > \cdots > \gamma_m$, the $L_2 - 1_\sigma$ formula has the following order of convergence*

$$\sum_{r=0}^{m} p_r \, {}_0D_t^{\gamma_r} h(t_{k+\sigma}) = \sum_{n=0}^{k} \left(\sum_{r=0}^{m} p_r \frac{\tau^{-\gamma_r}}{\Gamma(2-\gamma_r)} \hat{c}_n^{(k+1,\gamma_r)} \right) (h(t_{k-n+1}) - h(t_{k-n})) + \mathcal{O}(\tau^{3-\gamma_0}) \tag{14}$$

$$= \sum_{n=0}^{k} \hat{C}_{k-n}^{(k+1)} [h(t_{n+1}) - h(t_n)] + \mathcal{O}(\tau^{3-\gamma_0}). \tag{15}$$

Lemma 2.9 ([24]). *For any $h(t) \in C^3[0,T]$, we can obtain*

$$\partial_{\hat{t}} h(t_k) \cong \frac{1}{2\tau} [(2\sigma+1)h(t_{k+1}) - 4\sigma h(t_k) + (2\sigma-1)h(t_{k-1})] \tag{16}$$

$$= \frac{\partial h}{\partial t}(t_{k+\sigma}) + \mathcal{O}(\tau^2), \quad k \geq 1. \tag{17}$$

Lemma 2.10 ([24]). *Suppose that $\langle \cdot, \cdot \rangle_*$ is an inner product on $\hat{\mathcal{W}}_h$ and $\|\cdot\|_*$ is a norm deduced by the inner product. For any grid functions $w_0, w_1, \ldots, w_{k+1} \in \hat{\mathcal{W}}_h$, we have the following inequality*

$$\left\langle \sum_{n=0}^{k} \hat{C}_{k-n}^{(k+1)} \left[w^{n+1} - w^n \right], w^{k+\sigma} \right\rangle_* \geq \frac{1}{2} \sum_{n=0}^{k} \hat{C}_{k-n}^{(k+1)} \left[\left\| w^{n+1} \right\|_*^2 - \| w^n \|_*^2 \right], \tag{18}$$

$$\left\langle \partial_{\hat{t}} w^k, w^{k+\sigma} \right\rangle_* \geq \frac{1}{4\tau} \left(\mathcal{E}^{k+1} - \mathcal{E}^k \right), \tag{19}$$

where

$$\mathcal{E}^{k+1} = (2\sigma+1) \left\| w^{k+1} \right\|^2 - (2\sigma-1) \left\| w^k \right\|^2 + (2\sigma^2 + \sigma - 1) \left\| w^{k+1} - w^k \right\|^2, \quad k \geq 0.$$

Additionally, it holds that

$$\mathcal{E}^{k+1} \geq \frac{1}{\sigma} \left\| w^{k+1} \right\|^2, \quad k \geq 0.$$

Let us initiate by order reduction for system (6) by letting $\gamma_r = \alpha_r - 1$, $(0 \leq r \leq M)$ and $\mathcal{V}(x,t) = v_t(x,t)$, then

$$\frac{\partial^{\alpha_r} v(x,t)}{\partial t^{\alpha_r}} = \frac{\partial^{\gamma_r} \mathcal{V}(x,t)}{\partial t^{\gamma_r}},$$

and so

$$\partial_t \left(\frac{\partial}{\partial x} \left(q_1(x) \frac{\partial v}{\partial x} \right) \right) = \frac{\partial}{\partial x} \left(q_1(x) \frac{\partial \mathcal{V}}{\partial x} \right).$$

Subsequently, the equivalent system to (6) after order reduction can be formulated as

$$\sum_{r=0}^{m} p_r \frac{\partial^{\gamma_r} \mathcal{V}(x,t)}{\partial t^{\gamma_r}} = \frac{\partial}{\partial x} \left(q_1(x) \frac{\partial v}{\partial x} \right) + \tilde{f}(v(x,t), v(x, t-s), x, t), \quad 0 < t \leq T, \; 0 \leq x \leq L, \tag{20a}$$

$$\partial_t \left(\frac{\partial}{\partial x} \left(q_1(x) \frac{\partial v}{\partial x} \right) \right) = \frac{\partial}{\partial x} \left(q_1(x) \frac{\partial \mathcal{V}}{\partial x} \right), \tag{20b}$$

with the following initial and boundary conditions

$$v(x,t) = \tilde{r}(x,t), \quad 0 \leq x \leq L, \; t \in [-s, 0], \quad \mathcal{V}(x,0) = \hat{\varphi}(x), \quad 0 \leq x \leq L, \tag{20c}$$

$$v(0,t) = v(L,t) = 0, \quad 0 < t \leq T, \tag{20d}$$

$$\mathcal{V}(0,t) = \mathcal{V}(L,t) = 0, \quad 0 < t \leq T. \tag{20e}$$

2.2. Compact Difference Scheme Construction

Suppose that $v_t(x,t) = \mathcal{V}(x,t) \in C_{x,t}^{6,4}([0,T] \times [0,L])$, define the discretized functions

$$V_i^k = \mathcal{V}(x_i, t_k), \quad v_i^k = v(x_i, t_k), \; 0 \leq i \leq M, \; 0 \leq k \leq N.$$

Consider (20a) at $(x_i, t_{k+\sigma})$, and then we get

$$\sum_{r=0}^{m} p_r \frac{\partial^{\gamma_r} \mathcal{V}(x_i, t_{k+\sigma})}{\partial t^{\gamma_r}} = \frac{\partial}{\partial x}\left(q_1(x_i)\frac{\partial v(x_i, t_{k+\sigma})}{\partial x}\right) + \tilde{f}_i^{k+\sigma} \quad 0 < i < M, \ 0 \le k \le N-1, \quad (21)$$

such that

$$\tilde{f}_i^k = \tilde{f}(v(x_i, t_k), v(x_i, t_k - s), x_i, t_k).$$

From Lemma 2.10, we conclude that

$$\sum_{r=0}^{m} p_r \frac{\partial^{\gamma_r} \mathcal{V}(x_i, t_{k+\sigma})}{\partial t^{\gamma_r}} = \sum_{n=0}^{k} \hat{C}_{k-n}^{k+1}\left(V_i^{n+1} - V_i^n\right) + O\left(\tau^{3-\gamma_0}\right), \quad 0 < i < M, \ 0 \le k \le N-1, \quad (22)$$

and a direct expansion of Taylor type yields

$$\tilde{f}_i^{k+\sigma} = \tilde{F}_i^{k+\sigma} + O\left(\tau^2\right), \quad (23)$$

such that

$$\tilde{F}_i^{k+\sigma} = \tilde{f}\left((\sigma+1)V_i^k - \sigma V_i^{k-1}, \sigma V_i^{k+1-n_0} + (1-\sigma)V_i^{k-n_0}, x_i, t_{k+\sigma}\right). \quad (24)$$

Acting the averaging operator $\mathcal{A}$ on both sides of (21), noticing Lemma 2.10 and using Taylor expansion, we arrive at

$$\sum_{n=0}^{k} \hat{C}_{k-n}^{k+1}\left(\mathcal{A}V_i^{n+1} - \mathcal{A}V_i^n\right) = \mathcal{A}\left[\frac{\partial}{\partial x}\left(q_1(x_i)\frac{\partial v(x_i, t_{k+\sigma})}{\partial x}\right)\right] + \mathcal{A}\tilde{f}_i^{k+\sigma} \quad 0 < i < M, \ 0 \le k \le N-1, \quad (25)$$

Next, by using Lemma 2.5 and (23), we obtain

$$\sum_{n=0}^{k} \hat{C}_{k-n}^{k+1}\left(\mathcal{A}V_i^{n+1} - \mathcal{A}V_i^n\right) = \delta_x\left(\varkappa_2 \delta_x V\right)_i^{k+\sigma} + \mathcal{A}\tilde{F}_i^{k+\sigma} + \mathcal{S}_i^{k+\sigma}, \quad (26)$$

where a constant $\tilde{c}_0$ exists in order that

$$|\mathcal{S}_i^{k+\sigma}| \le \tilde{c}_0\left(\tau^2 + h^4\right), \quad 1 \le i \le M-1, \ 0 \le k \le N-1.$$

By considering (20b) at $(x_i, t_{1/2})$ and $(x_i, t_{k+\sigma})$, respectively, operating by $\mathcal{A}$ on both equations, we obtain by the aids of Taylor expansions and Lemmas 2.5 and 2.9, which

$$\delta_t\left(\delta_x\left(\varkappa_2 \delta_x V\right)_i^{1/2}\right) = \delta_x\left(\varkappa_2 \delta_x V\right)_i^{1/2} + s_i^{1/2}, \quad 1 \le i \le M-1, \quad (27)$$

$$\partial_{\hat{t}}\left(\delta_x\left(\varkappa_2 \delta_x V\right)_i^{k+\sigma}\right) = \delta_x\left(\varkappa_2 \delta_x V\right)_i^{k+\sigma} + s_i^{k+\sigma}, \quad 1 \le i \le M-1, \ 1 \le k \le N-1. \quad (28)$$

Moreover, there exists a constant $\tilde{c}_1 > 0$, such that

$$|s_i^{1/2}| \le \tilde{c}_1(\tau^2 + h^4), \quad 1 \le i \le M-1, \quad (29)$$

$$|s_i^{k+\sigma}| \le \tilde{c}_1(\tau^2 + h^4), \quad 1 \le i \le M-1, \ 1 \le k \le N-1. \quad (30)$$

By omitting the small terms in (26)–(28) and noticing the initial and boundary conditions, we construct a spatial fourth order difference scheme for problem (6), as follows

$$\sum_{n=0}^{k} \hat{C}_{k-n}^{k+1} \left(\mathcal{A} \mathcal{V}_i^{n+1} - \mathcal{A} \mathcal{V}_i^n \right) = \delta_x \left(\varkappa_2 \delta_x \nu \right)_i^{k+\sigma} + \mathcal{A} \tilde{\mathcal{F}}_i^{k+\sigma}, \quad 1 \leq i \leq M-1, \quad 0 \leq k \leq N-1, \tag{31a}$$

$$\delta_t \left(\delta_x (\varkappa_2 \delta_x \nu)_i^{1/2} \right) = \delta_x (\varkappa_2 \delta_x \mathcal{V})_i^{1/2}, \quad 1 \leq i \leq M-1, \tag{31b}$$

$$\partial_{\hat{t}} \left(\delta_x (\varkappa_2 \delta_x \nu)_i^{k+\sigma} \right) = \delta_x (\varkappa_2 \delta_x \mathcal{V})_i^{k+\sigma}, \quad 1 \leq i \leq M-1, \quad 1 \leq k \leq N-1, \tag{31c}$$

$$\nu_i^k = \rho_i^k, \quad 1 \leq i \leq M-1, \quad -n_0 \leq k \leq 0, \quad \mathcal{V}_i^0 = \hat{\psi}(x_i), \quad 1 \leq i \leq M-1, \tag{31d}$$

$$\nu_0^k = \nu_M^k = 0, \quad 0 \leq k \leq N, \tag{31e}$$

$$\mathcal{V}_0^k = \mathcal{V}_M^k = 0, \quad 0 \leq k \leq N. \tag{31f}$$

where

$$\tilde{\mathcal{F}}_i^{k+\sigma} = \tilde{f}\left((\sigma+1)\nu_i^k - \sigma \nu_i^{k-1}, \sigma \nu_i^{k+1-n_0} + (1-\sigma)\nu_i^{k-n_0}, x_i, t_{k+\sigma} \right). \tag{32}$$

3. The Stability and Convergence of the Constructed Difference Schemes

First, we will start with some technical lemmas, which will be helpful in the context of convergence and stability.

The nonlinear delay term $\tilde{f}(\mu, \nu, x, t)$ is sufficiently smooth and it satisfies the Lipschitz condition

$$|\tilde{f}(\mu_1, \nu, x, t) - \tilde{f}(\mu_2, \nu, x, t)| \leq c_1 |\mu_1 - \mu_2|, \quad \forall \mu_1, \mu_2 \in [0, L] \times [0, T], \tag{33}$$

$$|\tilde{f}(\mu, \nu_1, x, t) - \tilde{f}(\mu, \nu_2, x, t)| \leq c_1 |\nu_1 - \nu_2|, \quad \forall \nu_1, \nu_2 \in [0, L] \times [-\tau, T], \tag{34}$$

where c_1 and c_2 are two positive constants.

Lemma 3.1. *For any* $V_i^k, U_i^k \in \hat{\mathcal{W}}_h$, *if we define* $\varepsilon_i^k = V_i^k - U_i^k$, *and also*

$$\tilde{F}_i^{k+\sigma} = \tilde{f}\left((\sigma+1) U_i^k - \sigma U_i^{k-1}, \sigma U_i^{k+1-n_0} + (1-\sigma) U_i^{k-n_0}, x_i, t_{k+\sigma} \right), \tag{35}$$

the following estimate is satisfied

$$\left\| \tilde{F}_i^{k+\sigma} - \tilde{\mathcal{F}}_i^{k+\sigma} \right\|^2 \leq 4 \left[2c_1^2 \left(\left\| \varepsilon_i^k \right\|^2 + \left\| \varepsilon_i^{k-1} \right\|^2 \right) + c_2^2 \left(\left\| \varepsilon_i^{k+1-n_0} \right\|^2 + \left\| \varepsilon_i^{k-n_0} \right\|^2 \right) \right]. \tag{36}$$

Proof. Recalling $\tilde{F}_i^{k+\sigma}$ from (24) and under the assumptions of the nonlinear delay term (33), we can deduce the following estimate

$$|\tilde{F}_i^{k+\sigma} - \tilde{\mathcal{F}}_i^{k+\sigma}| \leq c_1 |(\sigma+1)\varepsilon_i^k - \sigma \varepsilon_i^{k-1}| + c_2 |\sigma \varepsilon_i^{k+1-n_0} + (1-\sigma)\varepsilon_i^{k-n_0}|, \quad 1 \leq i \leq M-1,$$

and so

$$\left\| \tilde{F}_i^{k+\sigma} - \tilde{\mathcal{F}}_i^{k+\sigma} \right\|^2 \leq h \sum_{i=1}^{M-1} \left(c_1 |(\sigma+1)\varepsilon_i^k - \sigma \varepsilon_i^{k-1}| + c_2 |\sigma \varepsilon_i^{k+1-n_0} + (1-\sigma)\varepsilon_i^{k-n_0}| \right)^2 \tag{37}$$

$$\leq 2c_1^2 h \sum_{i=1}^{M-1} \left[|(\sigma+1)\varepsilon_i^k - \sigma \varepsilon_i^{k-1}| \right]^2 + 2c_2^2 h \sum_{i=1}^{M-1} \left[|\sigma \varepsilon_i^{k+1-n_0} + (1-\sigma)\varepsilon_i^{k-n_0}| \right]^2 \tag{38}$$

$$\leq 4 \left[2c_1^2 \left(\left\| \varepsilon_i^k \right\|^2 + \left\| \varepsilon_i^{k-1} \right\|^2 \right) + c_2^2 \left(\left\| \varepsilon_i^{k+1-n_0} \right\|^2 + \left\| \varepsilon_i^{k-n_0} \right\|^2 \right) \right]. \tag{39}$$

□

For the convenience of our analysis, the following Grönwall inequality is recalled.

Lemma 3.2. Suppose that $\{H^k \mid k \geq 0\}$ is a non negative sequence that satisfies

$$H^{k+1} \leq A + B\tau \sum_{j=1}^{k} H^j, \quad k = 0, 1, \ldots,$$

then

$$H^{k+1} \leq A \exp(Bk\tau), \quad k = 0, 1, \ldots,$$

in which the positivity of the constants A and B must be taken into account.

Lemma 3.3. For any $p_i^k, q_i^k \in \hat{\mathcal{W}}_h$, the following estimates are satisfied

$$\left\langle \delta_t \left(\delta_x \left(\varkappa_2 \delta_x p \right)_i^{1/2} \right), -2\sigma p_i^1 \right\rangle = \frac{2\sigma}{\tau} \left[|p_i^1|_{1,\varkappa_2}^2 - \left\langle p_i^0, p_i^1 \right\rangle_{\varkappa_2} \right], \quad (40)$$

$$\left\langle \delta_x \left(\varkappa_2 \delta_x q \right)_i^{1/2}, -2\sigma p_i^1 \right\rangle = \sigma \left[\left\langle q_i^1, p_i^1 \right\rangle_{\varkappa_2} + \left\langle q_i^0, p_i^1 \right\rangle_{\varkappa_2} \right], \quad (41)$$

$$\left\langle \partial_{\hat{t}} \left(\delta_x \left(\varkappa_2 \delta_x p \right)_i^{k+\sigma} \right), -p_i^{k+\sigma} \right\rangle \geq \frac{1}{4\tau} \left(\tilde{\mathcal{E}}^{k+1} - \tilde{\mathcal{E}}^k \right), \quad (42)$$

$$\left\langle \delta_x \left(\varkappa_2 \delta_x q \right)_i^{k+\sigma}, -p_i^{k+\sigma} \right\rangle = \langle q_i^{k+\sigma}, p_i^{k+\sigma} \rangle_{\varkappa_2}. \quad (43)$$

where

$$\tilde{\mathcal{E}}^{k+1} = (2\sigma + 1)|w^{k+1}|_{1,\varkappa_2}^2 - (2\sigma - 1)|w^k|_{1,\varkappa_2}^2 + (2\sigma^2 + \sigma - 1)|w^{k+1} - w^k|_{1,\varkappa_2}^2, \quad k \geq 0. \quad (44)$$

Proof. Starting from the l.h.s of (40),

$$\left\langle \delta_t \left(\delta_x \left(\varkappa_2 \delta_x p \right)_i^{1/2} \right), -2\sigma p_i^1 \right\rangle = \frac{-2\sigma}{\tau} \left\langle \delta_x \left((\varkappa_2 \delta_x p)_i^1 \right) - \delta_x \left((\varkappa_2 \delta_x p)_i^0 \right), p_i^1 \right\rangle \quad (45)$$

$$= \frac{2\sigma}{\tau} h \sum_{i=0}^{M-1} \left[(\varkappa_2)_{i+1/2} \delta_x p_{i+1/2}^1 \delta_x p_{i+1/2}^1 - (\varkappa_2)_{i+1/2} \delta_x p_{i+1/2}^0 \delta_x p_{i+1/2}^1 \right], \quad (46)$$

then the r.h.s of (40) is achieved directly. Additionally, starting from the l.h.s of (41),

$$\left\langle \delta_x \left(\varkappa_2 \delta_x q \right)_i^{1/2}, -2\sigma p_i^1 \right\rangle = \frac{-2\sigma}{2} \left\langle \delta_x \left((\varkappa_2 \delta_x q)_i^1 \right) + \delta_x \left((\varkappa_2 \delta_x q)_i^0 \right), p_i^1 \right\rangle \quad (47)$$

$$= \sigma h \sum_{i=0}^{M-1} \left[(\varkappa_2)_{i+1/2} \delta_x q_{i+1/2}^1 \delta_x p_{i+1/2}^1 + (\varkappa_2)_{i+1/2} \delta_x q_{i+1/2}^0 \delta_x p_{i+1/2}^1 \right], \quad (48)$$

then the r.h.s of (41) is immediately held. Invoking the previous estimates and Lemma 2.10, we deduce

$$\left\langle \partial_{\hat{t}} \left(\delta_x \left(\varkappa_2 \delta_x p \right)_i^{k+\sigma} \right), -p_i^{k+\sigma} \right\rangle = \left\langle \partial_{\hat{t}} p_i^{k+\sigma}, p_i^{k+\sigma} \right\rangle_{\varkappa_2} \quad (49)$$

$$\geq \frac{1}{4\tau} \left(\tilde{\mathcal{E}}^{k+1} - \tilde{\mathcal{E}}^k \right), \quad (50)$$

where $\tilde{\mathcal{E}}^{k+1}$ is defined by (44) and so (42) is achieved. The estimate (43) is simply calculated. □

Now, we are in a position to combine the Lemmas 3.1–3.3 to prove the convergence and stability of the proposed compact difference scheme. To that end, Let

$$\rho_i^k = \mathcal{V}_i^k - \mathcal{V}_i^k, \quad e_i^k = V_i^k - v_i^k, \quad 0 \le i \le M, 0 \le k \le N.$$

Subtracting (31) from (26)–(28), (20c)–(20e), respectively, we obtain the error equations, as follows

$$\sum_{n=0}^{k} \hat{C}_{k-n}^{k+1} \left(\mathcal{A}\rho_i^{n+1} - \mathcal{A}\rho_i^n \right) = \delta_x \left(\varkappa_2 \delta_x e \right)_i^{k+\sigma} + \mathcal{A} \left(\tilde{F}_i^{k+\sigma} - \tilde{\mathcal{F}}_i^{k+\sigma} \right) + \mathcal{S}_i^{k+\sigma}, \quad 1 \le i \le M-1, \quad 0 \le k \le N-1, \quad (51a)$$

$$\delta_t \left(\delta_x \left(\varkappa_2 \delta_x e \right)_i^{1/2} \right) = \delta_x \left(\varkappa_2 \delta_x \rho \right)_i^{1/2} + s_i^{1/2}, \quad 1 \le i \le M-1, \quad (51b)$$

$$\partial_{\hat{t}} \left(\delta_x \left(\varkappa_2 \delta_x e \right)_i^{k+\sigma} \right) = \delta_x \left(\varkappa_2 \delta_x \rho \right)_i^{k+\sigma} + s_i^{k+\sigma}, \quad 1 \le i \le M-1, \quad 1 \le k \le N-1, \quad (51c)$$

$$e_i^k = 0, \quad 1 \le i \le M-1, \quad -n_0 \le k \le 0, \quad \rho_i^0 = 0, \quad 1 \le i \le M-1, \quad (51d)$$

$$e_0^k = e_M^k = 0, \quad 0 \le k \le N, \quad (51e)$$

$$\rho_0^k = \rho_M^k = 0, \quad 0 \le k \le N. \quad (51f)$$

Theorem 1. *Assume that $u(x,t) \in C_{x,t}^{6,4}([0,L] \times [-\tau, T])$ is the smooth solution of (5) and $\{V_i^k, v_i^k | 0 \le i \le M, -n_0 \le k \le N\}$ the numerical solution of the scheme (31). Subsequently, there exist positive constants h_0 and τ_0, independent of h and τ, such that, when $h \le h_0$ and $\tau \le \tau_0$, we have the error estimate*

$$\left\| e^k \right\|_\infty \le C_1 \left(\tau^2 + h^4 \right), \quad \tau \sum_{n=1}^{k} \|\rho^n\| \le C_1 \left(\tau^2 + h^4 \right), \quad -n_0 \le k \le N.$$

Proof. The proof will be preformed in two steps. Let us tackle the first one.
Step 1. When $k = 0$, the system (51) is as follows

$$\hat{C}_0^1 \left(\mathcal{A}\rho_i^1 - \mathcal{A}\rho_i^0 \right) = \sigma \delta_x \left(\varkappa_2 \delta_x e \right)_i^1 + (1-\sigma) \delta_x \left(\varkappa_2 \delta_x e \right)_i^0 + \mathcal{S}_i^\sigma, \quad 1 \le i \le M-1, \quad (52a)$$

$$\delta_t \left(\delta_x \left(\varkappa_2 \delta_x e \right)_i^{1/2} \right) = \delta_x \left(\varkappa_2 \delta_x \rho \right)_i^{1/2} + s_i^{1/2}, \quad 1 \le i \le M-1, \quad (52b)$$

$$e_i^k = 0, \quad 1 \le i \le M-1, \quad -n_0 \le k \le 0, \quad \rho_i^0 = 0, \quad 1 \le i \le M-1, \quad (52c)$$

$$e_0^0 = e_M^0 = 0, \quad (52d)$$

$$\rho_0^0 = \rho_M^0 = 0. \quad (52e)$$

Taking the inner product of (52a) with ρ^1, we obtain

$$\begin{aligned}
\hat{C}_0^1 \left\| \rho^1 \right\|_{\mathcal{A}}^2 &= \hat{C}_0^1 \langle \mathcal{A}\rho^0, \rho^1 \rangle + \langle \delta_x \left(\varkappa_2 \delta_x e \right)^\sigma, \rho^1 \rangle + \langle \mathcal{S}^\sigma, \rho^1 \rangle \\
&= \hat{C}_0^1 \langle \mathcal{A}\rho^0, \rho^1 \rangle + \langle \sigma \delta_x \left(\varkappa_2 \delta_x e \right)^1 + (1-\sigma) \delta_x \left(\varkappa_2 \delta_x e \right)^0, \rho^1 \rangle + \langle \mathcal{S}^\sigma, \rho^1 \rangle \\
&= \hat{C}_0^1 \langle \mathcal{A}\rho^0, \rho^1 \rangle - \sigma \langle e^1, \rho^1 \rangle_{\varkappa_2} + (1-\sigma) \langle \delta_x \left(\varkappa_2 \delta_x e \right)^0, \rho^1 \rangle + \langle \mathcal{S}^\sigma, \rho^1 \rangle,
\end{aligned} \quad (53)$$

Taking the inner product of (52b) with $-2\sigma e^1$ and invoking Lemma 3.3, we arrive at

$$\frac{2\sigma}{\tau} \left| e^1 \right|_{1,\varkappa_2}^2 = \frac{2\sigma}{\tau} \left\langle e^0, e^1 \right\rangle_{\varkappa_2} + \sigma \left(\left\langle e^1, \rho^1 \right\rangle_{\varkappa_2} + \left\langle \rho^0, e^1 \right\rangle_{\varkappa_2} \right) - 2\sigma \langle s^{1/2}, e^1 \rangle. \quad (54)$$

Adding (53) with (54) and using Young inequality and Lemma 2.6, we obtain

$$\frac{2}{3}\hat{C}_0^1 \left\|\rho^1\right\|^2 + \frac{2\sigma}{\tau}\left|e^1\right|_{1,\varkappa_2}^2 \leq \hat{C}_0^1 \langle \mathcal{A}\rho^0, \rho^1\rangle + (1-\sigma)\langle \delta_x(\varkappa_2 \delta_x e)^0, \rho^1\rangle + \langle \mathcal{S}^\sigma, \rho^1\rangle \tag{55}$$

$$+ \frac{2\sigma}{\tau}\left\langle e^0, e^1\right\rangle_{\varkappa_2} + \sigma\left\langle \rho^0, e^1\right\rangle_{\varkappa_2} - 2\sigma\langle s^{1/2}, e^1\rangle$$

$$\leq \left[\frac{2}{9}\hat{C}_0^1\|\rho^1\|^2 + \frac{9}{8}\hat{C}_0^1\|\rho^0\|^2\right] + \left[\frac{2\hat{C}_0^1}{9}\|\rho^1\|^2 + \frac{9(1-\sigma)}{8\hat{C}_0^1}\|\delta_x(\varkappa_2\delta_x e)^0\|^2\right]$$

$$+ \left[\frac{2\hat{C}_0^1}{9}\|\rho^1\|^2 + \frac{9}{8\hat{C}_0^1}\|\mathcal{S}^\sigma\|^2\right] + \left[\frac{2\sigma}{9\tau}\left|e^1\right|_{1,\varkappa_2}^2 + \frac{9\sigma}{2\tau}\left|e^0\right|_{1,\varkappa_2}^2\right]$$

$$+ \left[\frac{2\sigma}{9\tau}\left|e^1\right|_{1,\varkappa_2}^2 + \frac{9\tau\sigma}{8}\left|e^0\right|_{1,\varkappa_2}^2\right] + \left[\frac{2\sigma}{9\tau}\left|e^1\right|_{1,\varkappa_2}^2 + \frac{3\sigma\tau}{2}L^2\|s^{1/2}\|^2\right], \tag{56}$$

after a simplification, we get

$$\left|e^1\right|_{1,\varkappa_2}^2 \leq \frac{27\tau}{32\sigma}\hat{C}_0^1\|\rho^0\|^2 + \frac{27\tau(1-\sigma)}{32\sigma\hat{C}_0^1}\left\|\delta_x(\varkappa_2\delta_x e)^0\right\|^2 + \frac{27\tau}{32\sigma\hat{C}_0^1}\|\mathcal{S}^\sigma\|^2 + \frac{27}{8}\left|e^0\right|_{1,\varkappa_2}^2$$

$$+ \frac{27\tau^2}{32}\left|e^0\right|_{1,\varkappa_2}^2 + \frac{9\tau^2}{8}L^2\|s^{1/2}\|^2, \tag{57}$$

it follows from (52b) that

$$\delta_x(\varkappa_2\delta_x e)^1_i = \delta_x(\varkappa_2\delta_x e)^0_i + \tau\delta_x(\varkappa_2\delta_x \rho)^{1/2}_i + \tau s^{1/2}_i, \quad 1 \leq i \leq M-1. \tag{58}$$

Substituting (58) into (52a), we have

$$\hat{C}_0^1\left(\mathcal{A}\rho^1_i - \mathcal{A}\rho^0_i\right) = \left[\tau\sigma\delta_x(\varkappa_2\delta_x \rho)^{1/2}_i + \tau s^{1/2}_i\right] + \delta_x(\varkappa_2\delta_x e)^0_i + \mathcal{S}^\sigma_i, \quad 1 \leq i \leq M-1. \tag{59}$$

Taking the inner product of (59) with $\rho^{1/2}$, we obtain

$$\hat{C}_0^1\langle\left(\mathcal{A}\rho^1 - \mathcal{A}\rho^0\right), \rho^{1/2}\rangle = -\tau\sigma\left|\rho^{1/2}\right|_{1,\varkappa_2}^2 + \langle\tau\sigma s^{1/2}, \rho^{1/2}\rangle + \langle\delta_x(\varkappa_2\delta_x e)^0, \rho^{1/2}\rangle \tag{60}$$

$$+ \langle\mathcal{S}^\sigma, \rho^{1/2}\rangle, \tag{61}$$

By summation by parts and the Young inequality $ab \leq \frac{1}{2\theta}a^2 + \frac{\theta}{2}b^2$, with $\theta = \frac{3}{\hat{C}_0^1}$, this yields

$$\frac{\hat{C}_0^1}{2}\left(\left\|\rho^1\right\|_\mathcal{A}^2 - \left\|\rho^0\right\|_\mathcal{A}^2\right) = \langle\delta_x(\varkappa_2\delta_x e)^0, \rho^{1/2}\rangle - \tau\sigma\left|\rho^{1/2}\right|_{1,\varkappa_2}^2 + \langle\mathcal{S}^\sigma, \rho^{1/2}\rangle + \langle\tau\sigma s^{1/2}, \rho^{1/2}\rangle$$

$$\leq \left[\frac{3}{2\hat{C}_0^1}\left\|\delta_x(\varkappa_2\delta_x e)^0\right\|^2 + \frac{\hat{C}_0^1}{6}\|\rho^{1/2}\|^2\right] + \left[\frac{3}{2\hat{C}_0^1}\|\mathcal{S}^\sigma\|^2 + \frac{\hat{C}_0^1}{6}\|\rho^{1/2}\|^2\right]$$

$$+ \left[\frac{3\tau^2\sigma^2}{2\hat{C}_0^1}\|s^{1/2}\|^2 + \frac{\hat{C}_0^1}{6}\|\rho^{1/2}\|^2\right]$$

$$\leq \frac{\hat{C}_0^1}{4}\left(\|\rho^0\|^2 + \|\rho^1\|^2\right) + \frac{3}{2\hat{C}_0^1}\left\|\delta_x(\varkappa_2\delta_x e)^0\right\|^2 + \frac{3}{2\hat{C}_0^1}\|\mathcal{S}^\sigma\|^2 + \frac{3\tau^2\sigma^2}{2\hat{C}_0^1}\|s^{1/2}\|^2. \tag{62}$$

Subsequently, by invoking Lemma 2.6 and some simple manipulations, we get

$$\left\|\rho^1\right\|^2 \leq 9\|\rho^0\|^2 + \frac{18}{\left(\hat{C}_0^1\right)^2}\left\|\delta_x(\varkappa_2\delta_x e)^0\right\|^2 + \frac{18}{\left(\hat{C}_0^1\right)^2}\|\mathcal{S}^\sigma\|^2 + \frac{18\tau^2\sigma^2}{\left(\hat{C}_0^1\right)^2}\|s^{1/2}\|^2. \tag{63}$$

Step 2. When $k \geq 1$, we take the inner product of (51a) with $\rho^{k+\sigma}$ and obtain

$$\left\langle \sum_{n=0}^{k} \hat{C}_{k-n}^{k+1}\left(A\rho^{n+1} - A\rho^n\right), \rho^{k+\sigma} \right\rangle = \langle \delta_x\left(\varkappa_2 \delta_x e\right)^{k+\sigma}, \rho^{k+\sigma}\rangle$$
$$+ \langle \mathcal{A}\left(\tilde{F}^{k+\sigma} - \bar{\mathcal{F}}^{k+\sigma}\right), \rho^{k+\sigma}\rangle + \langle \mathcal{S}^{k+\sigma}, \rho^{k+\sigma}\rangle, \quad 1 \leq k \leq N-1. \quad (64)$$

By Lemmas 2.4 and 2.10, we deduce for $1 \leq k \leq N-1$,

$$\left\langle \sum_{n=0}^{k} \hat{C}_{k-n}^{k+1}\left(A\rho^{n+1} - A\rho^n\right), \rho^{k+\sigma} \right\rangle \geq \frac{1}{2} \sum_{n=0}^{k} \hat{C}_{k-n}^{k+1} \left[\|\rho^{n+1}\|_{\mathcal{A}}^2 - \|\rho^n\|_{\mathcal{A}}^2 \right]$$
$$= \frac{1}{2}\left(\sum_{n=1}^{k+1} \hat{C}_{k-n+1}^{k+1} \|\rho^n\|_{\mathcal{A}}^2 - \sum_{n=1}^{k} \hat{C}_{k-n}^{k} \|\rho^n\|_{\mathcal{A}}^2 - \hat{b}_k \|\rho^1\|_{\mathcal{A}}^2 - \hat{C}_k^{k+1} \|\rho^0\|_{\mathcal{A}}^2 \right). \quad (65)$$

Young inequality is used for any $\theta > 0$ to yield

$$\left| \langle \mathcal{S}^{k+\sigma}, \rho^{k+\sigma}\rangle \right| \leq \theta \|\rho^{k+\sigma}\|^2 + \frac{1}{4\theta} \|\mathcal{S}^{k+\sigma}\|^2. \quad (66)$$

Using Lemma 3.1 and also Young inequality, this gives

$$\langle \mathcal{A}\left(\tilde{F}^{k+\sigma} - \bar{\mathcal{F}}^{k+\sigma}\right), \rho^{k+\sigma}\rangle \leq \theta \|\rho^{k+\sigma}\|^2 + \frac{1}{4\theta} \|\tilde{F}^{k+\sigma} - \bar{\mathcal{F}}^{k+\sigma}\|_{\mathcal{A}}^2$$
$$\leq \frac{\theta}{2}\|\rho^{k+\sigma}\|^2 + \frac{1}{2\theta}\left[2c_1^2\left(\|\rho^k\|^2 + \|\rho^{k-1}\|^2\right) + c_2^2\left(\|\rho^{k+1-n_0}\|^2 + \|\rho^{k-n_0}\|^2\right) \right]. \quad (67)$$

Substituting (65)–(67) in (64), we get, for all $1 \leq k \leq N-1$,

$$\frac{1}{2}\left(\sum_{n=1}^{k+1} \hat{C}_{k-n+1}^{k+1} \|\rho^n\|_{\mathcal{A}}^2 - \sum_{n=1}^{k} \hat{C}_{k-n}^{k} \|\rho^n\|_{\mathcal{A}}^2 - \hat{b}_k \|\rho^1\|_{\mathcal{A}}^2 - \hat{C}_k^{k+1} \|\rho^0\|_{\mathcal{A}}^2 \right)$$
$$\leq \langle \delta_x\left(\varkappa_2 \delta_x e\right)^{k+\sigma}, \rho^{k+\sigma}\rangle + \theta \|\rho^{k+\sigma}\|^2 + \frac{1}{4\theta} \|\mathcal{S}^{k+\sigma}\|^2$$
$$+ \frac{\theta}{2} \|\rho^{k+\sigma}\|^2 + \frac{1}{2\theta}\left[2c_1^2\left(\|\rho^k\|^2 + \|\rho^{k-1}\|^2\right) + c_2^2\left(\|\rho^{k+1-n_0}\|^2 + \|\rho^{k-n_0}\|^2\right) \right]. \quad (68)$$

Taking the inner product of (51c) with $-e^{k+\sigma}$, we obtain

$$-\left\langle \partial_{\hat{t}}\left(\delta_x\left(\varkappa_2 \delta_x e\right)^{k+\sigma}\right), e^{k+\sigma}\right\rangle = -\left\langle \delta_x\left(\varkappa_2 \delta_x \rho\right)^{k+\sigma}, e^{k+\sigma}\right\rangle - \left\langle s^{k+\sigma}, e^{k+\sigma}\right\rangle, \quad 1 \leq k \leq N-1. \quad (69)$$

For the l.h.s of (69), after recalling Lemmas 2.10 and 3.3, we get

$$-\left\langle \partial_{\hat{t}}\left(\delta_x\left(\varkappa_2 \delta_x e\right)^{k+\sigma}\right), e^{k+\sigma}\right\rangle \geq \frac{1}{4\tau}\left(\tilde{\mathcal{E}}^{k+1} - \tilde{\mathcal{E}}^k\right), \quad 1 \leq k \leq N-1, \quad (70)$$

such that

$$\tilde{\mathcal{E}}^{k+1} = (2\sigma+1)|e^{k+1}|_{1,\varkappa_2}^2 - (2\sigma-1)|e^k|_{1,\varkappa_2}^2 + (2\sigma^2+\sigma-1)|e^{k+1} - e^k|_{1,\varkappa_2}^2, \quad k \geq 0,$$

and additionally,

$$\tilde{\mathcal{E}}^{k+1} \geq \frac{1}{\sigma}|e^{k+1}|_{1,\varkappa_2}^2, \quad k \geq 0. \quad (71)$$

By the Cauchy–Schwarz inequality, we have

$$\left|-\left\langle s^{k+\sigma}, e^{k+\sigma}\right\rangle\right| \leq \frac{1}{2}\left\|e^{k+\sigma}\right\|^2 + \frac{1}{2}\left\|s^{k+\sigma}\right\|^2, \quad 1 \leq k \leq N-1, \tag{72}$$

so plugging (70) and (72) into (69), this gives

$$\frac{1}{4\tau}\left(\tilde{\mathcal{E}}^{k+1} - \tilde{\mathcal{E}}^k\right) \leq -\left\langle \delta_x(\varkappa_2 \delta_x \rho)^{k+\sigma}, e^{k+\sigma}\right\rangle + \frac{1}{2}\left\|e^{k+\sigma}\right\|^2 + \frac{1}{2}\left\|s^{k+\sigma}\right\|^2, \quad 1 \leq k \leq N-1. \tag{73}$$

Adding (68) and (73), we obtain

$$\frac{1}{2}\left(\sum_{n=1}^{k+1} \hat{C}_{k-n+1}^{k+1}\|\rho^n\|_{\mathcal{A}}^2 - \sum_{n=1}^{k} \hat{C}_{k-n}^{k}\|\rho^n\|_{\mathcal{A}}^2 - \hat{b}_k\|\rho^1\|_{\mathcal{A}}^2 - \hat{C}_k^{k+1}\|\rho^0\|_{\mathcal{A}}^2\right) + \frac{1}{4\tau}\left(\tilde{\mathcal{E}}^{k+1} - \tilde{\mathcal{E}}^k\right)$$

$$\leq \frac{1}{2}\left\|e^{k+\sigma}\right\|^2 + \frac{1}{2}\left\|s^{k+\sigma}\right\|^2 + \theta\left\|\rho^{k+\sigma}\right\|^2 + \frac{1}{4\theta}\|\mathcal{S}^{k+\sigma}\|^2$$

$$+ \frac{\theta}{2}\left\|\rho^{k+\sigma}\right\|^2 + \frac{1}{2\theta}\left[2c_1^2\left(\|\rho^k\|^2 + \|\rho^{k-1}\|^2\right) + c_2^2\left(\|\rho^{k+1-n_0}\|^2 + \|\rho^{k-n_0}\|^2\right)\right].$$

Now, multiply both sides of (74) by 4τ, use Lemma 2.6 and do some arrangements, we then have

$$2\tau \sum_{n=1}^{k+1} \hat{C}_{k-n+1}^{k+1}\|\rho^n\|_{\mathcal{A}}^2 + \tilde{\mathcal{E}}^{k+1} \leq 2\tau\left(\sum_{n=1}^{k} \hat{C}_{k-n}^{k}\|\rho^n\|_{\mathcal{A}}^2\right) + \tilde{\mathcal{E}}^k + 2\tau\left(\hat{b}_k\|\rho^1\|^2 + \hat{C}_k^{k+1}\|\rho^0\|^2\right)$$

$$+ \frac{4\tau}{2}\left\|e^{k+\sigma}\right\|^2 + \frac{4\tau}{2}\|s^{k+\sigma}\|^2 + \frac{4\tau}{4\theta}\|\mathcal{S}^{k+\sigma}\|^2 + 2\tau(1+2\theta)\|\rho^{k+\sigma}\|^2 \tag{74}$$

$$+ \frac{4\tau}{2}\left[2c_1^2\left(\|\rho^k\|^2 + \|\rho^{k-1}\|^2\right) + c_2^2\left(\|\rho^{k+1-n_0}\|^2 + \|\rho^{k-n_0}\|^2\right)\right].$$

Denote $\mathcal{H}^{k+1} = 2\tau \sum_{n=1}^{k+1} \hat{C}_{k-n+1}^{k+1}\|\rho^n\|_{\mathcal{A}}^2 + \tilde{\mathcal{E}}^{k+1}$, we can write

$$\mathcal{H}^{k+1} \leq \mathcal{H}^k + 2\tau\left(\hat{b}_k\|\rho^1\|^2 + \hat{C}_k^{k+1}\|\rho^0\|^2\right) + \frac{4\tau}{2}\|e^{k+\sigma}\|^2 + \frac{4\tau}{2}\|s^{k+\sigma}\|^2 + \frac{4\tau}{4\theta}\|\mathcal{S}^{k+\sigma}\|^2$$

$$+ 8\tau\theta\|\rho^{k+\sigma}\|^2 + \frac{4\tau}{\theta}\left[2c_1^2\left(\|\rho^k\|^2 + \|\rho^{k-1}\|^2\right) + c_2^2\left(\|\rho^{k+1-n_0}\|^2 + \|\rho^{k-n_0}\|^2\right)\right]$$

$$\leq \mathcal{H}^1 + 2\tau\left(\sum_{n=1}^{k} \hat{b}_n\|\rho^1\|^2 + \sum_{n=1}^{k} \hat{C}_n^{k+1}\|\rho^0\|^2\right) + \frac{8\tau}{2}\sum_{n=1}^{k+1}\|e^n\|^2 + \frac{4\tau}{2}\sum_{n=1}^{k}\|s^{n+\sigma}\|^2 \tag{75}$$

$$+ \frac{4\tau}{4\theta}\sum_{n=1}^{k}\|\mathcal{S}^{n+\sigma}\|^2 + 4\tau(1+2\theta)\sum_{n=1}^{k+1}\|\rho^n\|^2 + \frac{4\tau}{2}\sum_{n=1}^{k}\left[2c_1^2\left(\|\rho^n\|^2 + \|\rho^{n-1}\|^2\right)\right.$$

$$\left. + c_2^2\left(\|\rho^{n+1-n_0}\|^2 + \|\rho^{n-n_0}\|^2\right)\right], \quad 1 \leq k \leq N-1.$$

From (71) and Lemma 2.4,

$$\mathcal{H}^{k+1} \geq \tau \sum_{r=0}^{m} p_r \frac{(1-\gamma_r)T^{-\gamma_r}}{\Gamma(2-\gamma_r)}(k+\sigma)^{1-\gamma_r}\sum_{n=1}^{k+1}\|\rho^n\|_{\mathcal{A}}^2 + \frac{1}{\sigma}|e^{k+1}|_{1,\varkappa_2}^2, \quad 1 \leq k \leq N-1, \tag{76}$$

$$\mathcal{H}^1 = 2\tau \hat{C}_0^1 \|\rho^1\|_{\mathcal{A}}^2 + \mathcal{E}^1 \leq 2\tau \hat{C}_0^1 \|\rho^1\|^2 + (2\sigma+1)|e^1|_{1,\varkappa_2}^2 + (2\sigma^2 + \sigma - 1)|e^1|_{1,\varkappa_2}^2. \tag{77}$$

Inserting the above two inequalities into (76) and considering

$$c := \max\{c_1, c_2\}, \quad A := \frac{2}{3}\sum_{r=0}^{m} p_r \frac{(1-\gamma_r)T^{-\gamma_r}}{\Gamma(2-\gamma_r)}(k+\sigma)^{1-\gamma_r},$$

yields

$$\frac{1}{\sigma}|e^{k+1}|_{1,\varkappa_2}^2 + \tau A \sum_{n=1}^{k+1}\|\rho^n\|^2 \le 2\tau\hat{C}_0^1\|\rho^1\|^2 + (2\sigma^2 + 3\sigma)|e^1|_{1,\varkappa_2}^2$$

$$+ 2\tau\left(\sum_{n=1}^{k}\hat{b}_n\|\rho^1\|^2 + \sum_{n=1}^{k}\hat{C}_n^{k+1}\|\rho^0\|^2\right) + 4\tau\sum_{n=1}^{k+1}\|e^n\|^2 + 2\tau\sum_{n=1}^{k}\|s^{n+\sigma}\|^2$$

$$+ \frac{\tau}{\theta}\sum_{n=1}^{k}\|S^{n+\sigma}\|^2 + 4\tau(1 + 2\theta + 3c^2)\sum_{n=1}^{k+1}\|\rho^n\|^2.$$

By choosing θ to achieve $A \ge 4(1 + 2\theta + 3c^2)$, invoking (57), (63), and denoting

$$\mathcal{G}_{k+1} := 2\tau\left(\hat{C}_0^1 + \sum_{n=1}^{k}\hat{b}_n\right)\left[9\|\rho^0\|^2 + \frac{18}{(\hat{C}_0^1)^2}\|\delta_x(\varkappa_2\delta_x e)^0\|^2 + \frac{18}{(\hat{C}_0^1)^2}\|S^\sigma\|^2 + \frac{18\tau^2\sigma^2}{(\hat{C}_0^1)^2}\|s^{1/2}\|^2\right]$$

$$+ (2\sigma^2 + 3\sigma)\left[\frac{27\tau}{32\sigma}\hat{C}_0^1\|\rho^0\|^2 + \frac{27\tau(1-\sigma)}{32\sigma\hat{C}_0^1}\|\delta_x(\varkappa_2\delta_x e)^0\|^2 + \frac{27\tau}{32\sigma\hat{C}_0^1}\|S^\sigma\|^2 + \frac{27}{8}|e^0|_{1,\varkappa_2}^2\right] \quad (78)$$

$$+ \frac{27\tau^2}{32}|e^0|_{1,\varkappa_2}^2 + \frac{9\tau^2}{8}L^2\|s^{1/2}\|^2\right] + 2\tau\sum_{n=1}^{k}\hat{C}_n^{k+1}\|\rho^0\|^2 + 2\tau\sum_{n=1}^{k}\|s^{n+\sigma}\|^2 + \frac{\tau}{\theta}\sum_{n=1}^{k}\|S^{n+\sigma}\|^2,$$

we obtain directly after following the assumptions (13),

$$|e^{k+1}|_1^2 \le \frac{1}{b_2}|e^{k+1}|_{1,\varkappa_2}^2 \le \frac{4\tau\sigma}{b_2}\sum_{n=1}^{k+1}\|e^n\|^2 + \frac{\sigma}{b_2}\mathcal{G}_{k+1}, \quad 1 \le k \le N-1, \quad (79)$$

invoking Lemma 2.6 gives

$$|e^{k+1}|_1^2 \le \frac{2\tau L^2\sigma}{3b_2}\sum_{n=1}^{k+1}|e^n|_1^2 + \frac{\sigma}{b_2}\mathcal{G}_{k+1}, \quad 1 \le k \le N-1, \quad (80)$$

applying Grönwall Lemma 3.2 yields

$$|e^{k+1}|_1^2 \le \frac{\sigma}{b_2}\exp\left(\frac{4L^2\sigma}{3b_2}\right)\mathcal{G}_{k+1}, \quad 1 \le k \le N-1. \quad (81)$$

Accordingly, the proof is completed. □

4. Almost Unconditional Stability

To discuss the stability of the compact difference scheme (31), we also use the discrete energy method. Let $\{\tilde{\mathcal{V}}_i^k, \tilde{v}_i^k | 0 \le i \le M, -n_0 \le k \le N\}$ be the solution of

$$\sum_{n=0}^{k} \hat{C}_{k-n}^{k+1} \left(\mathcal{A}\bar{\mathcal{V}}_i^{n+1} - \mathcal{A}\bar{\mathcal{V}}_i^n \right) = \delta_x \left(\varkappa_2 \delta_x \bar{v} \right)_i^{k+\sigma} + \mathcal{A}\hat{\mathbb{F}}_i^{k+\sigma}, \quad 1 \le i \le M-1, \quad 0 \le k \le N-1, \tag{82a}$$

$$\delta_t \left(\delta_x \left(\varkappa_2 \delta_x \bar{v} \right)_i^{1/2} \right) = \delta_x \left(\varkappa_2 \delta_x \bar{v} \right)_i^{1/2}, \quad 1 \le i \le M-1, \tag{82b}$$

$$\partial_{\hat{t}} \left(\delta_x \left(\varkappa_2 \delta_x \bar{v} \right)_i^{k+\sigma} \right) = \delta_x \left(\varkappa_2 \delta_x \bar{v} \right)_i^{k+\sigma}, \quad 1 \le i \le M-1, \quad 1 \le k \le N-1, \tag{82c}$$

$$\bar{v}_i^k = \hat{r}_i^k + \varrho_i^k, \quad 1 \le i \le M-1, \quad -n_0 \le k \le 0, \quad \bar{v}_i^0 = \hat{\psi}(x_i), \quad 1 \le i \le M-1, \tag{82d}$$

$$\bar{v}_0^k = \bar{v}_M^k = 0, \quad 0 \le k \le N, \tag{82e}$$

$$\bar{\mathcal{V}}_0^k = \bar{\mathcal{V}}_M^k = 0, \quad 0 \le k \le N, \tag{82f}$$

where

$$\hat{\mathbb{F}}_i^{k+\sigma} = \tilde{f}\left((\sigma+1) \bar{v}_i^k - \sigma \bar{v}_i^{k-1}, \sigma \bar{v}_i^{k+1-n_0} + (1-\sigma) \bar{v}_i^{k-n_0}, x_i, t_{k+\sigma} \right). \tag{83}$$

where ϱ_i^k denotes an initial perturbation term that is very small.

Theorem 2. *Let $\bar{\rho}_i^k = \bar{\mathcal{V}}_i^k - \mathcal{V}_i^k$, $\bar{e}_i^k = \bar{v}_i^k - v_i^k$, for $0 \le i \le M, -n_0 \le k \le N$. Subsequently, there exist constants c_4, c_5, h_0, τ_0 that fulfill*

$$\left\| \bar{e}^k \right\|_\infty \le c_4 \sum_{k=-n}^{0} \left\| \varrho^k \right\|, \quad 0 \le k \le N,$$

conditioned by

$$h \le h_0, \quad \tau \le \tau_0, \quad \max_{\substack{-n \le k \le 0 \\ 0 \le i \le M}} \left| \bar{e}_i^k \right| \le c_5.$$

Proof. The perturbation equations in terms of $\bar{\rho}_i^k$ and $\bar{e}_i^k$ come by subtracting (82) from (31) and similar to the proof of Theorem 1, the conclusion of stability holds immediately. □

5. Generalized Scheme for the Distributed Order Case

We are now in a position to consider the distributed order form of dmfCDWEs, which means that

$$\int_1^2 \omega(\alpha) \frac{\partial^\alpha u(x,t)}{\partial t^\alpha} d\alpha = \frac{\partial}{\partial x} \left(q_1(x) \frac{\partial u}{\partial x} \right) + q_2(x) \frac{\partial u}{\partial x} + f(u(x,t), u(x,t-s), x, t), \quad 0 < t \le T, \ 0 \le x \le L, \tag{84a}$$

with the following initial and boundary conditions

$$u(x,t) = d(x,t), \quad 0 \le x \le L, \quad t \in [-s, 0], \quad \frac{\partial u(x,0)}{\partial t} = \psi(x) = \lim_{t \to -0} \frac{\partial d(x,t)}{\partial t}, \tag{84b}$$

$$u(0,t) = \phi_0(t), \quad u(L,t) = \phi_L(t), \quad 0 < t \le T, \tag{84c}$$

Following the same manipulations illustrated before; starting from an exponential transformation technique, then a transformation to zero Dirichlet boundary conditions, we obtain the following system

$$\int_1^2 \omega(\alpha) \frac{\partial^\alpha v(x,t)}{\partial t^\alpha} d\alpha = \frac{\partial}{\partial x} \left(q_1(x) \frac{\partial v}{\partial x} \right) + \tilde{f}(v(x,t), v(x,t-s), x, t), \quad 0 < t \le T, \ 0 \le x \le L, \tag{85a}$$

with the initial and boundary conditions as

$$v(x,t) = \hat{r}(x,t), \quad 0 \le x \le L, \quad t \in [-s, 0], \quad \frac{\partial v(x,0)}{\partial t} = \hat{\psi}(x) = \lim_{t \to -0} \frac{\partial r(x,t)}{\partial t}, \tag{85b}$$

$$v(0,t) = v(L,t) = 0, \quad t > 0. \tag{85c}$$

A numerical quadrature rule can be adapted to transform the system (85) to dmfCDWEs. We recall Simpson's rule (also known as the three-point Newton–Cotes quadrature rule), a proof of which can be found in any descent textbook.

Lemma 5.1. *Consider an equidistant partition of the interval* $[1,2]$ *into* $2J$ *subintervals, let* $\Delta \alpha = \frac{1}{2J}$ *and denote* $\alpha_l = 1 + l\Delta \alpha$, $0 \leq l \leq 2J$. *Subsequently, the composite Simpson's rule reads*

$$\int_1^2 f(\alpha) d\alpha = \Delta \alpha \sum_{l=0}^{2J} \gamma_l f(\alpha_l) - \frac{(\Delta \alpha)^4}{180} f^{(4)}(\zeta), \quad \zeta \in [1,2], \tag{86}$$

where

$$\gamma_l = \begin{cases} \frac{1}{3}, & l = 0, 2J, \\ \frac{2}{3}, & l = 2, 4, \ldots, 2J-4, 2J-2, \\ \frac{4}{3}, & l = 1, 3, \ldots, 2J-3, 2J-1. \end{cases}$$

Define the function $G(\cdot\,; x_i, t_j) : \alpha \mapsto \omega(\alpha) \frac{\partial^\alpha v(x_i, t_j)}{\partial t^\alpha}$. Suppose that $G(\alpha) \in C^4([1,2])$, then by using Lemma 5.1, we approximate the distributed derivative as

$$\int_1^2 \omega(\alpha) \frac{\partial^\alpha v(x_i, t_{k+\sigma})}{\partial t^\alpha} d\alpha = \Delta \alpha \sum_{l=0}^{2J} \gamma_l \omega(\alpha_l) \, {}_0^C D_t^{\alpha_l} v(x_i, t_{k-1/2}) - \frac{(\Delta \alpha)^4}{180} G^{(4)}(\alpha; x_i, t_{k+\sigma}) \Big|_{\alpha = \zeta_i^k}$$

$$= \Delta \alpha \sum_{l=0}^{2J} \gamma_l \omega(\alpha_l) \, {}_0^C D_t^{\alpha_l} v(x_i, t_{k+\sigma}) + O\left((\Delta \alpha)^4\right), \tag{87}$$

for a $\zeta_i^k \in [1,2]$. Define

$$\hat{C}_n^{(k+1)} = \Delta \alpha \sum_{l=0}^{2J} \gamma_l \omega(\alpha_l) \frac{\tau^{-\gamma_l}}{\Gamma(2-\gamma_l)} c_n^{(k+1,\gamma_l)}, \quad n = 0, 1, \ldots, k,$$

then the constructed compact difference scheme has the following form

$$\sum_{n=0}^k \hat{C}_{k-n}^{k+1} \left(\mathcal{A} v_i^{n+1} - \mathcal{A} v_i^n \right) = \delta_x (\varkappa_2 \delta_x v)_i^{k+\sigma} + \mathcal{A} \hat{\mathcal{F}}_i^{k+\sigma}, \quad 1 \leq i \leq M-1, \quad 0 \leq k \leq N-1, \tag{88a}$$

$$\delta_t \left(\delta_x (\varkappa_2 \delta_x v)_i^{1/2} \right) = \delta_x (\varkappa_2 \delta_x \mathcal{V})_i^{1/2}, \quad 1 \leq i \leq M-1, \tag{88b}$$

$$\partial_{\hat{t}} \left(\delta_x (\varkappa_2 \delta_x v)_i^{k+\sigma} \right) = \delta_x (\varkappa_2 \delta_x \mathcal{V})_i^{k+\sigma}, \quad 1 \leq i \leq M-1, \quad 1 \leq k \leq N-1, \tag{88c}$$

$$v_i^k = \hat{r}_i^k, \quad 1 \leq i \leq M-1, \quad -n_0 \leq k \leq 0, \quad \mathcal{V}_i^0 = \hat{\psi}(x_i), \quad 1 \leq i \leq M-1, \tag{88d}$$

$$v_0^k = v_M^k = 0, \quad 0 \leq k \leq N, \tag{88e}$$

$$\mathcal{V}_0^k = \mathcal{V}_M^k = 0, \quad 0 \leq k \leq N. \tag{88f}$$

Remark 1. The local truncation error of the compact difference scheme (88) for the distributed order system (85) is of order $\mathcal{O}\left(\tau^2 + h^4 + (\Delta \alpha)^4\right)$. The convergence and stability estimates can be derived in the same manner as in Theorem 1 and Theorem 2.

6. Numerical Illustration

The purpose of the present section is to demonstrate the convergence rate of the method. We will consider the maximum absolute error between the exact solution $u(x_i, t_k)$ of the continuous problem and corresponding approximations u_i^k, which is given by

$$\epsilon_{\tau,h} = \max_{0 \leq i \leq M, 0 \leq k \leq N} |u(x_i, t_n) - u_i^n|. \tag{89}$$

Moreover, we define the standard rates

$$\rho_{\tau,h}^x = \log_2\left(\frac{\epsilon_{\tau,2h}}{\epsilon_{\tau,h}}\right), \& \&\rho_{\tau,h}^t = \log_2\left(\frac{\epsilon_{2\tau,h}}{\epsilon_{\tau,h}}\right). \tag{90}$$

We consider the following multiterm time fractional delay sup-diffusion problem

$$\sum_{r=0}^{2} p_r \frac{\partial^{\alpha_r} u(x,t)}{\partial t^{\alpha_r}} = \frac{(x+1)}{2}\frac{\partial^2 u}{\partial x^2} + (x+1)^2 \frac{\partial u}{\partial x} + f(u(x,t), u(x,t-0.2), x, t), \tag{91}$$

$$f(u(x,t), u(x,t-0.2), x, t) = -u^2(t,x) + u(t-0.2,x) + g(x,t), \quad \forall (x,t) \in [0,1] \times [0,1]. \tag{92}$$

Note that $g(x,t)$ is defined/derived, such that $u(x,t) = e^x x^2(1-x)^2 t^3$ is the exact solution. The exact solution determines the initial condition and boundary conditions. The difference scheme (31) is employed in order to obtain the numerical solution. First, the numerical accuracy of this scheme in time will be verified. Taking a sufficiently small step size h and varying step size τ, the numerical errors and numerical convergence orders are listed in the lower half of Table 1. The computational results presented in Table 1 confirm the second-order convergence of the difference scheme (31) in time.

Table 1. Absolute errors and standard convergence rates in space and time when approximating the solution u of (1) with ($\alpha_1 = 1.3$, $\alpha_2 = 1.5$, $\alpha_3 = 1.7$), while using the difference method (31). The parameters and conditions employed in this case correspond to those in Example 6.

		Spatial Analysis of Convergence			
		($p_0 = 1, p_1 = 1.25, p_2 = 2$)		($p_0 = 2, p_1 = 1.75, p_2 = 1$)	
τ	h	$\epsilon_{\tau,h}$	$\rho_{\tau,h}^x$	$\epsilon_{\tau,h}$	$\rho_{\tau,h}^x$
0.001	0.02×2^{-1}	1.877×10^{-3}	—	4.179×10^{-4}	—
	0.02×2^{-2}	1.388×10^{-4}	3.765	2.676×10^{-5}	3.965
	0.02×2^{-3}	9.733×10^{-6}	3.834	1.697×10^{-6}	3.979
	0.02×2^{-4}	6.306×10^{-7}	3.948	1.066×10^{-7}	3.992
0.0005	0.02×2^{-1}	2.480×10^{-4}	—	6.981×10^{-5}	—
	0.02×2^{-2}	1.601×10^{-5}	3.953	4.430×10^{-6}	3.978
	0.02×2^{-3}	1.016×10^{-6}	3.978	2.791×10^{-7}	3.988
	0.02×2^{-4}	6.399×10^{-8}	3.989	1.750×10^{-8}	3.995
		Temporal Analysis of Convergence			
		($p_0 = 1, p_1 = 1.25, p_2 = 2$)		($p_0 = 2, p_1 = 1.75, p_2 = 1$)	
h	τ	$\epsilon_{\tau,h}$	$\rho_{\tau,h}^t$	$\epsilon_{\tau,h}$	$\rho_{\tau,h}^t$
0.005	0.01×2^{-1}	3.357×10^{-4}	—	7.278×10^{-5}	—
	0.01×2^{-2}	9.152×10^{-5}	1.875	1.842×10^{-5}	1.984
	0.01×2^{-3}	2.395×10^{-5}	1.934	4.640×10^{-6}	1.989
	0.01×2^{-4}	7.450×10^{-6}	1.978	1.164×10^{-6}	1.995
0.001	0.01×2^{-1}	9.375×10^{-5}	—	5.378×10^{-6}	—
	0.01×2^{-2}	2.472×10^{-5}	1.923	1.359×10^{-6}	1.987
	0.01×2^{-3}	6.366×10^{-6}	1.964	3.409×10^{-7}	1.994
	0.01×2^{-4}	1.592×10^{-6}	1.993	8.534×10^{-8}	1.998

Next, the numerical accuracy of the difference scheme in space for solving this example is examined. The numerical results of this scheme for different step sizes in space are calculated and the numerical errors, as well as the numerical convergence orders are recorded in the upper half of Table 1. Again, from which, one can find that, in this case the fourth-order convergence is achieved.

7. Conclusions

A linearized difference scheme for solving a class of dmfCDWEs is constructed. With the help of an easy to execute and invertible exponential transformation, the considered problem can be converted into the delay variable coefficient fractional diffusion wave equation equivalently. Subsequently, we establish a fourth-order accurate numerical scheme that is based on a variable coefficient compact operator and with a temporal second order of convergence at a super-convergent point. The convergence and stability of the current numerical scheme are proved at length and a numerical example is finally added for the sake of demonstrating the theoretical findings.

Author Contributions: All authors contributed equally. All authors have read and agreed to the published version of the manuscript.

Funding: The first author wishes to acknowledge the support of RFBR Grant 19-01-00019.

Acknowledgments: The authors wish to thank the anonymous reviewers for their comments and criticism. All of their comments were taken into account in the revised version of the paper, resulting in a substantial improvement with respect to the original submission.

Conflicts of Interest: The authors declare no conflict of interest.

References

1. Metzler, R.; Klafter, J. The random walk's guide to anomalous diffusion: A fractional dynamics approach. *Phys. Rep.* **2000**, *339*, 1–77. [CrossRef]
2. Sun, Z.Z.; Wu, X. A fully discrete difference scheme for a diffusion-wave system. *Appl. Numer. Math.* **2006**, *56*, 193–209. [CrossRef]
3. Alikhanov, A. A new difference scheme for the fractional diffusion equation. *J. Comput. Phys.* **2015**, *280*, 424–438. [CrossRef]
4. Gafiychuk, V.; Datsko, B.Y. Pattern formation in a fractional reaction–diffusion system. *Phys. A Stat. Mech. Appl.* **2006**, *365*, 300–306. [CrossRef]
5. Tomovski, Ž.; Hilfer, R.; Srivastava, H. Fractional and operational calculus with generalized fractional derivative operators and Mittag–Leffler type functions. *Integral Transform. Spec. Funct.* **2010**, *21*, 797–814. [CrossRef]
6. Zhu, B.; Liu, L.; Wu, Y. Existence and uniqueness of global mild solutions for a class of nonlinear fractional reaction–diffusion equations with delay. *Comput. Math. Appl.* **2019**, *78*, 1811–1818. [CrossRef]
7. Hendy, A.S.; De Staelen, R.H.; Pimenov, V.G. A semi-linear delayed diffusion-wave system with distributed order in time. *Numer. Algorithms* **2018**, *77*, 885–903. [CrossRef]
8. Gao, G.H.; Alikhanov, A.A.; Sun, Z.Z. The temporal second order difference schemes based on the interpolation approximation for solving the time multi-term and distributed-order fractional sub-diffusion equations. *J. Sci. Comput.* **2017**, *73*, 93–121. [CrossRef]
9. Sun, Z.Z. An unconditionally stable and $O(\tau^2 + h^4)$ order L_∞ convergent difference scheme for linear parabolic equations with variable coefficients. *Numer. Methods Partial Differ. Equ. Int. J.* **2001**, *17*, 619–631. [CrossRef]
10. Pimenov, V.G.; Hendy, A.S.; De Staelen, R.H. On a class of non-linear delay distributed order fractional diffusion equations. *J. Comput. Appl. Math.* **2017**, *318*, 433–443. [CrossRef]
11. Hendy, A.S.; Macías-Díaz, J. A novel discrete Gronwall inequality in the analysis of difference schemes for time-fractional multi-delayed diffusion equations. *Commun. Nonlinear Sci. Numer. Simul.* **2019**, *73*, 110–119. [CrossRef]
12. Li, L.; Zhou, B.; Chen, X.; Wang, Z. Convergence and stability of compact finite difference method for nonlinear time fractional reaction–diffusion equations with delay. *Appl. Math. Comput.* **2018**, *337*, 144–152. [CrossRef]
13. Hendy, A.S.; Pimenov, V.G.; Macías-Díaz, J.E. Convergence and stability estimates in difference setting for time-fractional parabolic equations with functional delay. *Numer. Methods Partial Differ. Equ.* **2020**, *36*, 118–132. [CrossRef]
14. Bhrawy, A.; Zaky, M. Numerical simulation of multi-dimensional distributed-order generalized Schrödinger equations. *Nonlinear Dyn.* **2017**, *89*, 1415–1432. [CrossRef]

15. Hendy, A.S.; Zaky, M.A. Global consistency analysis of L1-Galerkin spectral schemes for coupled nonlinear space-time fractional Schrödinger equations. *Appl. Numer. Math.* **2020**, *156*, 276–302. [CrossRef]
16. Zaky, M.; Doha, E.; Tenreiro Machado, J. A spectral numerical method for solving distributed-order fractional initial value problems. *J. Comput. Nonlinear Dyn.* **2018**, *13*, 101007. [CrossRef]
17. Zaky, M.A.; Machado, J.T. Multi-dimensional spectral tau methods for distributed-order fractional diffusion equations. *Comput. Math. Appl.* **2020**, *79*, 476–488. [CrossRef]
18. Zheng, R.; Liu, F.; Jiang, X.; Turner, I.W. Finite difference/spectral methods for the two-dimensional distributed-order time-fractional cable equation. *Comput. Math. Appl.* **2020**, *80*, 1523–1537. [CrossRef]
19. Zhang, Q.; Ren, Y.; Lin, X.; Xu, Y. Uniform convergence of compact and BDF methods for the space fractional semilinear delay reaction–diffusion equations. *Appl. Math. Comput.* **2019**, *358*, 91–110. [CrossRef]
20. Çelik, C.; Duman, M. Crank–Nicolson method for the fractional diffusion equation with the Riesz fractional derivative. *J. Comput. Phys.* **2012**, *231*, 1743–1750. [CrossRef]
21. Zhao, X.; Sun, Z.Z.; Hao, Z.P. A fourth-order compact ADI scheme for two-dimensional nonlinear space fractional Schrodinger equation. *SIAM J. Sci. Comput.* **2014**, *36*, A2865–A2886. [CrossRef]
22. Zhang, Q.; Liu, L.; Zhang, C. Compact scheme for fractional diffusion-wave equation with spatial variable coefficient and delays. *Appl. Anal.* **2020**, 1–22. [CrossRef]
23. Zhang, Q.; Zhang, C. A new linearized compact multisplitting scheme for the nonlinear convection–reaction–diffusion equations with delay. *Commun. Nonlinear Sci. Numer. Simul.* **2013**, *18*, 3278–3288. [CrossRef]
24. Sun, H.; Sun, Z.Z.; Gao, G.H. Some temporal second order difference schemes for fractional wave equations. *Numer. Methods Partial Differ. Equ.* **2016**, *32*, 970–1001. [CrossRef]
25. Gao, G.H.; Sun, Z.Z. A compact finite difference scheme for the fractional sub-diffusion equations. *J. Comput. Phys.* **2011**, *230*, 586–595. [CrossRef]
26. Zhao, X.; Sun, Z.Z. Compact Crank–Nicolson schemes for a class of fractional Cattaneo equation in inhomogeneous medium. *J. Sci. Comput.* **2015**, *62*, 747–771. [CrossRef]

© 2020 by the authors. Licensee MDPI, Basel, Switzerland. This article is an open access article distributed under the terms and conditions of the Creative Commons Attribution (CC BY) license (http://creativecommons.org/licenses/by/4.0/).

Article

Delay Stability of n-Firm Cournot Oligopolies

Akio Matsumoto [1,*] and Ferenc Szidarovszky [2,*]

[1] Department of Economics, Chuo University, 742-1, Higashi-Nakano, Hachioji, Tokyo 192-0393, Japan
[2] Department of Mathematics, Corvinus University, Fövám tér 8, 1093 Budapest, Hungary
* Correspondence: akiom@tamacc.chuo-u.ac.jp (A.M.); szidarka@gmail.com (F.S.)

Received: 29 August 2020; Accepted: 14 September 2020; Published: 18 September 2020

Abstract: The dynamic behavior of n-firm oligopolies is examined without product differentiation and with linear price and cost functions. Continuous time scales are assumed with best response dynamics, in which case the equilibrium is asymptotically stable without delays. The firms are assumed to face both implementation and information delays. If the delays are equal, then the model is a single delay case, and the equilibrium is oscillatory stable if the delay is small, at the threshold stability is lost by Hopf bifurcation with cyclic behavior, and for larger delays, the trajectories show expanding cycles. In the case of the non-equal delays, the stability switching curves are constructed and the directions of stability switches are determined. In the case of growth rate dynamics, the local behavior of the trajectories is similar to that of the best response dynamics. Simulation studies verify and illustrate the theoretical findings.

Keywords: implementation delay; information delay; stability switching curve; Cournot oligopoly; growth rate dynamics

1. Introduction

Examining oligopoly models is a very frequently studied research area in mathematical economics. Based on the pioneering work of Cournot [1], many researchers were devoted to this interesting and challenging model and its variants and extensions. One frequently studied extension is obtained by considering the dynamic behavior of the firms. These models can be divided into several categories including linear and nonlinear models, discrete and continuous time scales, best response, and gradient adjustments. For discrete time scales Theocharis [2] showed that the equilibrium of n-firm linear oligopolies without product differentiation is asymptotically stable if $n = 2$, marginally stable if $n = 3$ and unstable if $n > 3$. For continuous time scales, McManus and Quandt [3] showed that the equilibrium is always asymptotically stable in the linear case regardless of the values of the positive speeds of adjustments. These classical results already indicated that the dynamic properties of the equilibrium strongly depends on the selection of time scales. Several generalizations and extensions were then introduced and studied in the literature. The early results up to the mid-70s are summarized in Okuguchi [4] and their multiproduct generalizations are presented in Okuguchi and Szidarovszky [5]. Different aspects of the classical Theocharis model were then examined by several authors including Canovas et al. [6], Hommes et al. [7], Lampart [8], Puu [9,10], Matsumoto and Szidarovszky [11] among others. Nonlinear models are discussed in Bischi et al. [12] and their extensions including delays are examined in Matsumoto and Szidarovszky [13].

In this paper, we reconsider the classical Theocharis model by examining the dynamic behavior of linear n-firm oligopolies without product differentiation and with the additional assumption that the firms face both implementation and information delays. As it is well known that in the linear case best response and gradient adjustment processes are equivalent with different speeds of adjustments, we deal only with best response dynamics. It is assumed that the firms face equal delays in both types. If the implementation and information delays are equal, then the model is equivalent with a single delay case mathematically. In this case, we show that the equilibrium is oscillatory asymptotically stable if the common delay is sufficiently small, at the threshold Hopf bifurcation occurs with cyclic and for larger delays expanding cyclic trajectories. If the delays are different, then a two-delay model is obtained. The stability switching curves are first constructed and then the directions of stability switches are determined. Growth rate dynamics result in nonlinear systems, their local linearizations around the equilibrium result in linear dynamics, that is equivalent to the best response case. So the local dynamics of the two systems are equivalent. Simulation studies verify and illustrate the theoretical findings of the paper. Even in the very special case of linear models, our analysis discovered several aspects of the dynamics which were not studied in the literature before. The importance of examining linear models is verified in addition to the fact that linearized nonlinear models have the same mathematical structures.

This paper develops as follows. Section 3 introduces the best response dynamics. First stability switching curves are constructed and then the case of equal delays is discussed in detail. Growth rate dynamics are introduced in Section 4. First, the stability switching curves are shown and then the directions of stability switches are determined. In both sections, numerical results and simulation studies verify and illustrate the theoretical results. Section 5 offers conclusions and outlines further research directions.

2. Model

The classical oligopoly model is presented reconsidering the classical results of Theocharis [2] and McManus and Quandt [3]. In the model, n firms are producing a homogeneous output. The price function is assumed to be linear,

$$p = a - b \sum_{j=1}^{n} x_j$$

where $a > 0$ is the maximum price, $b > 0$ is the slope of the price function and x_j is firm j's output. The production cost is also assumed to be linear with no fixed cost. The marginal cost of firm j is denoted by c_j, being positive. The profit function of firm i is defined by

$$\pi_i = \left(a - b \sum_{j=1}^{n} x_j \right) x_i - c_i x_i.$$

Under the Cournot competition, the firms decide how much to produce. As we focus only on interior solutions (If the optimal output level of a firm is zero, then the firm leaves the industry, so we can ignore such firms), the first-order condition of firm i for profit maximization is

$$\frac{\partial \pi_i}{\partial x_i} = a - 2bx_i - b \sum_{j \neq i}^{n} x_j - c_i = 0$$

and the second-order condition is satisfied,

$$\frac{\partial^2 \pi_i}{\partial x_i^2} = -2b < 0.$$

The best reply function is obtained through the first-order condition and depends on the choices of other firms,

$$x_i^* = \frac{a - c_i - b \sum_{j \neq i}^{n} x_j}{2b}.$$

Let us introduce a new notation,

$$\alpha_i = \frac{a - c_i}{2b}, \beta = \frac{1}{2} \text{ and } Q = \sum_{j=1}^{n} x_j$$

and make the conventional assumption:

Assumption 1. $c_i = c$ for all i and $a > c$.

Assumption 1 implies $\alpha_i = \alpha > 0$ for all i. As each firm makes an optimal choice at the Cournot equilibrium, its best reply function is written as

$$x_i^* = \frac{\alpha - \beta Q^*}{1 - \beta}.$$

The aggregate output of all firms is obtained by adding the individual outputs,

$$Q^* = \sum_{i=1}^{n} x_i^* = n \frac{\alpha - \beta Q^*}{1 - \beta}$$

that is solved for Q^* to have

$$Q^* = \frac{n\alpha}{1 + (n-1)\beta}.$$

Substituting Q^* into the best reply gives the individual output values at the Cournot equilibrium,

$$x_i^e = \frac{\alpha}{1 + (n-1)\beta} \text{ for } i = 1, 2, ..., n.$$

3. Best Reply Dynamics

Dynamic interpretation of the oligopoly model depends on how to define a learning process on how each firm observes its competitors' choices. Theocharis (1960) constructs the best reply dynamics with naive expectations in discrete time scales,

$$x_i(t+1) = \alpha - \beta \sum_{j \neq i}^{n} x_j(t)$$

where the adjustment to the optimal output in each period is perfect. His provocative result shows that the stability of the Cournot equilibrium is determined only by the number of the firms in an industry as mentioned in the Introduction. McManus and Quandt (1961) makes two reasonable modifications of Theocharis' assumptions: the discrete-time scales are replaced with continuous-time scales and the imperfect adjustment assumption is adopted in which the direction of output change is proportional to the discrepancy between the optimal and actual values,

$$\dot{x}_i(t) = k_i \left[\alpha - \beta \sum_{j \neq i}^{n} x_j(t) - x_i(t) \right] \text{ with } k_i > 0.$$

It is demonstrated that the Cournot equilibrium is always stable when the adjustment speeds are the same (i.e., $k_i = k$). Their result is in sharp contrast to Theocharis' result. We also note that this result remains true if all adjustment speeds are positive.

3.1. Stability Switching

In this study, we move one step forward from the McManus and Quandt model and introduce implementation delays (i.e., $\tau_1 > 0$) on the firm's own production and information delays ($\tau_2 > 0$) on the competitors' productions,

$$\dot{x}_i(t) = k \left[\alpha - \beta \sum_{j \neq i}^{n} x_j(t - \tau_2) - x_i(t - \tau_1) \right] \text{ for } i = 1, 2, ..., n. \tag{1}$$

Notice that dynamic system (1) has the Cournot equilibrium as the steady-state and its homogeneous part is

$$\dot{x}_i(t) = k\left[-x_i(t-\tau_1) - \beta \sum_{j\neq i}^{n} x_j(t-\tau_2)\right] \text{ for } i = 1, 2, ..., n. \quad (2)$$

The characteristic equation is

$$\varphi(\lambda) = \begin{pmatrix} \lambda + ke^{-\lambda\tau_1} & k\beta e^{-\lambda\tau_2} & \cdots & k\beta e^{-\lambda\tau_2} \\ k\beta e^{-\lambda\tau_2} & \lambda + ke^{-\lambda\tau_1} & \cdots & k\beta e^{-\lambda\tau_2} \\ \cdot & \cdot & \cdots & \cdots \\ k\beta e^{-\lambda\tau_2} & k\beta e^{-\lambda\tau_2} & \cdots & \lambda + ke^{-\lambda\tau_1} \end{pmatrix} = 0.$$

With new notation,

$$D = \text{diag}\left(\lambda + ke^{-\lambda\tau_1} - k\beta e^{-\lambda\tau_2}, ..., \lambda + ke^{-\lambda\tau_1} - k\beta e^{-\lambda\tau_2}\right)_{(n,n)}$$

$$a = \left(k\beta e^{-\lambda\tau_2}\right)_{(n,1)} \text{ and } b = (1)_{(n,1)},$$

the characteristic equation can be written as

$$\begin{aligned}\varphi(\lambda) &= \det\left(D + ab^T\right), \\ &= \det D \det\left(I + D^{-1}ab^T\right), \\ &= \det D\left[1 + b^T D^{-1} a\right].\end{aligned}$$

Hence

$$\begin{aligned}\varphi(\lambda) &= \left[\lambda + ke^{-\lambda\tau_1} - k\beta e^{-\lambda\tau_2}\right]^n \left[1 + \frac{nk\beta e^{-\lambda\tau_2}}{\lambda + ke^{-\lambda\tau_1} - k\beta e^{-\lambda\tau_2}}\right] \\ &= \left(\lambda + ke^{-\lambda\tau_1} - k\beta e^{-\lambda\tau_2}\right)^{n-1} \left(\lambda + ke^{-\lambda\tau_1} + k\beta(n-1)e^{-\lambda\tau_2}\right).\end{aligned}$$

It follows that we have two possibilities to solve $\varphi(\lambda) = 0$,

(i) $\lambda + ke^{-\lambda\tau_1} - k\beta e^{-\lambda\tau_2} = 0$,

(ii) $\lambda + ke^{-\lambda\tau_1} + k\beta(n-1)e^{-\lambda\tau_2} = 0$.

Without delays $\tau_1 = \tau_2 = 0$, the eigenvalues are negative,

$$\lambda_1 = -k(1-\beta) < 0 \text{ and } \lambda_2 = -k[1+\beta(n-1)] < 0,$$

implying that the equilibrium is asymptotically stable.

For positive delays, we follow the method discussed in Matsumoto and Szidarovszky [13] based on Gu et al. [14]. Consider equation (i) first. As $\lambda = 0$ does not solve equation (i), it can be rewritten as

$$1 + a_1(\lambda)e^{-\lambda\tau_1} + a_2(\lambda)e^{-\lambda\tau_2} = 0 \quad (3)$$

where

$$a_1(\lambda) = \frac{k}{\lambda} \text{ and } a_2(\lambda) = -\frac{k\beta}{\lambda}.$$

Equation (3) must have a pair of pure imaginary solutions when a stability switch occurs. Hence let $\lambda = i\omega$ with $\omega > 0$ (It is possible to take its conjugate with $\omega < 0$. Even so, we can arrive at the same result.) and we may consider the three terms in (3) as three vectors in the complex plane with the magnitudes 1, $|a_1(i\omega)|$ and $|a_2(i\omega)|$, respectively. Equation (3) means that if we put these vectors head to tail, they form a triangle with the internal angles θ_1 and θ_2 as illustrated in Figure 1.

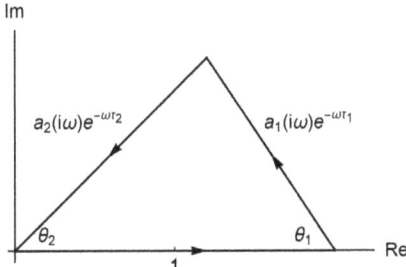

Figure 1. Triangle conditions.

These vectors form a triangle if and only if the sum of the lengths of any two adjacent line segments is not shorter than the length of the remaining line segment:

$$|a_1(i\omega)| + |a_2(i\omega)| \geq 1$$

and

$$-1 \leq |a_1(i\omega)| - |a_2(i\omega)| \leq 1.$$

For $\lambda = i\omega$,

$$a_1(i\omega) = -i\frac{k}{\omega} \text{ and } a_2(\lambda) = i\frac{k\beta}{\omega}$$

where the absolute values are

$$|a_1(i\omega)| = \frac{k}{\omega} \text{ and } |a_2(i\omega)| = \frac{k\beta}{\omega}$$

and the arguments are

$$\arg[a_1(i\omega)] = \frac{3\pi}{2} \text{ and } \arg[a_2(i\omega)] = \frac{\pi}{2}.$$

From the triangle conditions, we have the interval of ω for which $\lambda = i\omega$ can be a solution of equation (i) for some τ_1 and τ_2,

$$\omega \in I = \left[\frac{1}{2}k, \frac{3}{2}k\right].$$

The internal angles of θ_1 and θ_2 are calculated by the law of cosine as

$$\theta_1(\omega) = \cos^{-1}\left[\frac{4\omega^2 + 3k^2}{8k\omega}\right]$$

and

$$\theta_2(\omega) = \cos^{-1}\left[\frac{4\omega^2 - 3k^2}{4k\omega}\right].$$

For any $\omega \in I$, we may find all pairs of (τ_1, τ_2) satisfying (3) as follows:

$$\tau_1^{\pm}(\omega, \ell_1) = \frac{1}{\omega}\left[\frac{3}{2}\pi + (2\ell_1 - 1)\pi \pm \theta_1(\omega)\right] \quad (4)$$

and
$$\tau_2^{\mp}(\omega, \ell_2) = \frac{1}{\omega}\left[\frac{1}{2}\pi + (2\ell_2 - 1)\pi \mp \theta_2(\omega)\right]. \tag{5}$$

Since a symmetric triangle can be formed below the horizontal axis in Figure 1, four inner angles are defined, $\pm\theta_1(\omega)$ and $\mp\theta_2(\omega)$ (double-sign correspondence). By the definitions of the interior angles, we have the followings:

$$\arg\left(a_1(\omega)e^{-i\omega\tau_1}\right) + 2\ell_1\pi \pm \theta_1(\omega) = \pi$$

and

$$\arg\left(a_2(\omega)e^{-i\omega\tau_2}\right) + 2\ell_2\pi \mp \theta_2(\omega) = \pi$$

for $\ell_1 = 0, 1, 2, ...$ and $\ell_2 = 0, 1, 2, ...$ Solving these equations for τ_1 and τ_2 yields (4) and (5). So we have two sets of line segments,

$$C_1^+(\ell_1, \ell_2) = \left\{(\tau_1^+(\omega, \ell_1), \tau_2^-(\omega, \ell_2)) \mid \omega \in I, \, (\ell_1, \ell_2) \in \mathbf{Z}\right\}$$

and

$$C_1^-(\ell_1, \ell_2) = \left\{(\tau_1^-(\omega, \ell_1), \tau_2^+(\omega, \ell_2)) \mid \omega \in I, \, (\ell_1, \ell_2) \in \mathbf{Z}\right\}.$$

As ℓ_1 is the horizontal shift parameter and ℓ_2 is the vertical shift parameter, changing these values shifts these segments accordingly. Connecting these segments creates the stability switching curve (SSC, henceforth) under equation (i).

We now turn attention to equation (ii) that can be written as

$$1 + b_1(\lambda)e^{-\lambda\tau_1} + b_2(\lambda)e^{-\lambda\tau_2} = 0 \tag{6}$$

where

$$b_1(\lambda) = \frac{k}{\lambda} \text{ and } b_2(\lambda) = \frac{k\beta(n-1)}{\lambda}.$$

With $\lambda = i\omega$,

$$b_1(i\omega) = -i\frac{k}{\omega} \text{ and } b_2(i\omega) = -i\frac{k\beta(n-1)}{\omega},$$

their absolute values are

$$|b_1(i\omega)| = \frac{k}{\omega} \text{ and } |b_2(i\omega)| = \frac{k\beta(n-1)}{\omega},$$

and their arguments are

$$\arg(b_1(i\omega)) = \frac{3}{2}\pi \text{ and } \arg(b_2(i\omega)) = \frac{3}{2}\pi.$$

By the triangle conditions, the domains of ω are defined, respectively, by

$$I_2 = \left[\frac{1}{2}k, \frac{3}{2}k\right] \text{ if } n = 2$$

and

$$I_n = \left[\frac{n-3}{2}k, \frac{n+1}{2}k\right] \text{ if } n \geq 3.$$

As in the same way, the internal angles denoted as $\bar{\theta}_1$ and $\bar{\theta}_2$ generated under equation (ii) are obtained as

$$\bar{\theta}_1(\omega) = \cos^{-1}\left[\frac{4\omega^2 - k^2(n-3)(n+1)}{8k\omega}\right]$$

and

$$\bar{\theta}_2(\omega) = \cos^{-1}\left[\frac{4\omega^2 + k^2(n-3)(n+1)}{4k(n-1)\omega}\right].$$

For any $\omega \in I_2$ or I_n, we may find all pairs of (τ_1, τ_2) satisfying (6) as follows:

$$\bar{\tau}_1^{\pm}(\omega, m_1) = \frac{1}{\omega}\left[\frac{3}{2}\pi + (2m_1 - 1)\pi \pm \bar{\theta}_1(\omega)\right] \tag{7}$$

and

$$\bar{\tau}_2^{\mp}(\omega, m_2) = \frac{1}{\omega}\left[\frac{3}{2}\pi + (2m_2 - 1)\pi \mp \bar{\theta}_2(\omega)\right]. \tag{8}$$

As before, we have again two sets of line segments,

$$C_2^+(m_1, m_2) = \{(\bar{\tau}_1^+(\omega, m_1), \bar{\tau}_2^-(\omega, m_2)) \mid \omega \in I_2 \text{ or } I_n, (m_1, m_2) \in \mathbb{Z}\}$$

and

$$C_2^-(m_1, m_2) = \{(\bar{\tau}_1^-(\omega, m_1), \bar{\tau}_2^+(\omega, m_2)) \mid \omega \in I \text{ or } I_n, (m_1, m_2) \in \mathbb{Z}\}$$

which are shifted horizontally and vertically by changing the values of m_1 and m_2. Connecting these segments creates again the stability switching curves under equation (ii).

3.2. Equal Delays

Having found the delays' critical values, we may draw attention to the equal delay case before proceeding further with the different delay case. When the delays are equal, conditions (i) and (ii) are changed to

$$(i)'\ \lambda + ke^{-\lambda\tau} - k\beta e^{-\lambda\tau} = 0,$$

$$(ii)'\ \lambda + ke^{-\lambda\tau} + k\beta(n-1)e^{-\lambda\tau} = 0.$$

For $\lambda = i\omega$ with $\omega > 0$, equation (i)' is

$$i\omega + k(1-\beta)(\cos\omega\tau - i\sin\omega\tau) = 0.$$

Separating the real and imaginary parts gives the equations,

$$k(1-\beta)\cos\omega\tau = 0$$

$$k(1-\beta)\sin\omega\tau = \omega$$

from which

$$\cos\omega\tau = 0,\ \sin\omega\tau = 1 \text{ and } \omega = \frac{k}{2}.$$

Hence the critical values of τ for equation (i)' are determined as

$$\tau_\ell^* = \frac{2}{k}\left(\frac{\pi}{2} + 2\ell\pi\right) \text{ for } \ell = 0, 1, 2, \dots \tag{9}$$

Similarly, for equation (ii)', we have

$$\cos\omega\tau = 0,\ \sin\omega\tau = 1 \text{ and } \omega = \frac{k(n+1)}{2}.$$

Hence the critical values of τ are determined as

$$\tau_m^*(n) = \frac{2}{k(n+1)}\left(\frac{\pi}{2} + 2m\pi\right) \text{ for } m = 0, 1, 2, \dots \tag{10}$$

It is confirmed that

$$\tau_0^*(n) < \tau_0^* \text{ for any } n \geq 2.$$

Therefore stability switching occurs when $\tau = \tau_0^*(n)$. To check the direction of stability switches, we select τ as the bifurcation parameter and consider the eigenvalues as functions of τ, $\lambda = \lambda(\tau)$. Then we differentiate equation (ii)' with respect to τ,

$$\lambda' + k(1 + (n-1)\beta)e^{\lambda \tau}\left(-\lambda'\tau - \lambda\right) = 0$$

and solving this for λ' gives

$$\lambda' = \frac{-\lambda^2}{1 + \lambda \tau}.$$

The sign of the real part for $\lambda = i\omega$ is positive,

$$\text{Re}\left[(\lambda')_{\lambda = i\omega}\right] = \frac{\omega^2}{1 + (\omega \tau)^2} > 0.$$

As equations (i)' is obtained from (ii)' with $n = 0$, this derivation also applies to equation (i)'. Hence we have the following result when the delays are equal:

Theorem 1. *The Cournot equilibrium is locally asymptotically stable for $\tau < \tau_0^*(n)$, loses its stability at $\tau = \tau_0^*(n)$ and stability cannot be regained for $\tau > \tau_0^*(n)$ where*

$$\tau_0^*(n) = \frac{\pi}{k(n+1)}.$$

Theorem 1 is numerically confirmed when

$$\alpha = 10, \ k = 0.5.$$

We perform simulations with three different values of n, $n = 2$, $n = 3$, and $n = 4$. The simulations are done with Mathematica, ver. 12.1. The corresponding critical values of τ are

$$\tau_0^*(2) = \frac{2\pi}{3}, \ \tau_0^*(3) = \frac{\pi}{2} \text{ and } \tau_0^*(4) = \frac{2\pi}{5}$$

which imply that the stability region becomes smaller as n increases. This is also clear from the form of $\tau_0^*(n)$ in Theorem 1. In each simulation below, we take $\tau = \tau_0^*(n) - 0.2$ for the red convergent curve and $\tau = \tau_0^*(n) + 0.1$ for the divergent green curve and assume constant functions for $t \leq 0$. In duopoly, the initial functions are defined as

$$\varphi_1(t) = x_1^e - 3 \text{ and } \varphi_2(t) = x_2^e - 2 \text{ for } t \leq 0.$$

In tiropoly and quartopoly, the appropriate functions are similarly defined and the initial values are selected from the neighborhood of the equilibrium point. Although it is clear that the simulation results strongly depend on the model's specification, we can see the followings from those simulations illustrated in Figure 2A–C:

(1) Theorem 1 is numerically confirmed for $n = 2, 3, 4$; it is seen that the Cournot equilibrium is stable for $\tau < \tau_0^*(n)$, loses stability, and bifurcates to a cyclic oscillation for $\tau = \tau_0^*(n)$.
(2) The trajectories are oscillatory because only complex roots can solve the characteristic equations.
(3) It is further confirmed that the trajectories are oscillatory expanding for $\tau > \tau_0^*(n)$ and thus sooner or later become negative, losing economic meaning.
(4) The time at which the negative production takes place the first time becomes smaller as n increases. Indeed, the green curve first crosses the horizontal axis at $t \simeq 32.012$ in triopoly in Figure 2B and at $t \simeq 25.423$ in quartopoly in Figure 2C. Although it is not illustrated in Figure 2A, the trajectory becomes negative at $t \simeq 59.641$ in duopoly.

Results (3) and (4) are inevitable because the best reply functions are linear and the resultant dynamical system does not have enough nonlinearities to prevent the trajectories from becoming negative. We also have essentially the same results in the case of different delays.

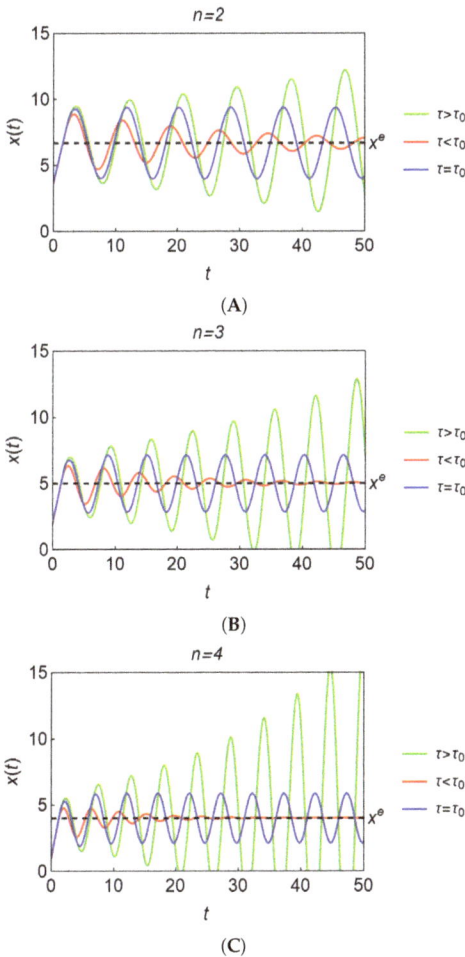

Figure 2. (**A**) Oscillatory dynamics in duopoly. (**B**) Oscillatory dynamics in triopoly. (**C**) Oscillatory dynamics in quartopoly.

4. Growth Rate Dynamics

In this section, we make one modification to the delay best reply dynamical system, (1), and pursue the possibility of bounded dynamics when the system includes some nonlinearities. In particular, the growth rate adjustment is assumed in which the growth rate of output is controlled by the difference between the optimal output and the actual output,

$$\frac{\dot{x}_i(t)}{x_i(t)} = k\left[\alpha - x_i(t-\tau_1) - \beta \sum_{j\neq i}^{n} x_j(t-\tau_2)\right] \text{ for } i = 1, 2, ..., n. \tag{11}$$

System (11) has the same stationary point as system (1). The homogeneous part of its linearized version is

$$\dot{x}_i(t) = K\left[-x_i(t-\tau_1) - \beta \sum_{j \neq i}^n x_j(t-\tau_2)\right] \text{ for } i = 1, 2, ..., n, \quad (12)$$

where

$$K = kx^e.$$

Comparing (12) with (2) reveals that only the adjustment parameters are different. Thus, the formulas for the critical delays in (4), (5), (7) and (8) obtained in the best reply dynamic system can be applied to the growth rate dynamical system (12) if k is replaced with K.

The remaining part of this section is divided into two. The stability switching curves under the growth rate dynamics are constructed and numerical simulations are performed in the first subsection. The stability index is examined to provide theoretical backgrounds with the directions of stability switches for the numerical results in the second part.

4.1. Stability Switching Curves

It is assumed henceforth that K replaces k. Then the pairs of $(\tau_1^+(\omega, \ell_1), \tau_2^-(\omega, \ell_2))$ and $(\tau_1^-(\omega, \ell_1), \tau_2^+(\omega, \ell_2))$ in (4) and (5) satisfy the following characteristic equation,

$$\lambda + Ke^{-\lambda \tau_1} - K\beta e^{-\lambda \tau_2} = 0 \quad (13)$$

where the definitions of θ_1 and θ_2 should be changed to

$$\theta_1(\omega) = \cos^{-1}\left[\frac{4\omega^2 + 3K^2}{8K\omega}\right], \quad (14)$$

$$\theta_2(\omega) = \cos^{-1}\left[\frac{4\omega^2 - 3K^2}{4K\omega}\right] \quad (15)$$

and the interval ω is redefined by

$$I = \left[\frac{1}{2}K, \frac{3}{2}K\right].$$

We then have two sets of line segments in the first quadrant of the (τ_1, τ_2) plane,

$$L_1^+(\ell_1, \ell_2) = \left\{(\tau_1^+(\omega, \ell_1), \tau_2^-(\omega, \ell_2)) \mid \omega \in I, (\ell_1, \ell_2) \in \mathbf{Z}\right\} \quad (16)$$

and

$$L_1^-(\ell_1, \ell_2) = \left\{(\tau_1^-(\omega, \ell_1), \tau_2^+(\omega, \ell_2)) \mid \omega \in I, (\ell_1, \ell_2) \in \mathbf{Z}\right\} \quad (17)$$

similar to the case of best reply dynamics. Lemma 1 characterizes the relations of the segments $L_1^+(\ell_1, \ell_2)$ and $L_1^-(\ell_1, \ell_2)$ for the extreme values of ω in interval I.

Lemma 1. $L_1^+(\ell_1, \ell_2 + 1) = L_1^-(\ell_1, \ell_2)$ holds for the initial point of I, $\omega = K/2$, and $L_1^-(\ell_1, \ell_2) = L_1^+(\ell_1, \ell_2)$ holds for the terminal point of I, $\omega = 3K/2$.

Proof. Substituting $\omega = K/2$ into (14) and (15) gives

$$\theta_1(K/2) = \cos^{-1}(1) = 0 \text{ and } \theta_2(K/2) = \cos^{-1}(-1) = \pi$$

implying that

$$\tau_1^{\pm}(K/2, \ell_1) = \frac{2}{K}\left(\frac{3}{2}\pi + (2\ell_1 - 1)\pi\right)$$

and
$$\tau_2^+(K/2, \ell_2 + 1) = \tau_2^-(K/2, \ell_2) = \frac{2}{K}\left(\frac{1}{2}\pi + 2\ell_2\pi\right).$$

Hence $L_1^+(\ell_1, \ell_2 + 1) = L_1^-(\ell_1, \ell_2)$ at the initial point of I. In the same way, for $\omega = 3K/2$,
$$\theta_1(3K/2) = \cos^{-1}(1) = 0 \text{ and } \theta_2(3K/2) = \cos^{-1}(1) = 0$$

implying that
$$\tau_1^+(3K/2, \ell_1) = \tau_1^-(3K/2, \ell_1) = \frac{2}{3K}\left(\frac{3}{2}\pi + (2\ell_1 - 1)\pi\right)$$

and
$$\tau_2^-(3K/2, \ell_2) = \tau_2^+(3K/2, \ell_2) = \frac{2}{3K}\left(\frac{1}{2}\pi + (2\ell_2 - 1)\pi\right).$$

Hence $L_1^+(\ell_1, \ell_2) = L_1^-(\ell_1, \ell_2)$ at the terminal point of I. This completes the proof. □

Pairs of $(\bar{\tau}_1^+(m_1), \bar{\tau}_2^-(m_1))$ and $(\bar{\tau}_1^-(m_2), \bar{\tau}_2^+(m_2))$ from (7) and (8) satisfy the characteristic equation,
$$\lambda + Ke^{-\lambda\tau_1} + K\beta(n-1)e^{-\lambda\tau_2} = 0 \tag{18}$$

where the definitions of $\bar{\theta}_1$ and $\bar{\theta}_2$ should be changed to
$$\bar{\theta}_1(\omega) = \cos^{-1}\left[\frac{4\omega^2 - K^2(n-3)(n+1)}{8K\omega}\right] \tag{19}$$

and
$$\bar{\theta}_2(\omega) = \cos^{-1}\left[\frac{4\omega^2 + K^2(n-3)(n+1)}{4K(n-1)\omega}\right] \tag{20}$$

and the interval for ω is defined, respectively, by
$$I_2 = \left[\frac{1}{2}K, \frac{3}{2}K\right] \text{ if } n = 2$$

and
$$I_n = \left[\frac{n-3}{2}K, \frac{n+1}{2}K\right] \text{ if } n \geq 3.$$

We also have two line segments of (τ_1, τ_2),
$$L_2^+(m_1, m_2) = \{(\bar{\tau}_1^+(\omega, m_1), \bar{\tau}_2^-(\omega, m_2)) \mid \omega \in I_2 \text{ or } I_n, (m_1, m_2) \in \mathbb{Z}\} \tag{21}$$

and
$$L_2^-(m_1, m_2) = \{(\bar{\tau}_1^-(\omega, m_1), \bar{\tau}_2^+(\omega, m_2)) \mid \omega \in I \text{ or } I_n, (m_1, m_2) \in \mathbb{Z}\} \tag{22}$$

similarly to the case of best reply dynamics. Similarly to Lemma 1, we have the followings:

Lemma 2. *In the case of $n = 2$, $L_2^+(m_1, m_2 + 1) = L_2^-(m_1, m_2)$ holds for the initial point of I_2, $\omega = K/2$, and $L_2^-(m_1, m_2) = L_2^+(m_1, m_2)$ holds for the terminal point of I_2, $\omega = 3K/2$.*

Notice that for $n = 3$,
$$\lim_{\omega \to 0} \bar{\tau}_1^\pm(\omega, m_1) = \infty \text{ and } \lim_{\omega \to 0} \bar{\tau}_2^\pm(\omega, m_2) = \infty.$$

The equality of the segments does not hold at the initial point of I_3 but only at the terminal point which can be proved similarly to Lemma 1.

Lemma 3. *In the case of $n = 3$, $L_2^+(m_1, m_2) = L_2^-(m_1, m_2)$ holds for the terminal point of I_3, $\omega = 2K$.*

If $n \geq 4$, then the following result holds.

Lemma 4. *In the case of $n \geq 4$, $L_2^+(m_1, m_2) = L_2^-(m_1 + 1, m_2)$ holds for the initial point of I_n, $\omega = (n-1)K/2$ and $L_2^+(m_1, m_2) = L_2^-(m_1, m_2)$ holds for the terminal point $\omega = (n+1)K/2$.*

In the following, we will construct stability switching curves. To this end, we specify the parameters' values as $\alpha = 10$ and $k = 0.1$. In Figure 3, the dotted red loci are described by $L_1^-(\ell_1, \ell_2)$ with $\ell_1 = 0$ and $\ell_2 = 0, 1$ and the dotted blue locus by $L_1^+(0, 1)$. The black point a' is the initial point of $L_1^+(0, 1)$ and $L_1^-(0, 0)$ and its coordinates are

$$a' = \left(\frac{3}{2}\pi, \frac{3}{2}\pi\right)$$

at which $L_1^+(0, 1) = L_1^-(0, 0)$ holds by Lemma 1. The black point b' is the terminal point of $L_1^+(0, 1)$ and $L_1^-(0, 1)$ and its coordinates are

$$b' = \left(\frac{1}{2}\pi, \frac{3}{2}\pi\right)$$

at which $L_1^+(0, 1) = L_1^-(0, 1)$ holds by Lemma 1. The blue and red solid curves are described by $L_2^+(0, 0)$ and $L_2^-(0, 0)$. They are connected at point a,

$$a = \left(\frac{1}{2}\pi, \frac{1}{2}\pi\right)$$

at which $L_2^+(0, 0) = L_2^-(0, 0)$ by Lemma 2.

The dotted and solid curves are smoothly connected as is seen in Figure 3. As a result, the (τ_1, τ_2) region is divided into two subregions by the stability switching curve connecting the left-most parts among the segments of $L_2^\pm(0, 0)$, $L_1^\pm(0, 1)$, and $L_1^-(0, 1)$. As the Cournot equilibrium is stable when there are no delays, it is stable in the region including the origin and left to the connecting curve.

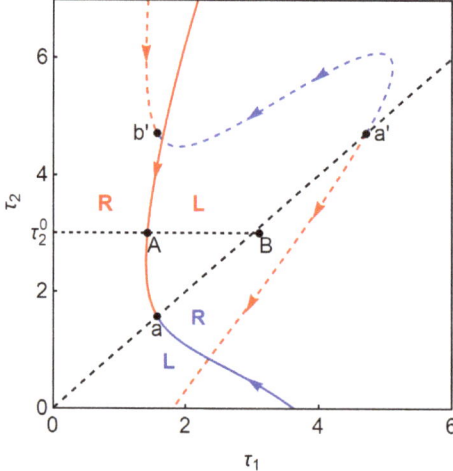

Figure 3. Stability switching curve (SSC) with $n = 2$.

We want to investigate the influence of τ_1 and τ_2. Two simulations in the case of $n = 2$ are performed with initial functions,

$$\varphi_1(t) = x_1^e - 2 \text{ and } \varphi_2(t) = x_2^e + 1 \text{ for } t \leq 0.$$

The first simulation result along the diagonal is presented in Figure 4A. The delays increase from $\tau_i = 0$ to $\tau_i = 3.4$ with an increment of 0.003 along the diagonal. The Cournot equilibrium is asymptotically stable for smaller delays and becomes unstable through a Hopf bifurcation at

$$\tau_i^a = \tau_0^*(2) = \frac{1}{2}\pi \text{ for } i = 1, 2,$$

producing a limit cycle that further bifurcates to a multi-periodic cycle for larger delays. The second result with the different two delays is given in Figure 4B. The value of τ_1 increases from $\tau_1 = \tau_1^A (\simeq 1.423)$ to $\tau_1 = \tau_1^B (= 3.4)$ along the dotted horizontal line at $\tau_2 = 3$. More precisely, the bifurcation diagrams with two delays are constructed in the following procedure with *Mathematica*, version 12.1. The value of τ_2 is fixed at 3, and the value of τ_1 is increased from $\tau_{min} = \tau_1^A$ to $\tau_{max} = \tau_1^B$ with an increment $(\tau_{max} - \tau_{min})/1000$. For each value of τ_1, dynamic system (11) runs for $0 \leq t \leq T (= 2000)$, and the data for $t \leq T - 100$ are discarded to get rid of the initial disturbance. The local maxima and minima out of the remaining data are plotted against this τ_1 value. Then the value of τ_1 is increased and then the same procedure is repeated until τ_1 arrives at τ_{max}. The following bifurcation diagrams are obtained in the same way. The resulting bifurcation diagram shows that the dynamic system experience similar dynamics. The stability of the equilibrium point is confirmed for the zero delay and holds for $\tau_1 < \tau_1^A$ and $\tau_2 = 3$. In both diagrams (and the following diagrams), notation $\tilde{x}^e = \log [x^e]$ is used.

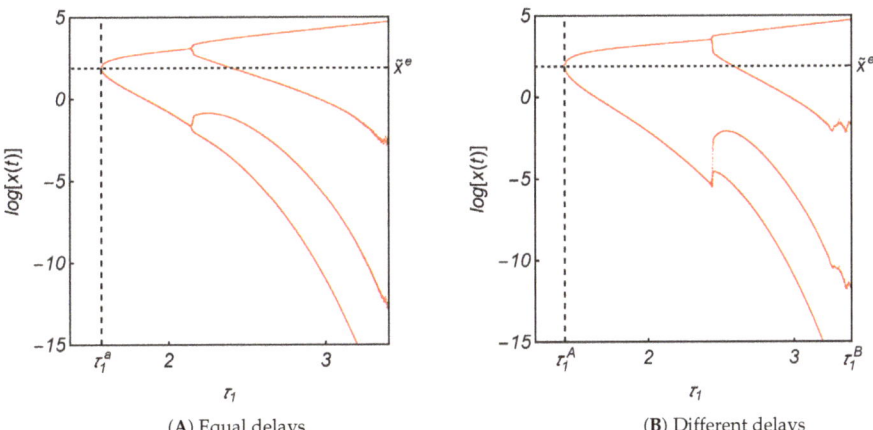

(A) Equal delays (B) Different delays

Figure 4. Bifurcation diagrams with $n = 2$.

We now increase the number of firms to 3. Figure 5A shows the stability switching curves. The line segments of $L_2^+(0,0)$ (i.e., the solid blue curve) and $L_2^-(0,0)$ (i.e., the solid red curve) take the L-shaped profile and rotate counter-clockwise at point a to the extent that the solid red curve is located furthermost to the left. By Lemma 4, both line segments head to point a, the terminal point as ω increases to $2K$. We simulate the model (11) along the diagonal (i.e., $\tau_1 = \tau_2$) and the dotted horizontal line at $\tau_2 = 3$ (i.e., $\tau_1 \neq \tau_2$) in Figure 5A. As we find qualitatively no big differences between these simulation results as in Figure 4A,B, we depict only the bifurcation diagram with different delays in Figure 5B. It is seen that alá "period-doubling bifurcation" occurs in which the Cournot equilibrium is asymptotically stable for $\tau_1 < \tau_1^A (\simeq 1.136)$, loses stability at $\tau_1 = \tau_1^A$ and bifurcates to a limit cycle from which new limit cycles emerge having a doubled period of the cycle as τ_1 increases from τ_1^A. We also see that further increasing τ_1 gives rise to complicated dynamics that suddenly shrinks to a limit cycle with multiple local maxima and minima at some critical point.

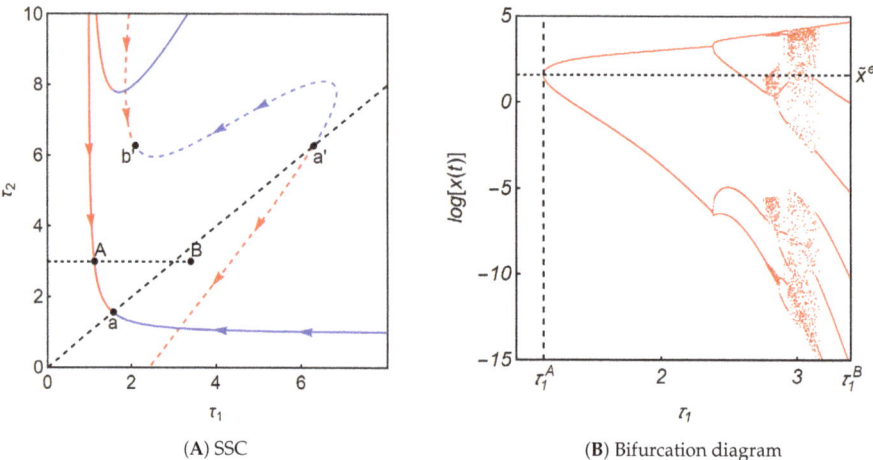

Figure 5. Dynamic properties of Equation (11) with $n = 3$.

In the case of $n = 4$, as is seen in Figure 6A, the solid red and blue segments rotate counter-clockwise further at point a, leading to that the red segment crosses the vertical axis. In Figure 6B, we see that the bifurcation diagram gets more complicated and various dynamics can emerge.

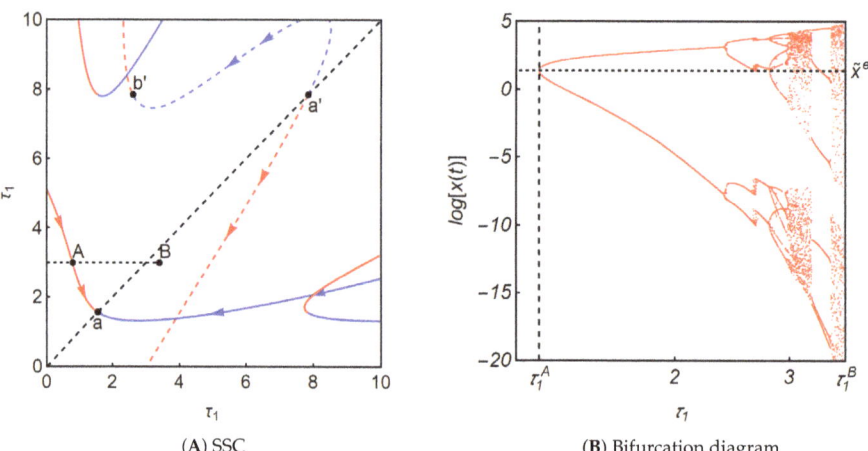

Figure 6. Dynamic properties of Equation (11) with $n = 4$.

Lastly, we simulate system (11) with $n = 9$. The shape of the stability switching curve is different from those with smaller n. In Figure 7A, the positive-sloping dotted line is the diagonal, the dotted-red line is $L_2^-(0,0)$ as before and the black dots are the starting or ending points of the segments. A remarkable difference is that the solid red-blue segments consist of the wave-shaped curve. Accordingly, the bifurcation diagram is obtained along the horizontal dotted line at $\tau_2 = 2$ and exhibits a different route to chaos. The stability of the Cournot equilibrium is lost at $\tau_1 = \tau_1^A$

($\simeq 0.646$), regained at $\tau_1 = \tau_1^B$ ($\simeq 5.441$), and then lost again at $\tau_1 = \tau_1^C$ ($\simeq 7.306$). Unstable oscillatory trajectories get complicated for $\tau_1 > \tau_1^D$ ($\simeq 7.697$). It is known that time delays destabilize dynamic systems. This simulation, however, indicates that time delays can also stabilize the systems.

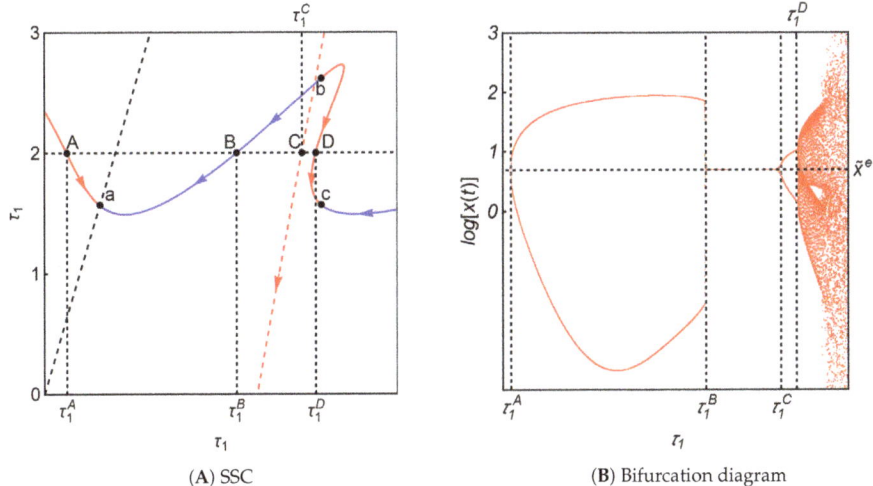

(A) SSC (B) Bifurcation diagram

Figure 7. Dynamic properties of Equation (11) with $n = 9$.

Dynamic system (11) examines the birth of complicated dynamics through a period-doubling bifurcation and the occurrence of stability loss and gain. Needless to say, time delays play prominent roles. In addition, taking account of the fact that only the firm's number is different in those numerical studies, the larger number could influence the system's dynamics by increasing the degree of interactions among the firms.

4.2. Stability Index

We compute the stability index to provide a theoretical background for finding directions of stability switches. First, we denote the second and third vectors of (3) by Q_1 and Q_2,

$$Q_1 = a_1(i\omega)e^{-i\omega\tau_1} = -i\frac{K}{\omega}(\cos\omega\tau_1 - i\sin\omega\tau_1)$$

and

$$Q_2 = a_2(i\omega)e^{-i\omega\tau_2} = i\frac{K\beta}{\omega}(\cos\omega\tau_2 - i\sin\omega\tau_2).$$

Having Q_1 and Q_2, we further denote the real and imaginary parts by the followings:

$$R_1 = \operatorname{Re} Q_1 = -\frac{K}{\omega}\sin\omega\tau_1 \text{ and } I_1 = \operatorname{Im} Q_1 = -\frac{K}{\omega}\cos\omega\tau_1$$

and

$$R_2 = \operatorname{Re} Q_2 = \frac{K\beta}{\omega}\sin\omega\tau_2 \text{ and } I_2 = \operatorname{Im} Q_2 = \frac{K\beta}{\omega}\cos\omega\tau_2.$$

Finally, the stability index is defined as follows:

$$\begin{aligned} S &= R_2 I_1 - R_1 I_2 \\ &= \frac{K^2\beta}{\omega^2}(\sin\omega\tau_1\cos\omega\tau_2 - \cos\omega\tau_1\sin\omega\tau_2), \end{aligned}$$

hence
$$S = \frac{K^2\beta}{\omega^2} \sin\left[\omega\left(\tau_1 - \tau_2\right)\right]. \tag{23}$$

In the same way, we denote the second and third vectors of (6) by $\bar{Q}_1$ and $\bar{Q}_2$,
$$\bar{Q}_1 = b_1(i\omega)e^{-i\omega\tau_1} = -i\frac{K}{\omega}\left(\cos\omega\tau_1 - i\sin\omega\tau_1\right)$$
and
$$\bar{Q}_2 = b_2(i\omega)e^{-i\omega\tau_2} = -i\frac{K\beta(n-1)}{\omega}\left(\cos\omega\tau_2 - i\sin\omega\tau_2\right).$$

The real and imaginary parts are the followings:
$$\bar{R}_1 = \operatorname{Re}\bar{Q}_1 = -\frac{K}{\omega}\sin\omega\tau_1 \text{ and } \bar{I}_1 = \operatorname{Im}\bar{Q}_1 = -\frac{K}{\omega}\cos\omega\tau_1$$
and
$$\bar{R}_2 = \operatorname{Re}\bar{Q}_2 = -\frac{K\beta(n-1)}{\omega}\sin\omega\tau_2 \text{ and } \bar{I}_2 = \operatorname{Im}\bar{Q}_2 = -\frac{K\beta(n-1)}{\omega}\cos\omega\tau_2,$$

moreover, the stability index is as follows:
$$\begin{aligned}\bar{S} &= \bar{R}_2\bar{I}_1 - \bar{R}_1\bar{I}_2 \\ &= -\frac{K^2\beta(n-1)}{\omega^2}\left(\sin\omega\tau_1\cos\omega\tau_2 - \cos\omega\tau_1\sin\omega\tau_2\right).\end{aligned}$$

Hence
$$\bar{S} = -\frac{K^2\beta(n-1)}{\omega^2}\sin\left[\omega\left(\tau_1 - \tau_2\right)\right]. \tag{24}$$

We call the direction of the curve that corresponds to increasing ω the *positive direction*. We also call the region on the left-hand side *the region on the left* when we head in the positive direction of the curve. *Region on the right* is defined similarly. Concerning the stability changes, we have the following result from Matsumoto and Szidarovszky (2018) that is based on Gu et al. (2005):

Theorem 2. *Let (τ_1, τ_2) be a point on the stability switching curves, when $i\omega$ is a simple pure complex eigenvalue. Assume we look toward increasing values of ω on the curve, and a point (τ_1, τ_2) moves from the region on the right to the region on the left. A pair of eigenvalues crosses the imaginary axis to the right if $S > 0$ or $\bar{S} > 0$. If $S < 0$ or $\bar{S} < 0$, then crossing is in the opposite direction.*

The condition of the theorem is satisfied if all $i\omega$ egenvalues are single. It can be proved that the multiple eigenvalues, if any, are isolated from each other, so do the corresponding points on the stability switching curve. Hence at these points, the directions of stability switching are the same as those in the points of their neighborhoods.

We now compute the stability index on the solid red segment of the stability switching curve in Figure 3. The red segment is a locus of the following points,
$$L_2^-(0,0) = \left\{\left(\bar{\tau}_1^-(\omega,0), \bar{\tau}_2^+(\omega,0)\right) \mid \omega \in \left[\frac{1}{2}K, \frac{3}{2}K\right]\right\}$$

From (7) and (8), we have
$$\begin{aligned}\omega\left(\bar{\tau}_1^-(\omega,0) - \bar{\tau}_2^+(\omega,0)\right) &= \left[\frac{3}{2}\pi - \pi - \bar{\theta}_1(\omega)\right] - \left[\frac{3\pi}{2} - \pi + \bar{\theta}_2(\omega)\right] \\ &= -\left[\bar{\theta}_1(\omega) + \bar{\theta}_2(\omega)\right]\end{aligned}$$

implying
$$\sin[\omega(\tau_1 - \tau_2)] = -\sin[\bar{\theta}_1(\omega) + \bar{\theta}_2(\omega)] < 0$$
when $\theta_1 + \theta_2 < \pi$. If $\theta_1 + \theta_2 = \pi$, then the triangle reduces to a line such that
$$|a_1(i\omega)| - |a_2(i\omega)| = \pm 1.$$

That is, in Equation (13),
$$\frac{K}{\omega} - \frac{K\beta}{\omega} = \frac{K}{2\omega} = 1$$
showing that $\omega = \pi/2$ being the left endpoint of interval I, given for ω, which gives the common starting point of two line segments. In the case of Equation (18),
$$\frac{K}{\omega} - \frac{K\beta(n-1)}{\omega} = \frac{K}{\omega}\left(\frac{3-n}{2}\right).$$

If $n = 2$, this equals $+1$ if $\omega = K/2$, which is the initial point of I. If $n = 3$, then this expression is always zero, so cannot be $+1$ or -1. If $n > 3$, then this expression can be only -1, when $\omega = K(n-3)/2$, which is the left endpoint of interval I_n which gives again the common starting point of two line segments. In these points, the direction of stability switching is the same as that in the two connecting segments. So in the rest of the discussion, we will assume that $\theta_1 + \theta_2 < \pi$. Hence the stability index $\bar{S}$ is positive on the solid red segments of the stability switching curve. In Figure 3, the arrows on the solid red segment indicate the positive direction and the red R and L mean the right and left regions along the red segment. As (τ_1, τ_2) moves from the R-region to the L-region and $\bar{S} > 0$, Theorem 2 implies that a solution pair of (18) crosses the imaginary axis to the right. That is, stability is lost. As seen in Figure 4B, the stability is lost at point A with $\tau_1 = \tau_1^A$ when τ_1 increases along the horizontal dotted line at $\tau_2^0 = 3$.

Similarly, we can compute the stability index on the solid blue segment,
$$L_2^+(0,0) = \left\{(\bar{\tau}_1^+(\omega,0), \bar{\tau}_2^-(\omega,0)) \mid \omega \in \left[\frac{1}{2}K, \frac{3}{2}K\right]\right\}.$$

From (7) and (8) with K,
$$\begin{aligned}\omega(\bar{\tau}_1^+(\omega,0) - \bar{\tau}_2^-(\omega,0)) &= \left[\frac{3}{2}\pi - \pi + \bar{\theta}_1(\omega)\right] - \left[\frac{3\pi}{2} - \pi - \bar{\theta}_2(\omega)\right] \\ &= \bar{\theta}_1(\omega) + \bar{\theta}_2(\omega).\end{aligned}$$

Then
$$\sin[\omega(\bar{\tau}_1^+(\omega,0) - \bar{\tau}_2^-(\omega,0))] = \sin[\bar{\theta}_1(\omega) + \bar{\theta}_2(\omega)] > 0.$$
The stability index $\bar{S}$ is negative,
$$\bar{S} = -\frac{K^2\beta(n-1)}{\omega^2}\sin[\bar{\theta}_1(\omega) + \bar{\theta}_2(\omega)] < 0.$$

The blue L and R denote the right-region and the left-region with respect to the solid blue segment. Hence the stability is lost when a pair of (τ_1, τ_2) crosses the blue segment from the L-region to the R-region.

Consider the stability switching on the dotted red segment located in the upper-left corner of Figure 3. The segment is described by
$$L_1^-(0,1) = \left\{(\tau_1^-(\omega,0), \tau_2^+(\omega,1)) \mid \omega \in \left[\frac{1}{2}K, \frac{3}{2}K\right]\right\}.$$

Then
$$\omega\left(\tau_1^-(\omega,0) - \tau_2^+(\omega,1)\right) = -\pi - (\theta_1(\omega) + \theta_2(\omega)).$$

The stability index is positive

$$S = \frac{K^2\beta}{\omega} \sin\left[\theta_1(\omega) + \theta_2(\omega)\right] > 0$$

showing that crossing these segments from R to L, stability is lost.

In the lower part of Figure 3, there is a small segment of $L_1^-(0,0)$ where

$$L_1^-(0,0) = \left\{ \tau_1^-(\omega,0),\ \tau_2^+(\omega,0) \mid \omega \in \left[\frac{1}{2}K,\ \frac{3}{2}K\right] \right\},$$

so

$$\begin{aligned}
\omega\left(\tau_1^-(\omega,0) - \tau_2^+(\omega,0)\right) &= \left[\frac{3}{2}\pi - \pi - \theta_1(\omega)\right] - \left[\frac{1}{2}\pi + \pi - \theta_2(\omega)\right] \\
&= \pi - [\theta_1(\omega) + \theta_2(\omega)].
\end{aligned}$$

Then
$$\sin\left[\omega\left(\tau_1^-(\omega,0) - \tau_2^+(\omega,0)\right)\right] = \sin\left[\tilde{\theta}_1(\omega) + \tilde{\theta}_2(\omega)\right] > 0$$

meaning that crossing this segment from the stable region, at least one eigenvalue changes the sign of its real part from negative to positive, implying stability loss.

5. Concluding Remarks

In this paper, n-firm dynamic oligopolies were examined without product differentiation and with linear price and cost functions. Continuous time scales were assumed reconsidering the classical dynamic model of McManus and Quandt (1961) with the best response dynamics. Without delays, the equilibrium is always asymptotically stable without delays regardless of the values of the positive adjustment speeds. We examined how this stability is lost when the firms face implementation and information delays. For the sake of mathematical simplicity, it was assumed that the firms have the same marginal costs and identical delays in both types. If these delays are equal, then a single-delay model is obtained. If the delay is sufficiently small, then the equilibrium is oscillatory stable, at the threshold, the trajectories show cyclic behavior and for larger delays, the cycles become expanding. If the delays are different, then in the resulting two-delay case the stability switching curves were first constructed and then the directions of the stability switches were determined. Growth rate dynamics brought nonlinearities into the model, but their linearized version is identical with best response dynamics, so shows similar local dynamics. Numerical results and simulation studies verify and illustrate the theoretical findings.

This research can be continued in two different ways. One is the consideration of different model modifications such as product differentiation, multi-product models, oligopsonies, labor-managed, and rent seeking oligopolies, including market saturation to mention only a few. The other research direction could be to examine nonlinear models, the local dynamics are similar to that of linear models, however with very different global dynamic behavior.

Author Contributions: Conceptualization, methodology, A.M. and F.S.; software, A.M.; validation, A.M. and F.S.; formal analysis, A.M.; writing—original draft preparation, A.M.; writing—review and editing, F.S.; visualization, A.M.; All authors have read and agreed to the published version of the manuscript.

Funding: This research received the financial support from the Japan Society for the Promotion of Science (Grant-in-Aid for Scientific Research (C) 20K01566).

Acknowledgments: The authors would like to thank the anonymous reviewers for their careful reading and valuable comments.

Conflicts of Interest: The authors declare no conflict of interest.

References

1. Cournot, A. *Recherches sur les Principes Mathématiques de la Théorie des Richessess*; Hachette: Paris, France, 1833.
2. Theocharis, T.D. On the stability of the Cournot solution on the oligopoly problem. *Rev. Econ. Stud.* **1960**, *27*, 133–134. [CrossRef]
3. McManus, M.; Quandt, R. Comments on the stability of the Cournot oligopoly model. *Rev. Econ. Stud.* **1961**, *28*, 136–139. [CrossRef]
4. Okuguchi, K. *Expectations and Stability in Oligopoly Models*; Springer: Berlin, Germany, 1976.
5. Okuguchi, K.; Szidarovszky, F. *The Theory of Oligopoly with Multi-Product Firms*, 2nd ed.; Springer: Berlin, Germany, 1999.
6. Ćanovas, J.; Puu, T.; Ruiz, M. The Cournot-Theocharis problem reconsidered. *Chaos Solitions Fractals* **2008**, *37*, 1025–1039. [CrossRef]
7. Hommes, C.; Ochea, M.; Tuinstra, J. Evolutionary competition between adjustment processes in Cournot oligopoly: Instability and complex dynamics. *Dyn. Games Appl.* **2018**, *8*, 822–843. [CrossRef]
8. Lampart, M. Stability of the Cournot equilibrium for a Cournot oligopoly model with n-competitors. *Chaos Solitions Fractals* **2012**, *45*, 1081–1085. [CrossRef]
9. Puu, T. On the stability of Cournot equilibrium when the number of competitors increases. *J. Econ. Behav. Organ.* **2008**, *66*, 445–456. [CrossRef]
10. Puu, T. Rational expectations and the Cournot-Theocharis problem. *Discret. Dyn. Nat. Soc.* **2006**, *2006*, 32103. [CrossRef]
11. Matsumoto, A.; Szidarovszky, F. Theocharis problem reconsidered in differentiated oligopoly. *Discret. Dyn. Nat. Soc.* **2014**, *2014*, 630351. [CrossRef]
12. Bischi, G.-I.; Chiarella, C.; Kopel, M.; Szidarovszky, F. *Nonlinear Oligopolies: Stability and Bifurcations*; Springer: Berlin, Germany, 2010.
13. Matsumoto, A.; Szidarovszky, F. *Dynamic Oligopolies with Time Delays*; Springer Nature: Berlin/Heidelberg, Germany, 2018.
14. Gu, K.; Nicolescu, S.-I.; Chen, I. On the stability crossing curves for general systems with two delays. *J. Math. Appl.* **2005**, *3311*, 231–253. [CrossRef]

© 2020 by the authors. Licensee MDPI, Basel, Switzerland. This article is an open access article distributed under the terms and conditions of the Creative Commons Attribution (CC BY) license (http://creativecommons.org/licenses/by/4.0/).

Article

The Convergence Analysis of a Numerical Method for a Structured Consumer-Resource Model with Delay in the Resource Evolution Rate

Luis M. Abia [1], Óscar Angulo [2,*], Juan C. López-Marcos [1] and Miguel A. López-Marcos [1]

[1] Departamento de Matemática Aplicada, Facultad de Ciencias, Universidad de Valladolid e IMUVa, 47011 Valladolid, Spain; abia@mac.uva.es (L.M.A.); lopezmar@mac.uva.es (J.C.L.-M.); malm@mac.uva.es (M.A.L.-M.)
[2] Departamento de Matemática Aplicada, ETS de Ingenieros de Telecomunicación, Universidad de Valladolid e IMUVa, 47011 Valladolid, Spain
* Correspondence: oscar@mat.uva.es; Tel.: +34-983-423000 (ext. 5835); Fax: +34-983-423-661

Received: 05 August 2020; 24 August 2020; Published: 27 August 2020

Abstract: In this paper, we go through the development of a new numerical method to obtain the solution to a size-structured population model that describes the evolution of a consumer feeding on a dynamical resource that reacts to the environment with a lag-time response. The problem involves the coupling of the partial differential equation that represents the population evolution and an ordinary differential equation with a constant delay that describes the evolution of the resource. The numerical treatment of this problem has not been considered before when a delay is included in the resource evolution rate. We analyzed the numerical scheme and proved a second-order rate of convergence by assuming enough regularity of the solution. We numerically confirmed the theoretical results with an academic test problem.

Keywords: size-structured population; consumer-resource model; delay differential equation; numerical methods; characteristics method; convergence analysis

MSC: 92D25; 92D40; 65M25; 65M12; 35B40

1. Introduction

We consider a model that describes the evolution of a size-structured consumer in an environment inhabited by a single unstructured resource. We assume that the resource responds to the environment with a constant time delay. The model is composed of two nonlinear coupled problems. On the one hand, a size-structured population model governed by a first-order hyperbolic equation with a nonlocal and nonlinear boundary condition, which represents the evolution over time of the consumer:

$$\begin{cases} u_t + (g(x,s(t),t)\,u)_x = -\mu(x,s(t),t)\,u, & x_0 < x < x_M(t), \quad t > 0, \\ g(x_0,s(t),t)\,u(x_0,t) = \int_{x_0}^{x_M(t)} \alpha(x,s(t),t)\,u(x,t)\,dx, & t > 0, \\ u(x,0) = u_0(x), & x_0 \leq x \leq x_M^0. \end{cases} \quad (1)$$

Variables x and t represent the size and the time, respectively. Size is the variable which structures the individuals in the consumer population, x_0 denotes the newborn individual's size (the size at birth) that is assumed to be constant and positive and $x_M(t)$ represents the maximum size of individuals at time t. Then, $x_M(0) = x_M^0$ is the initial maximum size. The dependent variables u and s are the density of individuals of size x and the amount of the resource available at time t, respectively. The so-called vital rates, the mortality, the fertility and the growth rates, are given by μ, α and g, respectively.

The mortality and the fertility are nonnegative functions and the growth rate has no-sign restriction, although we assume that $g(x_0, \cdot, \cdot) > 0$, which means that each individual increases in size at birth. Vital rates depend on the structuring variable, the time and the amount of the resource available, to take into account the influences of these factors on the dynamics of the population. The size of the individuals changes according to the differential equation

$$x'(t) = g(x(t), s(t), t), \quad t > 0.$$

As we allow a negative growth rate g, we are considering the case in which an individual in the population could shrink in size under a food shortage environmental condition. In particular, the maximum size, $x_M(t)$, is not fixed in the model and evolves following the corresponding characteristic curve of (1)

$$\begin{cases} x'_M(t) = g(x_M(t), s(t), t), & t > 0, \\ x_M(0) = x_M^0. \end{cases} \quad (2)$$

On the other hand, the unstructured resource $s(t)$, $t \geq 0$, which provides feeding for the individuals of the population, evolves with time according to the following initial value problem for a delay ordinary differential equation:

$$\begin{cases} s'(t) = f(s(t), s(t-\tau), I(t), t), & t > 0, \\ s(t) = s_0(t), & t \in [-\tau, 0]. \end{cases} \quad (3)$$

The rate of change in the evolution of the resource is given by a function f that includes dependence on time and on the consumer population through the nonlocal term

$$I(t) = \int_{x_0}^{x_M(t)} \gamma(x, s(t), t)\, u(x, t)\, dx, \quad t \geq 0, \quad (4)$$

where γ is a function that represents the individual rate of consumption for individuals of a determined size. It also depends on the amount of the resource available at times t and $t - \tau$, $s(t)$ and $s(t-\tau)$ respectively. The memory effect, $s(t-\tau)$, can be seen as a deferred influence on the environment of the resource affordability. A common biological situation for this phenomenon occurs when the resource population is close to the carrying capacity of the environment or near extinction, which can make the population react with a certain delay [1]. Another situation in which this deferred influence occurs is when the resource is formed only by the adult individuals of the population, and the delay is due to the maturation time [2]. The solution of the model (1), (3) is determined once we know the initial conditions $u(x, 0) = u_0(x)$, $x_0 \leq x \leq x_M^0$, and $s(t) = s_0(t)$, $t \in [-\tau, 0]$.

Although a large number of papers on consumer models have been considered in past years, not so many address the case in which one of the populations is structured. Consumer-resource models have been studied from the very initial work of Kooijman and Metz [2]. These authors presented a mathematical model for the development of an ectothermic population (*Daphnia magna*, water flea) in which the amount of food they are supplied with represents a regulatory mechanism for the population density. The model was length-structured and also included a stage of maturity: a differentiation among juveniles and adults. This work is considered as the origin of the modern dynamic energy budget theory. The well-posedness of the model equations and the continuously dependence on model data was studied by Thieme [3]. The dynamical properties of the model were explored numerically in [4], and later, analytically in [5–7]. However the description of the evolution of the resource as a delay differential equation within a physiological structured consumer-resource model has not been developed theoretically yet.

The theoretical treatment of this model is not easy; therefore, its numerical analysis is a valid tool and sometimes the only one affordable. The integration of the model without delay has been developed by means of the Excalator Boxcar Train (EBT) [8] and a characteristics method [9]. This last work included the convergence proof of the method while the convergence of the popular EBT method was recently considered [10,11]. However, the numerical treatment of the coupled model with a resource evolving according to a delay differential equation remained unexplored until our study, so our goal was to provide a numerical method to perform the integration of such a model and its convergence analysis.

The remainder of the paper is organized as follows. Section 2 is devoted to the introduction of the numerical method to integrate problem (1), (3). We employ the technique of integration along the characteristic curves in order to obtain the new numerical scheme. In Section 3, we carry out the convergence analysis of the scheme. It is based on a theoretical framework that involves the properties of consistency and stability of the numerical method. We also pay attention to the properties required by the numerical quadrature rule used in the integration. Finally, in Section 4, we present a test done in order to numerically confirm the theoretical order of convergence.

2. Numerical Method

We begin with the description of the numerical scheme. We must employ the integration along the characteristic curves; therefore, we should remind the reader of the definition of $x(t; t^*, x_*)$, $x_* \in [x_0, x_M(t)]$, $t^* > 0$, as the solution of

$$\begin{cases} \dfrac{d}{dt} x(t; t^*, x_*) = g(x(t; t^*, x_*), s(t), t), & t \geq t^*, \\ x(t^*; t^*, x_*) = x_*, \end{cases} \quad (5)$$

that represents the characteristic curve that at time t^* starts at x_*. Now, we consider $w(t; t^*, x_*) = u(x(t; t^*, x_*), t)$, for each $x_* \in [x_0, x_M(t)]$, $t^* > 0$. It satisfies the following initial value problem

$$\begin{cases} \dfrac{d}{dt} w(t; t^*, x_*) = -\mu^*(x(t; t^*, x_*), s(t), t)\, w(t; t^*, x_*), & t > t^*, \\ w(t^*; t^*, x_*) = u(x_*, t^*), \end{cases} \quad (6)$$

where $\mu^*(x, s(t), t) = \mu(x, s(t), t) + g_x(x, s(t), t)$, whose solution is given by the following formula

$$u(x(t; t^*, x_*), t) = u(x_*, t^*) \exp\left\{-\int_{t^*}^{t} \mu^*\left(x(\omega; t^*, x_*), s(\omega), \omega\right) d\omega\right\}, \quad t \geq t^*. \quad (7)$$

For the numerical method, we discretize this expression, together with the boundary condition and the initial data in (1), and the initial value problem in (3).

The integration is carried out on a finite time interval $[0, T]$. Then, given $J, L \in \mathbb{N}$, we define the discretization parameters $h = (x_M^0 - x_0)/J$, $k = \tau/L$, and the number of discrete time levels $N = [T/k]$, which are given by $t^n = nk$, $-L \leq n \leq N$. We begin the description of the numerical method with the initial values, which, in this case, are the initial size discretization $\mathbf{X}^0$, $X_j^0 = x_0 + jh$, $0 \leq j \leq J$, the initial condition on the initial grid, $\mathbf{U}^0$, $U_j^0 = u_0(X_j^0)$, $0 \leq j \leq J$, and the resource, that has to be initialized on the interval $[-\tau, 0]$, with the values $S^n = s_0(t^n)$, $-L \leq n \leq 0$.

We compute, for $0 \leq n \leq N-1$, the approximation at the general level t^{n+1}: $\{\mathbf{X}^{n+1}, S^{n+1}, \mathbf{U}^{n+1}\}$, $\mathbf{X}^{n+1} = (X_0^{n+1}, X_1^{n+1}, \ldots, X_{J+n+1}^{n+1})$, $\mathbf{U}^{n+1} = (U_0^{n+1}, U_1^{n+1}, \ldots, U_{J+n+1}^{n+1})$ when the values at the previous time steps are known: $\{\mathbf{X}^m, S^m, \mathbf{U}^m\}$, $\mathbf{X}^m = (X_0^m, X_1^m, \ldots, X_{J+m}^m)$, $\mathbf{U}^m = (U_0^m, U_1^m, \ldots, U_{J+m}^m)$, $0 \leq m \leq n$. We should point out that we design a second-order method based on the modified Euler scheme and on the trapezoidal quadrature rule. This means that we need an intermediate stage in which we compute a first-order approximation: $\{\mathbf{X}^{n+1,*}, S^{n+1,*}, \mathbf{U}^{n+1,*}\}$, $\mathbf{X}^{n+1,*} = (X_0^{n+1,*}, X_1^{n+1,*}, \ldots, X_{J+n+1}^{n+1,*})$, $\mathbf{U}^{n+1,*} = (U_0^{n+1,*}, U_1^{n+1,*}, \ldots, U_{J+n+1}^{n+1,*})$, given by the equations

$$\begin{cases} X_0^{n+1,*} = x_0, \quad X_{j+1}^{n+1,*} = X_j^n + k\, g(X_j^n, S^n, t^n), \quad 0 \le j \le J+n, \\ S^{n+1,*} = S^n + k\, f(S^n, S^{n-L}, \mathcal{Q}(\mathbf{X}^n, \gamma^n(\mathbf{X}, S) \cdot \mathbf{U}^n), t^n), \\ U_{j+1}^{n+1,*} = U_j^n \exp\{-k\,\mu^*(X_j^n, S^n, t^n)\}, \quad 0 \le j \le J+n, \\ g(X_0^{n+1,*}, S^{n+1,*}, t^{n+1})\, U_0^{n+1,*} = \mathcal{Q}(\mathbf{X}^{n+1,*}, \boldsymbol{\alpha}^{n+1,*}(\mathbf{X}, S) \cdot \mathbf{U}^{n+1,*}). \end{cases} \quad (8)$$

As soon as we have computed the intermedium quantities $\{\mathbf{X}^{n+1,*}, S^{n+1,*}, \mathbf{U}^{n+1,*}\}$ we can obtain the approximations at the advanced time level with

$$\begin{cases} X_0^{n+1} = x_0, \quad X_{j+1}^{n+1} = X_j^n + \dfrac{k}{2}\left(g(X_j^n, S^n, t^n) + g(X_{j+1}^{n+1,*}, S^{n+1,*}, t^{n+1})\right), \quad 0 \le j \le J+n, \\ S^{n+1} = S^n + \dfrac{k}{2}\Big(f(S^n, S^{n-L}, \mathcal{Q}(\mathbf{X}^n, \gamma^n(\mathbf{X}, S) \cdot \mathbf{U}^n), t^n) \\ \qquad\qquad + f(S^{n+1,*}, S^{n+1-L}, \mathcal{Q}(\mathbf{X}^{n+1,*}, \gamma^{n+1,*}(\mathbf{X}, S) \cdot \mathbf{U}^{n+1,*}), t^{n+1})\Big), \\ U_{j+1}^{n+1} = U_j^n \exp\left\{-\dfrac{k}{2}\left(\mu^*(X_j^n, S^n, t^n) + \mu^*(X_{j+1}^{n+1,*}, S^{n+1,*}, t^{n+1})\right)\right\}, \quad 0 \le j \le J+n, \\ g(X_0^{n+1}, S^{n+1}, t^{n+1})\, U_0^{n+1} = \mathcal{Q}(\mathbf{X}^{n+1}, \boldsymbol{\alpha}^{n+1}(\mathbf{X}, S) \cdot \mathbf{U}^{n+1}). \end{cases} \quad (9)$$

In (8) and (9), at each time level, $\gamma^n(\mathbf{X}, S)$ and $\boldsymbol{\alpha}^n(\mathbf{X}, S)$ represent vectors with components $\gamma_j^n(\mathbf{X}, S) = \gamma(X_j^n, S^n, t^n)$ and $\boldsymbol{\alpha}_j^n(\mathbf{X}, S) = \alpha(X_j^n, S^n, t^n)$, $0 \le j \le J+n$, respectively, and $\gamma^{n,*}(\mathbf{X}, S)$ and $\boldsymbol{\alpha}^{n,*}(\mathbf{X}, S)$ vectors with components $\gamma_j^{n,*}(\mathbf{X}, S) = \gamma(X_j^{n,*}, S^{n,*}, t^n)$ and $\boldsymbol{\alpha}_j^{n,*}(\mathbf{X}, S) = \alpha(X_j^{n,*}, S^{n,*}, t^n)$, $0 \le j \le J+n+1$, respectively. The notation $\gamma^n(\mathbf{X}, S) \cdot \mathbf{U}^n$, $\boldsymbol{\alpha}^n(\mathbf{X}, S) \cdot \mathbf{U}^n$, $\gamma^{n,*}(\mathbf{X}, S) \cdot \mathbf{U}^{n,*}$ and $\boldsymbol{\alpha}^{n,*}(\mathbf{X}, S) \cdot \mathbf{U}^{n,*}$ represent the componentwise products of the corresponding vectors, and $\mathcal{Q}$ is an appropriate second-order quadrature rule, as given below. We observed that, in this procedure, the total number of nodes in the grid increases in one at each time step due to the new node that fluxes from the left boundary into the domain. This feature led us to name this kind of scheme the aggregation grid nodes method (AGN) [12]. The algorithm is explicit and totally straightforward and can be resumed as in the pseudocode given in Box 1.

Box 1. Pseudocode of the algorithm AGN (and SGN).

```
% Choose L, J integers
k = τ/L;  h = (x_M^0 − x_0)/J
N = [T/k]
% Initialization
do j=0,J
        X_j^0 = x_0 + jh   % Initial grid points
        U_j^0 = u_0(X_j^0)
end
do n=-L,0
        S^n = s_0(t^n)
end
% Advancing the solution on time, step by step
do n=0, N-1
        Formulae in (8)   % First stage to X^{n+1,*}, S^{n+1,*}, U^{n+1,*}
        Formulae in (9)   % Second stage to X^{n+1}, S^{n+1}, U^{n+1}
% Optional (difference between AGN and SGN methods)
        Formulae in (10)  % Selection procedure in case of SGN method
end
```

In the numerical experimentation, we use a more efficient procedure in order to keep the number of nodes constant (and consequently, to not increase the computational cost) at each time step. Toward that goal, we eliminate a chosen node from the grid at each time step. The selection procedure we use in the method depends on the dynamics of the grid, as it was introduced in [13]. We refer to this kind of scheme as the selection grid nodes method (SGN) [12]. The node picked out is the first X_l^{n+1} that satisfies

$$|X_{l+1}^{n+1} - X_{l-1}^{n+1}| = \min_{1 \leq j \leq J} |X_{j+1}^{n+1} - X_{j-1}^{n+1}|. \tag{10}$$

Once it is elected, we rename the nodes to keep the index labels from $j = 0$ to $j = J$.

Finally, we pay attention to the nonlocal terms; they are discretized by means of a composite quadrature rule, based on the trapezoidal formula, on the grid points $\mathbf{X} = (X_0, X_1, \ldots, X_\aleph)$, $\aleph \in \mathbb{N}$,

$$\mathcal{Q}(\mathbf{X}, \mathbf{V}) = \sum_{l=0}^{\aleph-1} \frac{X_{l+1} - X_l}{2} (V_l + V_{l+1}), \quad \text{where} \quad \mathbf{V} = (V_0, V_1, \ldots, V_\aleph).$$

3. Convergence Analysis

The convergence analysis here follows the discretization framework developed in [14]. In this framework, suitable discrete normed spaces and operators are introduced to formulate the equations of the numerical method. Then, appropriate properties of consistency and nonlinear stability are established. The numerical method is an adaptation of the one given in [9], and so is the analysis. However, for the sake of completeness, we provide the details of the numerical method formulation, and for the sake of simplicity, we present only those parts of the convergence analysis proofs associated with estimates derived from the discretization of the delay term in the differential equation. That means that the step by step recurrences for the errors at each time level increase the order of the difference equation as the time discretization parameter tends to zero. For the theoretical analysis, we consider the AGN method.

With this aim, we fix $T > 0$ and assume that the problem (1), (3) has a unique solution $u(x, t)$, $x \in [x_0, x_M(t)]$, and $s(t)$, $t \in [-\tau, T]$, with the following regularity assumptions:

Hypothesis 1 (H1). *$u \in \mathcal{C}^2([x_0, x_M(t)] \times [0, T])$ and is nonnegative.*

Hypothesis 2 (H2). *$s \in \mathcal{C}^2([-\tau, T])$ and is nonnegative.*

Additionally, we assume that there exists a compact neighborhood D of $\{s(t), 0 \leq t \leq T\}$, such that

Hypothesis 3 (H3). *$\gamma \in \mathcal{C}^2([x_0, x_M(t)] \times D \times [0, T])$ and is nonnegative.*

Hypothesis 4 (H4). *$\mu \in \mathcal{C}^2([x_0, x_M(t)] \times D \times [0, T])$ and is nonnegative.*

Hypothesis 5 (H5). *$\alpha \in \mathcal{C}^2([x_0, x_M(t)] \times D \times [0, T])$ and is nonnegative.*

Hypothesis 6 (H6). *$g \in \mathcal{C}^3([x_0, x_M(t)] \times D \times [0, T])$ and there exists a positive constant C such that $g(x_0, s, t) \geq C$, $s, t \geq 0$. In addition, the characteristic curves $x(t; t^*, x_*)$ defined by (5), are continuous and differentiable with respect to the initial values $(t^*, x_*) \in [0, T] \times [x_0, x_M(t)]$.*

Finally, we suppose that there are compact neighborhoods, D_f of $\{s(t), -\tau \leq t \leq T\}$, and D_I of

$$\left\{ \int_{x_0}^{x_M(t)} \gamma(x, s(t), t) u(x, t) \, dx, \, 0 \leq t \leq T \right\}, \text{ such that}$$

Hypothesis 7 (H7). *$f \in \mathcal{C}^2(D_f \times D_f \times D_I \times [0, T])$ and is nonnegative.*

Now, we choose $L \in \mathbb{N}$ and assume that the time discretization parameter, k, takes values in the set
$$K = \{k : k = \tau/(\nu L), \nu \in \mathbb{N}\}.$$

Then for $k \in K$, we set $N = [T/k]$ and choose $J \in \mathbb{N}$ such that $h = (x_M^0 - x_0)/J$, and the ratio $r := k/h$ is a positive constant fixed throughout the analysis. Then, x_M^0 is always the last node in the size grid. For each $k \in K$, we define the space
$$\mathcal{A}_k = \prod_{n=0}^{N} \left(\mathbb{R}^{J+n} \times \mathbb{R}^{J+n+1} \right) \times \mathbb{R}^{L+N+1},$$

where vectors $(\mathbf{y}^0, \mathbf{V}^0, \ldots, \mathbf{y}^N, \mathbf{V}^N, \mathbf{a}) \in \mathcal{A}_k$ are used to describe the following approximations: to the inner and right-hand boundary grid nodes, given by $\mathbf{y}^n \in \mathbb{R}^{J+n}$, and to the theoretical solution at the complete grid, $\mathbf{V}^n \in \mathbb{R}^{J+n+1}$, for each time level t^n, $0 \leq n \leq N$; and to the theoretical solution to the delay differential equation at time levels t^n, $-L \leq n \leq N$, provided by $\mathbf{a} \in \mathbb{R}^{L+N+1}$. We also consider the space
$$\mathcal{B}_k = \left(\mathbb{R}^J \times \mathbb{R}^{J+1} \times \mathbb{R}^{L+1} \right) \times \mathbb{R}^N \times \prod_{n=1}^{N} \left(\mathbb{R}^{J+n} \times \mathbb{R}^{J+n} \right) \times \mathbb{R}^N,$$

where each component of vector $\left(\mathbf{Y}^0, \mathbf{P}^0, \mathbf{A}^0, \mathbf{P}_0, \mathbf{Y}^1, \mathbf{P}^1, \ldots, \mathbf{Y}^N, \mathbf{P}^N, \mathbf{A} \right) \in \mathcal{B}_k$ describes the residuals of the discrete solution when it is substituted on the discrete equations that define the numerical method: we discriminate $\mathbf{Y}^0$, $\mathbf{P}^0$, $\mathbf{A}^0 = (A^{-L}, A^{-L+1}, \ldots, A^0)$, given by the approximations to the initial conditions; $\mathbf{P}_0$, provided by the solution at the left boundary node at each time level; $\mathbf{Y}^n$, $\mathbf{P}^n$, $0 \leq n \leq N$, given by the grid nodes and solution to them; and $\mathbf{A} = (A^1, A^2, \ldots, A^N)$, due to the solution to the delay differential equation, as we show below. Both spaces, $\mathcal{A}_k$ and $\mathcal{B}_k$, have the same dimension.

In order to measure the size of the errors, we define
$$\|\boldsymbol{\eta}\|_\infty = \max_{1 \leq j \leq \aleph} |\eta_j|, \quad \boldsymbol{\eta} = (\eta_1, \eta_2, \ldots, \eta_\aleph) \in \mathbb{R}^\aleph,$$
$$\|\mathbf{V}^n\|_1 = \sum_{j=0}^{J+n} h |V_j^n|, \quad \mathbf{V}^n = (V_0^n, V_1^n, \ldots, V_{J+n}^n) \in \mathbb{R}^{J+n+1}.$$

Thus, we endow the spaces $\mathcal{A}_k$ and $\mathcal{B}_k$ with the following norms. If $(\mathbf{y}^0, \mathbf{V}^0, \ldots, \mathbf{y}^N, \mathbf{V}^N, \mathbf{a}) \in \mathcal{A}_k$,
$$\left\| \left(\mathbf{y}^0, \mathbf{V}^0, \ldots, \mathbf{y}^N, \mathbf{V}^N, \mathbf{a} \right) \right\|_{\mathcal{A}_k} = \max \left(\max_{0 \leq n \leq N} \|\mathbf{y}^n\|_\infty, \max_{0 \leq n \leq N} \|\mathbf{V}^n\|_\infty, \|\mathbf{a}\|_\infty \right).$$

On the other hand, if $\left(\mathbf{Y}^0, \mathbf{P}^0, \mathbf{A}^0, \mathbf{P}_0, \mathbf{Y}^1, \mathbf{P}^1, \ldots, \mathbf{Y}^N, \mathbf{P}^N, \mathbf{A} \right) \in \mathcal{B}_k$,
$$\left\| \left(\mathbf{Y}^0, \mathbf{P}^0, \mathbf{A}^0, \mathbf{P}_0, \mathbf{Y}^1, \mathbf{P}^1, \ldots, \mathbf{Y}^N, \mathbf{P}^N, \mathbf{A} \right) \right\|_{\mathcal{B}_k}$$
$$= \|\mathbf{Y}^0\|_\infty + \|\mathbf{P}^0\|_\infty + \|\mathbf{A}^0\|_\infty + \|\mathbf{P}_0\|_\infty + \sum_{n=1}^{N} k \|\mathbf{Y}^n\|_\infty + \sum_{n=1}^{N} k \|\mathbf{P}^n\|_\infty + \sum_{n=1}^{N} k |A^n|.$$

Now, for each $k \in K$, we define
$$\mathbf{x}^n = (x_1^n, \ldots, x_{J+n}^n) \in \mathbb{R}^{J+n}, x_j^0 = x_0 + jh, 1 \leq j \leq J; x_j^n = x(t_n; t_{n-1}, x_{j-1}^{n-1}), 1 \leq j \leq J+n, 1 \leq n \leq N; \quad (11)$$

and we denote $x_0^n = x_0$, $n \geq 0$. Recall that $x(t; t^*, x_*)$ represents the theoretical solution to problem (5), $t^* \in [0, T]$, $x^* \in [x_0, x_M(t)]$. In addition, if u represents the theoretical solution to (1) we define

$$\mathbf{u}^n = (u_0^n, u_1^n, \ldots, u_{J+n}^n) \in \mathbb{R}^{J+n+1}, \quad u_j^n = u(x_j^n, t_n), \quad 0 \leq j \leq J+n, \quad 0 \leq n \leq N. \tag{12}$$

Finally, if s is the theoretical solution to (3) then we define

$$\mathbf{s}_k = (s^{-L}, s^{-L+1}, \ldots, s^N), \quad s^n = s(t^n), \quad -L \leq n \leq N. \tag{13}$$

Therefore, $\tilde{\mathbf{u}}_k = (\mathbf{x}^0, \mathbf{u}^0, \mathbf{x}^1, \mathbf{u}^1, \ldots, \mathbf{x}^N, \mathbf{u}^N, \mathbf{s}_k) \in \mathcal{A}_k$.

In the following, we introduce the discretization operator. Let R be a positive constant and let we denote by $B_{\mathcal{A}_k}(\tilde{\mathbf{u}}_k, R k^p) \subset \mathcal{A}_k$ the open ball with center $\tilde{\mathbf{u}}_k$ and radius $R k^p$, $1 < p < 2$. Then, we define

$$\Phi_k : B_{\mathcal{A}_k}(\tilde{\mathbf{u}}_k, R k^p) \to \mathcal{B}_k,$$

$$\Phi_k\left(\mathbf{y}^0, \mathbf{V}^0, \ldots, \mathbf{y}^N, \mathbf{V}^N, \mathbf{a}\right) = \left(\mathbf{Y}^0, \mathbf{P}^0, \mathbf{A}^0, \mathbf{P}_0, \mathbf{Y}^1, \mathbf{P}^1, \ldots, \mathbf{Y}^N, \mathbf{P}^N, \mathbf{A}\right), \tag{14}$$

given by the following equations, first

$$\mathbf{Y}^0 = \mathbf{y}^0 - \mathbf{X}^0 \in \mathbb{R}^J, \tag{15}$$

$$\mathbf{P}^0 = \mathbf{V}^0 - \mathbf{U}^0 \in \mathbb{R}^{J+1}, \tag{16}$$

$$\mathbf{A}^0 = \mathbf{a}^0 - \mathbf{S}^0 \in \mathbb{R}^{L+1}, \tag{17}$$

where $\mathbf{a}^0 = (a^{-L}, a^{-L+1}, \ldots, a^0)$. Vectors $\mathbf{X}^0$ and $\mathbf{U}^0$ represent approximations at $t = 0$, respectively, to the initial grid nodes and to the theoretical solution at these nodes. Vector $\mathbf{S}^0$ represents an approximation to the initial resource in the interval $[-\tau, 0]$. Second,

$$P_0^{n+1} = V_0^{n+1} - \frac{\mathcal{Q}\left(\mathbf{y}^{n+1}, \boldsymbol{\alpha}^{n+1}(\mathbf{y}, \mathbf{a}) \cdot \mathbf{V}^{n+1}\right)}{g(x_0, a^{n+1}, t^{n+1})}, \tag{18}$$

$$Y_{j+1}^{n+1} = \frac{1}{k}\left\{y_{j+1}^{n+1} - y_j^n - \frac{k}{2}\left(g(y_j^n, a^n, t^n) + g(y_{j+1}^{n+1,*}, a^{n+1,*}, t^{n+1})\right)\right\}, \tag{19}$$

$$P_{j+1}^{n+1} = \frac{1}{k}\left\{V_{j+1}^{n+1} - V_j^n \exp\left(-\frac{k}{2}\left(\mu^*\left(y_j^n, a^n, t^n\right) + \mu^*\left(y_{j+1}^{n+1,*}, a^{n+1,*}, t^{n+1}\right)\right)\right)\right\}, \tag{20}$$

$0 \leq j \leq J + n - 1$,

$$A^{n+1} = \frac{1}{k}\left\{a^{n+1} - a^n - \frac{k}{2}\left(f\left(a^n, a^{n-L}, \mathcal{Q}(\mathbf{y}^n, \boldsymbol{\gamma}^n(\mathbf{y}, \mathbf{a}) \cdot \mathbf{V}^n), t^n\right)\right.\right.$$
$$\left.\left. + f\left(a^{n+1,*}, a^{n-L+1}, \mathcal{Q}(\mathbf{y}^{n+1,*}, \boldsymbol{\gamma}^{n+1,*}(\mathbf{y}, \mathbf{a}) \cdot \mathbf{V}^{n+1,*}), t^{n+1}\right)\right)\right\}, \tag{21}$$

$0 \leq n \leq N - 1$. Where, with the notation introduced in Section 2,

$$y_{j+1}^{n+1,*} = y_j^n + k\, g(y_j^n, a^n, t^n), \tag{22}$$

$$V_{j+1}^{n+1,*} = V_j^n \exp\left(-k\, \mu^*\left(y_j^n, a^n, t^n\right)\right), \tag{23}$$

$0 \leq j \leq J + n - 1$,

$$V_0^{n+1,*} = \frac{\mathcal{Q}(\mathbf{y}^{n+1,*}, \boldsymbol{\alpha}^{n+1,*}(\mathbf{y}, \mathbf{a}) \cdot \mathbf{V}^{n+1,*})}{g(x_0, a^{n+1,*}, t_{n+1})}, \tag{24}$$

$$a^{n+1,*} = a^n + k f\left(a^n, a^{n-L}, \mathcal{Q}(\mathbf{y}^n, \gamma^n(\mathbf{y}, a) \cdot \mathbf{V}^n), t^n\right), \tag{25}$$

$0 \leq n \leq N - 1$. We rewrite the quadrature rule employed in (18)–(25) in the next general form $\mathcal{Q}(\mathbf{X}, \mathbf{V}) = \sum_{l=0}^{J+n} q_l(\mathbf{X}) V_l$. Written in this way, we highlight that the number of nodes considered at each time level t^n is $J + n + 1$, counting both the boundary node $X_0^n = x_0^n = x_0, 0 \leq n \leq N$ and the interior grid nodes $X_j^n, 1 \leq j \leq J + n, 0 \leq n \leq N$. This notation is also valid even when we consider quadrature rules whose nodes are, at each time level, a subgrid of $\mathbf{X}^n, 0 \leq n \leq N$, by asumming that the corresponding weights $q_l(\mathbf{X})$ of some of the nodes are zero.

Note that Φ_k takes into account all the grid nodes and their corresponding solution values at each time level, and it employs quadrature rules possibly based on a subgrid. If $\tilde{\mathbf{U}}_k = (\mathbf{X}^0, \mathbf{U}^0, \mathbf{X}^1, \mathbf{U}^1, \ldots, \mathbf{X}^N, \mathbf{U}^N, \mathbf{s}_k) \in B_{A_k}(\tilde{\mathbf{u}}_k, R k^p)$, satisfies

$$\Phi_k(\tilde{\mathbf{U}}_k) = \mathbf{0} \in \mathcal{B}_k, \tag{26}$$

the nodes, $\mathbf{X}^n$ and $0 \leq n \leq N$ and the corresponding values of the solution, $\mathbf{U}^n, 0 \leq n \leq N$, at such nodes are numerical solutions to the scheme defined by (9) when the composite trapezoidal quadrature rule is used on a subgrid at each time step. On the other hand, the numerical solution to the scheme defined by (9) satisfies (26).

Henceforth, C will denote a positive constant, independent of k, h ($k = r h$), j ($0 \leq j \leq J + n$) and n ($-L \leq n \leq N$), and C may have different values at different places.

As in [12], it is possible to establish the following property that we called (SG) for the selection procedure given by (10): if $\{x_{j_l^n}^n\}_{l=0}^{M(n)}$, denotes a subgrid of the grid $\{x_j^n\}_{j=0}^{J+n+1}$ at the time level t^n, $0 \leq n \leq N$,

(SG) There exists a positive constant C such that, for h sufficiently small, $x_{j_{l+1}^n}^n - x_{j_l^n}^n \leq C h$, $0 \leq l \leq M(n) - 1$, $x_{j_0^n}^n = x_0$, $x_{j_{M(n)}^n}^n = x_{J+n}^n$, $0 \leq n \leq N$.

The property (SG) condenses the essential information about the adaptive selection procedure to allow the proof, under the hypotheses (H1)–(H7), of the following general properties of the composite trapezoidal quadrature $\mathcal{Q}(\mathbf{X}, \mathbf{V})$ based on the subgrids,

(P1) $|I(t^n) - \mathcal{Q}(\mathbf{x}^n, \gamma^n(\mathbf{x}, s) \cdot \mathbf{u}^n)| \leq C h^2$, when $h \to 0, 0 \leq n \leq N$.

(P2) $\left| \int_{x_0}^{x_M(t^n)} \alpha(x, s(t^n), t^n) u(x, t^n) \, dx - \mathcal{Q}(\mathbf{x}^n, \boldsymbol{\alpha}^n(\mathbf{x}, s) \cdot \mathbf{u}^n) \right| \leq C h^2$, when $h \to 0, 0 \leq n \leq N$.

(P3) $|q_j(\mathbf{x}^n)| \leq q h$, where q is a positive constant independent of h, k, j ($0 \leq j \leq J + n$) and n ($-L \leq n \leq N$), for $0 \leq j \leq J + n, 0 \leq n \leq N$.

(P4) Let R and p be positive constants with $1 < p < 2$. The quadrature weights q_j are Lipschitz continuous functions on $B_\infty(\mathbf{x}^n, R k^p), 0 \leq j \leq J + n, 0 \leq n \leq N$.

(P5) Let R and p be positive constants with $1 < p < 2$. If $\mathbf{y}^n, \mathbf{z}^n \in B_\infty(\mathbf{x}^n, R k^p)$, $\mathbf{V}^n \in B_\infty(\mathbf{u}^n, R k^p)$ and $a^n \in B_\infty(s^n, R k^p)$; then

$$\left| \sum_{i=0}^{J+n} (q_i(\mathbf{y}^n) - q_i(\mathbf{z}^n)) \, \gamma(z_i^n, a^n, t^n) \, V_i^n \right| \leq C \|\mathbf{y}^n - \mathbf{z}^n\|_\infty,$$

when $k \to 0, 0 \leq n \leq N$.

(P6) Let R and p be positive constants with $1 < p < 2$. If $\mathbf{y}^n, \mathbf{z}^n \in B_\infty(\mathbf{x}^n, R k^p)$, $\mathbf{V}^n \in B_\infty(\mathbf{u}^n, R k^p)$ and $a^n \in B_\infty(s^n, R k^p)$; then

$$\left| \sum_{i=0}^{J+n} (q_i(\mathbf{y}^n) - q_i(\mathbf{z}^n)) \, \alpha(z_i^n, a^n, t^n) \, V_i^n \right| \leq C \|\mathbf{y}^n - \mathbf{z}^n\|_\infty,$$

when $h \to 0, 0 \leq n \leq N$.

Then, the properties (P1)–(P6) allow us to establish that the nonlinear discrete operators describing Φ_k are well defined for $k \in K$ small enough, as we formulate in the following theorem (we refer to [9] about details of the proof).

Theorem 1. *Assume that hypotheses (H1)–(H7) hold and that the quadrature rules used in (18)–(25) satisfy the properties (P1)–(P6). If*

$$\left(\mathbf{X}^0, \mathbf{V}^0, \ldots, \mathbf{X}^N, \mathbf{V}^N, \mathbf{S}\right) \in B_{\mathcal{A}_k}(\tilde{\mathbf{u}}_k, R\, k^p),$$

where R is a fixed positive constant and $1 < p < 2$, then, for k sufficiently small,

$$\mathcal{Q}(\mathbf{X}^n, \gamma^n(\mathbf{X}, S) \cdot \mathbf{V}^n) \in D_I, \tag{27}$$

$0 \leq n \leq N$. *Furthermore, there exists a positive constant R', independent of k, such that, as $k \to 0$, $\mathbf{X}^{n,*} \in B_\infty(\mathbf{x}^n, R'\, k^p)$, $S^{n,*} \in B_\infty(s^n, R'\, k^p)$ and $\mathbf{V}^{n,*} \in B_\infty(\mathbf{u}^n, R'\, k^p)$, and*

$$\mathcal{Q}(\mathbf{X}^{n,*}, \gamma^{n,*}(\mathbf{X}, S) \cdot \mathbf{V}^{n,*}) \in D_I, \tag{28}$$

$1 \leq n \leq N$.

Now, we define the local discretization error as

$$\Phi_k(\tilde{\mathbf{u}}_k) = (Z^0, \mathbf{L}^0, \sigma^0, \mathbf{L}_0, Z^1, \mathbf{L}^1, \ldots, Z^N, \mathbf{L}^N, \sigma) \in \mathcal{B}_k,$$

and we say that the discretization (14) is consistent if, as $k \to 0$,

$$\lim \|\Phi_k(\tilde{\mathbf{u}}_k)\|_{\mathcal{B}_k} = 0.$$

The following theorem establishes the consistency of the numerical scheme defined by Equations (26).

Theorem 2 (Consistency). *Assume that hypotheses (H1)–(H7) hold and that the considered quadrature rules satisfy properties (P1)–(P6). Then, as $k \to 0$, the local discretization error satisfies*

$$\|\Phi_k(\tilde{\mathbf{u}}_k)\|_{\mathcal{B}_k} = \|\mathbf{u}^0 - \mathbf{U}^0\|_\infty + \|\mathbf{x}^0 - \mathbf{X}^0\|_\infty + \max_{-L \leq n \leq 0} |s^n - S^n| + O(h^2 + k^2). \tag{29}$$

Proof. The only novelty with respect to the proof of the consistency in [9] is to establish bounds for the truncation errors in the numerical approximations to the solution of the delay differential equation that drives the dynamics of the resource. Therefore, the regularity hypotheses (H1)–(H7), properties (P1), (P2) of the quadrature rule and error bounds for the explicit Euler method allow us to bound

$$\begin{aligned}
|s^n - s^{n,*}| &\leq \left|s(t^n) - s^{n-1} - k f\left(s^{n-1}, s^{n-1-L}, I(t^{n-1}), t^{n-1}\right)\right| \\
&\quad + k\left|f\left(s^{n-1}, s^{n-1-L}, I(t^{n-1}), t^{n-1}\right) - f\left(s^{n-1}, s^{n-1-L}, \mathcal{Q}(\mathbf{x}^{n-1}, \gamma^{n-1}(\mathbf{x},s) \cdot \mathbf{u}^{n-1}, t^{n-1})\right)\right| \\
&\leq C\left(k^2 + k\left|I(t^{n-1}) - \mathcal{Q}(\mathbf{x}^{n-1}, \gamma^{n-1}(\mathbf{x},s) \cdot \mathbf{u}^{n-1})\right|\right) \\
&\leq C\left(k^2 + k h^2\right). \tag{30}
\end{aligned}$$

As was made in [9], for h sufficiently small, we also can attain,

$$\left|\mathcal{Q}\left(\mathbf{x}^n, \gamma^n(\mathbf{x},s)\, \mathbf{u}^n\right) - \mathcal{Q}\left(\mathbf{x}^{n,*}, \gamma^{n,*}(\mathbf{x},s)\, \mathbf{u}^{n,*}\right)\right| \leq C\left(h^2 + k^2\right), \tag{31}$$

and bounds on the truncation errors corresponding to the numerical approximations to the nodal grids, $\mathbf{Z}^n, 1 \leq n \leq N$,

$$|Z_j^n| \leq C(k^2 + h^2 k), \quad 1 \leq j \leq J+n, \quad 1 \leq n \leq N. \tag{32}$$

Next, (21), the regularity hypotheses (H1)–(H7), the property (P1), inequalities (30) and (31) and the error bound of the modified Euler scheme employed, allow us to achieve

$$\begin{aligned}
|\sigma^n| &\leq \frac{1}{k}\Bigg\{\left|s^n - s^{n-1} - \frac{k}{2}\left(f\left(s^{n-1}, s^{n-1-L}, I(t^{n-1}), t^{n-1}\right) + f\left(s^n, s^{n-L}, I(t^n), t^n\right)\right)\right| \\
&\quad + \frac{k}{2}\left|f\left(s^{n-1}, s^{n-1-L}, I(t^{n-1}), t^{n-1}\right) - f\left(s^{n-1}, s^{n-1-L}, \mathcal{Q}(\mathbf{x}^{n-1}, \gamma^{n-1}(\mathbf{x},s) \cdot \mathbf{u}^{n-1}), t^{n-1}\right)\right| \\
&\quad + \frac{k}{2}\left|f\left(s^n, s^{n-L}, I(t^n), t^n\right) - f\left(s^{n,*}, s^{n-L}, I(t^n), t^n\right)\right| \\
&\quad + \frac{k}{2}\left|f\left(s^{n,*}, s^{n-L}, I(t^n), t^n\right) - f\left(s^{n,*}, s^{n-L}, \mathcal{Q}(\mathbf{x}^{n,*}, \gamma^{n,*}(\mathbf{x},s) \cdot \mathbf{u}^{n,*}), t^n\right)\right|\Bigg\} \\
&\leq C\Big\{k^2 + |s^n - s^{n,*}| + \left|I(t^{n-1}) - \mathcal{Q}(\mathbf{x}^{n-1}, \gamma^{n-1}(\mathbf{x},s) \cdot \mathbf{u}^{n-1})\right| \\
&\quad + |I(t^n) - \mathcal{Q}(\mathbf{x}^n, \gamma^n(\mathbf{x},s) \cdot \mathbf{u}^n)| + |\mathcal{Q}(\mathbf{x}^n, \gamma^n(\mathbf{x},s) \cdot \mathbf{u}^n) - \mathcal{Q}(\mathbf{x}^{n,*}, \gamma^{n,*}(\mathbf{x},s) \cdot \mathbf{u}^{n,*})|\Big\} \\
&\leq C\left(k^2 + h^2\right),
\end{aligned} \tag{33}$$

$1 \leq n \leq N$. Finally, the bounds for the truncation errors produced by $U_j^n, 0 \leq j \leq J+n, 1 \leq n \leq N$, are derived as in [9],

$$|L_j^n| \leq C(k^2 + h^2), \quad 0 \leq j \leq J+n, \quad 1 \leq n \leq N. \tag{34}$$

Therefore, (29) follows from (32)–(34). □

Another piece of notion that plays an important role in the analysis of the numerical method is the stability with k-dependent thresholds. For $k \in K$, let R_k be a real number (the stability threshold) with $0 < R_k < \infty$. We say that the discretization (26) is stable for $\tilde{u}_k$ restricted to the thresholds R_k, if there exist two positive constants $k_0 \in K$ and $\mathfrak{S}$ (the stability constant) such that, for any $k \in K$ with $k \leq k_0$, the open ball $B_{\mathcal{A}_k}(\tilde{u}_k, R_k)$ is contained in the domain of $\Phi_{k''}$, and for all $\tilde{V}_k, \tilde{W}_k$ in that ball,

$$\|\tilde{V}_k - \tilde{W}_k\|_{\mathcal{A}_k} \leq \mathfrak{S}\|\Phi_k(\tilde{V}_k) - \Phi_k(\tilde{W}_k)\|_{\mathcal{B}_k}.$$

We introduce the following auxiliary result, proved in [9] where the same quadrature rule and the same procedure of selection of the subgrid at each time step were used.

Proposition 1. *Assume that hypotheses (H1)–(H7) hold and that the considered quadrature rules satisfy properties (P1)–(P6). Let be $\mathbf{y}^n, \mathbf{z}^n \in B_\infty(\mathbf{x}^n, R k^p)$, $\mathbf{V}^n, \mathbf{W}^n \in B_\infty(\mathbf{u}^n, R k^p)$ and $a^n, b^n \in B_\infty(s^n, R k^p)$, where R is a positive constant and $1 < p < 2$. Then, as $k \to 0$,*

$$|\mathcal{Q}(\mathbf{y}^n, \gamma^n(\mathbf{y}, a) \cdot \mathbf{V}^n) - \mathcal{Q}(\mathbf{z}^n, \gamma^n(\mathbf{z}, b) \cdot \mathbf{W}^n)| \leq C\left(\|\mathbf{V}^n - \mathbf{W}^n\|_1 + \|\mathbf{y}^n - \mathbf{z}^n\|_\infty + |a^n - b^n|\right), \tag{35}$$

$$|\mathcal{Q}(\mathbf{y}^{n,*}, \gamma^{n,*}(\mathbf{y}, a) \cdot \mathbf{V}^{n,*}) - \mathcal{Q}(\mathbf{z}^{n,*}, \gamma^{n,*}(\mathbf{z}, b) \cdot \mathbf{W}^{n,*})| \leq C\left(\|\mathbf{V}^{n,*} - \mathbf{W}^{n,*}\|_1 + \|\mathbf{y}^{n,*} - \mathbf{z}^{n,*}\|_\infty + |a^{n,*} - b^{n,*}|\right), \tag{36}$$

$1 \leq n \leq N$.

Now, we introduce the theorem that establishes the *stability* of the discretization defined by Equations (26).

Theorem 3 (Stability). *Assume that hypotheses (H1)–(H7) hold and that the considered quadrature rules satisfy properties (P1)–(P6). Then, the discretization is stable for $\tilde{u}_k$ with $R_k = R k^p$, $1 < p < 2$.*

Proof. We denote

$$\Phi_k\left(\mathbf{y}^0, \mathbf{V}^0, \mathbf{y}^1, \mathbf{V}^1, \ldots, \mathbf{y}^N, \mathbf{V}^N, \mathbf{a}\right) = \left(\mathbf{Y}^0, \mathbf{P}^0, \mathbf{A}^0, \mathbf{P}_0, \mathbf{Y}^1, \mathbf{P}^1, \ldots, \mathbf{Y}^N, \mathbf{P}^N, \mathbf{A}\right),$$

$$\Phi_k\left(\mathbf{z}^0, \mathbf{W}^0, \mathbf{z}^1, \mathbf{W}^1, \ldots, \mathbf{z}^N, \mathbf{W}^N, \mathbf{b}\right) = \left(\mathbf{Z}^0, \mathbf{R}^0, \mathbf{B}^0, \mathbf{R}_0, \mathbf{Z}^1, \mathbf{R}^1, \ldots, \mathbf{Z}^N, \mathbf{R}^N, \mathbf{B}\right),$$

for $(\mathbf{y}^0, \mathbf{V}^0, \mathbf{y}^1, \mathbf{V}^1, \ldots, \mathbf{y}^N, \mathbf{V}^N, \mathbf{a}), (\mathbf{z}^0, \mathbf{W}^0, \mathbf{z}^1, \mathbf{W}^1, \ldots, \mathbf{z}^N, \mathbf{W}^N, \mathbf{b}) \in B_{\mathcal{A}_k}(\tilde{\mathbf{u}}_k, R_k)$. Now, we set

$$\mathbf{E}^n = \mathbf{V}^n - \mathbf{W}^n \in \mathbb{R}^{J+n+1}, \Delta^n = \mathbf{y}^n - \mathbf{z}^n \in \mathbb{R}^{J+n}, 0 \leq n \leq N; \sigma^n = b^n - a^n \in \mathbb{R}, -L \leq n \leq N,$$

$$\mathbf{E}^{n,*} = \mathbf{V}^{n,*} - \mathbf{W}^{n,*} \in \mathbb{R}^{J+n+1}, \Delta^{n,*} = \mathbf{y}^{n,*} - \mathbf{z}^{n,*} \in \mathbb{R}^{J+n}, \sigma^{n,*} = b^{n,*} - a^{n,*} \in \mathbb{R},$$

$1 \leq n \leq N$. From (22) and hypothesis (H6), we obtain

$$\|\Delta^{n,*}\|_\infty \leq (1 + Ck)\|\Delta^{n-1}\|_\infty + Ck\,|\sigma^{n-1}|, \tag{37}$$

$1 \leq n \leq N$. Next, from (25), by means of hypothesis (H7) and inequality (35), we arrive at

$$\begin{aligned}
|\sigma^{n,*}| &\leq |\sigma^{n-1}| + k\left|f\left(a^{n-1}, a^{n-1-L}, Q\left(\mathbf{y}^{n-1}, \gamma^{n-1}(\mathbf{y}, a) \cdot \mathbf{V}^{n-1}\right), t^{n-1}\right)\right.\\
&\quad \left. -f\left(b^{n-1}, b^{n-1-L}, Q\left(\mathbf{z}^{n-1}, \gamma^{n-1}(\mathbf{z}, b) \cdot \mathbf{W}^{n-1}\right), t^{n-1}\right)\right|\\
&\leq (1 + Ck)|\sigma^{n-1}| + Ck|\sigma^{n-1-L}| + Ck\left\{\|\Delta^{n-1}\|_\infty + \|\mathbf{E}^{n-1}\|_1\right\},
\end{aligned} \tag{38}$$

$1 \leq n \leq N$. Now, from (23), hypotheses (H4), (H6) and the boundedness of $\|\mathbf{W}^n\|_\infty$, we have, for k sufficiently small,

$$|E_j^{n,*}| \leq (1 + Ck)\|\mathbf{E}^{n-1}\|_1 + Ck\left\{|\sigma^{n-1}| + \|\Delta^{n-1}\|_\infty\right\}, \tag{39}$$

$1 \leq j \leq J + n$. Additionally, (24), hypothesis (H6) and inequalities (36)–(38), allow us to obtain, for k sufficiently small,

$$|E_0^{n,*}| \leq C\left\{|\sigma^{n-1}| + |\sigma^{n-1-L}| + \|\Delta^{n-1}\|_\infty + \|\mathbf{E}^{n-1}\|_1 + \|\mathbf{E}^{n,*}\|_1\right\}, \tag{40}$$

$1 \leq n \leq N$. Then, we multiply $|E_j^{n,*}|$ by h and sum in j, $0 \leq j \leq J + n$, to achieve a bound of $\|\mathbf{E}^{n,*}\|_1$, $1 \leq n \leq N$. Therefore, from (39), (40) and that $k = rh$, we derive, for h sufficiently small,

$$\begin{aligned}
\|\mathbf{E}^{n,*}\|_1 &\leq Ch\left\{|\sigma^{n-1}| + |\sigma^{n-1-L}| + \|\Delta^{n-1}\|_\infty + \|\mathbf{E}^{n-1}\|_1 + \|\mathbf{E}^{n,*}\|_1\right\}\\
&\quad + h\sum_{j=1}^{J+n}\left\{(1 + Ck)\|\mathbf{E}^{n-1}\|_1 + Ck\left(|\sigma^{n-1}| + \|\Delta^{n-1}\|_\infty\right)\right\}\\
&\leq Ch\|\mathbf{E}^{n,*}\|_1 + C\left\{|\sigma^{n-1}| + |\sigma^{n-1-L}| + \|\Delta^{n-1}\|_\infty + \|\mathbf{E}^{n-1}\|_1\right\},
\end{aligned} \tag{41}$$

$1 \leq n \leq N$, and for h sufficiently small,

$$\|\mathbf{E}^{n,*}\|_1 \leq C\left\{|\sigma^{n-1}| + |\sigma^{n-1-L}| + \|\Delta^{n-1}\|_\infty + \|\mathbf{E}^{n-1}\|_1\right\}, \tag{42}$$

$1 \leq n \leq N$. Next, (19), hypotheses (H6)–(H7), (37) and (38) enable us to arrive at

$$\begin{aligned} |\Delta_j^n| &\leq (1+Ck)|\Delta_{j-1}^{n-1}| + Ck\left\{|\sigma^{n-1}| + \left|\Delta_j^{n,*}\right| + |\sigma^{n,*}|\right\} + k\left|Y_j^n - Z_j^n\right| \\ &\leq (1+Ck)|\Delta_{j-1}^{n-1}| + Ck\left\{\left\|\Delta^{n-1}\right\|_\infty + \left\|\mathbf{E}^{n-1}\right\|_1 + |\sigma^{n-1}| + \left|\sigma^{n-1-L}\right|\right\} + k\left|Y_j^n - Z_j^n\right|, \end{aligned} \quad (43)$$

$1 \leq j \leq J+n$, $1 \leq n \leq N$. Thus, from (43), when $N \geq n > j \geq 1$, we have

$$\begin{aligned} |\Delta_j^n| &\leq Ck\sum_{l=0}^{j-1}(1+Ck)^l\left(\|\mathbf{E}^{n-1-l}\|_1 + \|\Delta^{n-1-l}\|_\infty + |\sigma^{n-1-l}| + |\sigma^{n-1-l-L}|\right) \\ &\quad + k\sum_{l=0}^{j-1}(1+Ck)^l|Y_{j-l}^{n-l} - Z_{j-l}^{n-l}|, \end{aligned} \quad (44)$$

and when $J+n \geq j \geq n \geq 1$, it follows that

$$\begin{aligned} |\Delta_j^n| &\leq (1+Ck)^n|\Delta_{j-n}^0| + Ck\sum_{l=0}^{n-1}(1+Ck)^l\left(\|\mathbf{E}^{n-1-l}\|_1 + \|\Delta^{n-1-l}\|_\infty + |\sigma^{n-1-l}| + |\sigma^{n-1-l-L}|\right) \\ &\quad + k\sum_{l=0}^{n-1}(1+Ck)^l|Y_{j-l}^{n-l} - Z_{j-l}^{n-l}|. \end{aligned} \quad (45)$$

Then, by means of (44) and (45), we can conclude that

$$\|\Delta^n\|_\infty \leq C\left\{\|\Delta^0\|_\infty + \sum_{m=0}^{n-1}k\|\mathbf{E}^m\|_1 + \sum_{m=0}^{n-1}k\|\Delta^m\|_\infty + \sum_{m=-L}^{n-1}k|\sigma^m| + \sum_{m=1}^{n}k\|\mathbf{Y}^m - \mathbf{Z}^m\|_\infty\right\}, \quad (46)$$

$1 \leq n \leq N$. On the other hand, from (21), (H7) and (35)–(38) and (42), for k small enough, we obtain that

$$\begin{aligned} |\sigma^n| &\leq (1+Ck)|\sigma^{n-1}| + Ck\left\{|\sigma^{n,*}| + |\sigma^{n-L}| + |\mathcal{Q}(\mathbf{y}^{n,*}, \gamma^{n,*}(\mathbf{y},a)\cdot\mathbf{V}^{n,*}) - \mathcal{Q}(\mathbf{z}^{n,*}, \gamma^{n,*}(\mathbf{z},b)\cdot\mathbf{W}^{n,*})|\right. \\ &\quad \left. + \left|\mathcal{Q}(\mathbf{y}^{n-1}, \gamma^{n-1}(\mathbf{y},a)\cdot\mathbf{V}^{n-1}) - \mathcal{Q}(\mathbf{z}^{n-1}, \gamma^{n-1}(\mathbf{z},b)\cdot\mathbf{W}^{n-1})\right| + \left|\sigma^{n-L-1}\right|\right\} + k|A^n - B^n| \\ &\leq (1+Ck)|\sigma^{n-1}| + Ck\left\{\left\|\Delta^{n-1}\right\|_\infty + \left\|\mathbf{E}^{n-1}\right\|_1\right\} \\ &\quad + Ck\left\{\|\Delta^{n,*}\|_\infty + \|\mathbf{E}^{n,*}\|_1 + |\sigma^{n,*}| + |\sigma^{n-L-1}| + |\sigma^{n-L}|\right\} + k|A^n - B^n| \\ &\leq (1+Ck)|\sigma^{n-1}| + Ck\left\{\left|\sigma^{n-L-1}\right| + \left|\sigma^{n-L}\right| + \left\|\Delta^{n-1}\right\|_\infty + \left\|\mathbf{E}^{n-1}\right\|_1\right\} + k|A^n - B^n|, \end{aligned} \quad (47)$$

$1 \leq n \leq N$. Thus,

$$\begin{aligned} |\sigma^n| &\leq (1+Ck)^n|\sigma^0| + k\sum_{l=0}^{n-1}(1+Ck)^l|A^{n-l} - B^{n-l}| \\ &\quad + Ck\sum_{l=0}^{n-1}(1+Ck)^l\left\{\left|\sigma^{n-l-L-1}\right| + \left|\sigma^{n-l-L}\right| + \left\|\Delta^{n-l-1}\right\|_\infty + \left\|\mathbf{E}^{n-l-1}\right\|_1\right\}, \end{aligned} \quad (48)$$

$1 \leq n \leq N$. Therefore,

$$|\sigma^n| \leq C\left\{|\sigma^0| + \sum_{m=-L}^{n-L}k|\sigma^m| + \sum_{m=1}^{n}k|A^m - B^m| + \sum_{m=0}^{n-1}k\|\Delta^m\|_\infty + \sum_{m=0}^{n-1}k\|\mathbf{E}^m\|_1\right\}, \quad (49)$$

$1 \leq n \leq N$. Finally, when we study the residuals caused by the approximation to the solution, from (20), hypotheses (H4), (H6), (H7), the inequalities (35), (37), (38) and $\|\mathbf{W}^{n-1}\|_\infty \leq C$, we have, for k sufficiently small,

$$|E_j^n| \leq k|P_j^n - R_j^n| + (1 + Ck)|E_{j-1}^{n-1}| + Ck\left\{\left|\Delta_{j-1}^{n-1}\right| + \left|\sigma^{n-1}\right| + \left|\Delta_j^{n,*}\right| + \left|\sigma^{n,*}\right|\right\}$$
$$\leq k|P_j^n - R_j^n| + (1 + Ck)|E_{j-1}^{n-1}| + Ck\left\{\left\|\Delta^{n-1}\right\|_\infty + \left\|\mathbf{E}^{n-1}\right\|_1 + \left|\sigma^{n-1}\right| + \left|\sigma^{n-1-L}\right|\right\}, \quad (50)$$

$1 \leq j \leq J + n$, $1 \leq n \leq N$. Now, we derive the stability estimate for the boundary node from (18). We employ hypotheses (H5) and (H6), the properties (P3) and (P6), and that $\|\mathbf{W}^n\|_\infty \leq C$, to achieve

$$|E_0^n| \leq |P_0^n - R_0^n| + C\left\{|\sigma^n| + \|\Delta^n\|_\infty + \|\mathbf{E}^n\|_1\right\}, \quad (51)$$

$1 \leq n \leq N$. Thus, from (50), when $N \geq n > j \geq 1$, we obtain

$$|E_j^n| \leq (1 + Ck)^j |E_0^{n-j}| + k \sum_{l=0}^{j-1} (1 + Ck)^l |P_{j-l}^{n-l} - R_{j-l}^{n-l}|$$
$$+ Ck \sum_{l=0}^{j-1} (1 + Ck)^l \left\{\|\mathbf{E}^{n-1-l}\|_1 + \|\Delta^{n-1-l}\|_\infty + |\sigma^{n-1-l}| + |\sigma^{n-1-l-L}|\right\}. \quad (52)$$

Additionally, when $J + n \geq j \geq n \geq 1$, it follows that

$$|E_j^n| \leq (1 + Ck)^n |E_{j-n}^0| + k \sum_{l=0}^{n-1} (1 + Ck)^l |P_{j-l}^{n-l} - R_{j-l}^{n-l}|$$
$$+ Ck \sum_{l=0}^{n-1} (1 + Ck)^l \left\{\|\mathbf{E}^{n-1-l}\|_1 + \|\Delta^{n-1-l}\|_\infty + |\sigma^{n-1-l}| + |\sigma^{n-1-l-L}|\right\}. \quad (53)$$

Thus, we can conclude that

$$|E_j^n| \leq C\left\{\|\mathbf{E}^0\|_1 + \sum_{m=0}^{n-1} k\|\mathbf{E}^m\|_1 + \sum_{m=0}^{n-1} k\|\Delta^m\|_\infty + \sum_{m=-L}^{n-1} k|\sigma^m| + \sum_{m=1}^{n} k\|\mathbf{P}^m - \mathbf{R}^m\|_\infty\right\}. \quad (54)$$

As we are interested in a maximum norm bound of $\mathbf{E}^n$, next, we proceed to achieve a bound for the term $\|\mathbf{E}^n\|_1$, $1 \leq n \leq N$. Then, for k small enough,

$$\|\mathbf{E}^n\|_1 \leq C\left\{\|\mathbf{E}^0\|_1 + \sum_{m=1}^{n} k\|\mathbf{E}^m\|_1 + \sum_{m=0}^{n} k\|\Delta^m\|_\infty + \sum_{m=-L}^{n} k|\sigma^m| + \sum_{m=1}^{n} k\|\mathbf{P}^m - \mathbf{R}^m\|_\infty + \|P_0 - R_0\|_\infty\right\}, \quad (55)$$

$1 \leq n \leq N$. Thus, by means of the discrete Gronwall lemma,

$$\|\mathbf{E}^n\|_1 \leq C\left\{\|\mathbf{E}^0\|_1 + \sum_{m=0}^{n} k\|\Delta^m\|_\infty + \sum_{m=-L}^{n} k|\sigma^m| + \sum_{m=1}^{n} k\|\mathbf{P}^m - \mathbf{R}^m\|_\infty + \|P_0 - R_0\|_\infty\right\}, \quad (56)$$

$1 \leq n \leq N$. Next, we substitute (56) into (46), and by means of the discrete Gronwall Lemma, for k sufficiently small, it follows that

$$\|\Delta^n\|_\infty \leq C\left\{\|\Delta^0\|_\infty + \|\mathbf{E}^0\|_1 + \sum_{m=-L}^{n-1} k|\sigma^m| + \|P_0 - R_0\|_\infty + \sum_{m=1}^{n-1} k\|\mathbf{P}^m - \mathbf{R}^m\|_\infty + \sum_{m=1}^{n} k\|\mathbf{Y}^m - \mathbf{Z}^m\|_\infty\right\}, \quad (57)$$

$1 \leq n \leq N$. Now, we substitute (56) and (57) in (49), and apply again the discrete Gronwall Lemma, to attain

$$|\sigma^n| \leq C \left\{ \|\Delta^0\|_\infty + \|E^0\|_1 + \|\sigma^0\|_\infty + \sum_{m=1}^{n} k\,|A^m - B^m| + \|P_0 - R_0\|_\infty \right.$$
$$\left. + \sum_{m=1}^{n-1} k\,\|P^m - R^m\|_\infty + \sum_{m=1}^{n-1} k\,\|Y^m - Z^m\|_\infty \right\}, \tag{58}$$

$1 \leq n \leq N$. Then, we substitute (58) in (56) and (57) to obtain

$$\|\Delta^n\|_\infty \leq C \left\{ \|\Delta^0\|_\infty + \|E^0\|_1 + \|\sigma^0\|_\infty + \|P_0 - R_0\|_\infty + \sum_{m=1}^{n-1} k\,|A^m - B^m| \right.$$
$$\left. + \sum_{m=1}^{n-1} k\,\|P^m - R^m\|_\infty + \sum_{m=1}^{n} k\,\|Y^m - Z^m\|_\infty \right\}, \tag{59}$$

and

$$\|E^n\|_1 \leq C \left\{ \|\Delta^0\|_\infty + \|E^0\|_1 + \|\sigma^0\|_\infty + \|P_0 - R_0\|_\infty + \sum_{m=1}^{n} k\,|A^m - B^m| \right.$$
$$\left. + \sum_{m=1}^{n} k\,\|P^m - R^m\|_\infty + \sum_{m=1}^{n} k\,\|Y^m - Z^m\|_\infty \right\}, \tag{60}$$

$1 \leq n \leq N$. Finally, we substitute (58)–(60) in (51) and (54) to arrive at

$$\|E^n\|_\infty \leq C \left\{ \|\Delta^0\|_\infty + \|E^0\|_1 + \|\sigma^0\|_\infty + \|P_0 - R_0\|_\infty + \sum_{m=1}^{n} k\,|A^m - B^m| \right.$$
$$\left. + \sum_{m=1}^{n} k\,\|P^m - R^m\|_\infty + \sum_{m=1}^{n} k\,\|Y^m - Z^m\|_\infty \right\}, \tag{61}$$

$1 \leq n \leq N$. Thus, due to (58), (59) and (61) we have

$$\left\| \left(\Delta^0, E^0, \ldots, \Delta^N, E^N, \sigma \right) \right\|_{\mathcal{A}_k}$$
$$\leq C \|(\Delta^0, E^0, \sigma^0, P_0 - R_0, Y^1 - Z^1, P^1 - R^1, \ldots, Y^N - Z^N, P^N - R^N, A - B)\|_{\mathcal{B}_k}.$$

□

We finish the analysis of the numerical method with the convergence. The global discretization error is defined as

$$\tilde{e}_k = \tilde{u}_k - \tilde{U}_k \in \mathcal{A}_k,$$

We say that the discretization (26) is convergent if there exists $k_0 \in K$ such that, for each $k \in K$ with $k \leq k_0$, (26) has a solution $\tilde{U}_k$ for which, as $k \to 0$,

$$\lim \|\tilde{u}_k - \tilde{U}_k\|_{\mathcal{A}_k} = \lim \|\tilde{e}_k\|_{\mathcal{A}_k} = 0.$$

In our analysis, we use the following result of the general discretization framework introduced in [14].

Theorem 4. *Assume that* (26) *is consistent and stable with thresholds* R_k. *If* Φ_k *is continuous in* $B(\tilde{u}_k, R_k)$ *and* $\|l_k\|_{\mathcal{B}_k} = o(R_k)$ *as* $k \to 0$, *then:*

i) For k sufficiently small, the discrete Equations (26) possess a unique solution in $B(\tilde{\mathbf{u}}_k, R_k)$.

ii) As $k \to 0$, $\|\tilde{\mathbf{e}}_k\|_{\mathcal{A}_k} = O(\|\mathbf{l}_k\|_{\mathcal{B}_k})$.

Finally, we propose the next theorem which establishes the *convergence* of the numerical method defined by Equations (26).

Theorem 5. *Assume that hypotheses (H1)–(H7) hold and that the considered quadrature rules satisfy properties (P1)–(P6). Then, for k sufficiently small, the numerical method defined by Equations (26) has a unique solution $\tilde{\mathbf{U}}_k \in B(\tilde{\mathbf{u}}_k, R_k)$ and*

$$\|\tilde{\mathbf{U}}_k - \tilde{\mathbf{u}}_k\|_{\mathcal{A}_k} \leq C \left(\|\mathbf{x}^0 - \mathbf{X}^0\|_\infty + \|\mathbf{u}^0 - \mathbf{U}^0\|_\infty + \|\mathbf{s}^0 - \mathbf{S}^0\|_\infty + O(h^2 + k^2) \right). \tag{62}$$

The proof of Theorem 5 is immediately derived by means of consistency (Theorem 2), stability (Theorem 3) and Theorem 4. Specifically, if $\mathbf{X}^0 = \mathbf{x}^0$, $\mathbf{U}^0 = \mathbf{u}^0$ and $\mathbf{S}^0 = \mathbf{s}^0$, the proposed numerical scheme is second-order accurate.

Next, we also establish an error bound for the differences between the theoretical solution u evaluated at the numerical values of the grid nodes, and the approximation obtained with the numerical method.

Theorem 6. *Assume that hypotheses (H1)–(H7) hold and that the considered quadrature rules satisfy properties (P1)–(P6). Then, for k sufficiently small, let*

$$\mathbf{u}_k^\dagger = \left(\mathbf{u}^{0,\dagger}, \mathbf{u}^{1,\dagger}, \ldots, \mathbf{u}^{N,\dagger} \right) \in \prod_{n=0}^{N} \mathbb{R}^{J+n},$$

be defined by $\mathbf{u}^{n,\dagger} = \left(u(X_0^n, t^n), u(X_1^n, t^n), \ldots, u(X_{J+n}^n, t^n) \right) \in \mathbb{R}^{J+n}$, $0 \leq n \leq N$, *where* X_j^n, $0 \leq j \leq J+n$, $0 \leq n \leq N$, *are the grid nodes given in (26). Then, the solution* $\tilde{\mathbf{U}}_k$ *satisfies*

$$\max_{0 \leq n \leq N} \left\{ \max \left\{ \|\mathbf{u}^{n,\dagger} - \mathbf{U}^n\|_\infty, |s(t^n) - S^n| \right\} \right\} \leq C \left(\|\mathbf{x}^0 - \mathbf{X}^0\|_\infty + \|\mathbf{u}^0 - \mathbf{U}^0\|_\infty + \|\mathbf{s}^0 - \mathbf{S}^0\|_\infty + O(h^2 + k^2) \right). \tag{63}$$

Proof. We only have to consider the triangle inequality

$$|u(X_j^n, t^n) - U_j^n| \leq |u(X_j^n, t^n) - u(x_j^n, t^n)| + |u(x_j^n, t^n) - U_j^n|,$$

$0 \leq j \leq J+n$, $0 \leq n \leq N$. The smoothness hypothesis (H1) on u is enough to derive that the first term on the right hand side of the inequality is $O(|X_j^n - x_j^n|) = O(h^2 + k^2)$, as proved in Theorem 5. Additionally, the second-order rate of convergence proved in such a Theorem shows that the second term is $O(h^2 + k^2)$. □

The last convergence theorem remains true even if the method employs a cuadrature rule at each time-step in which only a selection of nodes are involved, whenever the subgrid of these cuadrature nodes satisfies the (SG) property. In the case of the selection procedure given by (10), the convergence can be demonstrated in two stages. First, as proven in [12], we can establish that the selection procedure (10) creates subgrids with the property (SG) if the procedure is applied to a grid whose nodes are in the neighborhood of the theoretical ones within a radius $R k^p$. This is just the case, as the convergence Theorem 6 establishes, step by step, if the solutions at each fixed time level are obtained by the discrete operator (14).

4. Numerical Results

We carried out different simulations with the aim of exploring the behavior of the numerical method for the structured consumer population with a delayed resource model (1), (3). We tried to corroborate the convergence theoretical results with an academical test. It consisted of a test problem with meaningful nonlinearities both from mathematical and biological points of view in which the delay differential equation for the resource employed $\tau = 1$. We used the following size-specific growth, fertility and mortality rates

$$g(x,z,t) = \frac{\lambda}{2} \frac{1+z}{z} \left(\left(\frac{z}{1+z} \right)^2 - x^2 \right) + \frac{x\rho}{1+z} \left(\frac{29}{30} - \frac{z}{\kappa} \right),$$

$$\alpha(x,z,t) = \frac{3}{2} \lambda \frac{1 + \left(\frac{z}{\Lambda \left(\frac{29}{30} - \frac{z}{\kappa} \right)} \right)^{\frac{-30\lambda}{29\rho}}}{1 + 2 \left(\frac{z}{\Lambda \left(\frac{29}{30} - \frac{z}{\kappa} \right)} \right)^{\frac{-30\lambda}{29\rho}}},$$

$$\mu(x,z,t) = \lambda \frac{1+z}{z} \left(\frac{z}{1+z} + 2x \right) - \frac{3\rho}{1+z} \left(\frac{29}{30} - \frac{z}{\kappa} \right).$$

The weight function in (4) was $\gamma(x,z,t) = x^2$, and finally, the function that drove the dynamics of the resource (3) was chosen to be

$$f(z(t), z(t-1), \zeta, t) = \rho z(t) \left(1 - \frac{z(t-1)}{\kappa} \right) - \rho \zeta \left(\kappa + 30 \left(z(t) - z(t-1) \right) \right) \frac{(1+z(t))^5}{\kappa (z(t))^4 (1+4e^{-\lambda t})}.$$

With this choice of data functions, the problem (1), (3) has the following solution:

$$u(x,t) = \left(\frac{s(t)}{1+s(t)} - x \right)^2 + e^{-\lambda t} \left(\left(\frac{s(t)}{1+s(t)} \right)^2 - x^2 \right),$$

$$s(t) = \frac{29}{30} \frac{\Lambda e^{29\rho t/30}}{1 + \Lambda e^{29\rho t/30}/\kappa},$$

where $\Lambda = 24$; $\kappa = 5$; and ρ and λ are parameters that must be fixed. In the experiment, we employed $\rho = 0.1$ and $\lambda = 0.3$.

The knowledge of the exact solution to the problem allowed us to compare it with the numerical solution and to compute the error caused by the numerical approximation. Then, given h and k the discretization parameters in size and time, and the corresponding numerical solution with $(\mathbf{X}^0, \mathbf{U}^0, \mathbf{X}^1, \mathbf{U}^1, \ldots, \mathbf{X}^N, \mathbf{U}^N, \mathbf{S})$, we computed

$$\hat{e}_{h,k}^\dagger = \max \left\{ \max_{0 \leq n \leq N} \left\{ \max_{0 \leq j \leq J} |u(X_j^n, t^n) - U_j^n| \right\}, \max_{0 \leq n \leq N} \{|s(t^n) - S^n|\} \right\}. \quad (64)$$

Note that the exact positions of the grid nodes, x_j^n, $0 \leq j \leq J$, $0 \leq n \leq N$, are unknown, so we compared them with the solution on the computed grid X_j^n, $0 \leq j \leq J$, $0 \leq n \leq N$. In Figure 1, we show the evolution of the computed error in both consumer and resource populations. We observe that the error is produced mainly at the birth (minimum size) and increases with time.

We can also obtain the experimental order of convergence by means of the following well-known formula:

$$\widehat{order}_{2h,2k} = \frac{\log \left(\hat{e}_{2h,2k}^\dagger / \hat{e}_{h,k}^\dagger \right)}{\log 2}.$$

We can show the convergence of our method and that it is second-order accurate by means of Table 1. In the test, the initial size interval was taken as $[0, x_M^0]$, with $x_M^0 = 0.8$, and the numerical integration was carried out on the time interval $[0, T]$, with $T = 10$. Each column and each row of the table represent a computation with different values of the time and size discretization parameters, respectively. The upper number of each entry in columns two to six of said table represents the error computed from (64) and the lower quantity is the experimental order of convergence. We also remark that the ratio $r := k/h$ was kept fixed as soon as we considered approximations along each diagonal of the table. We realize that the results in Table 1 clearly confirm the expected (theoretically) second-order convergence for these approximations. Different values of parameters ρ and λ confirm the same behavior of the error.

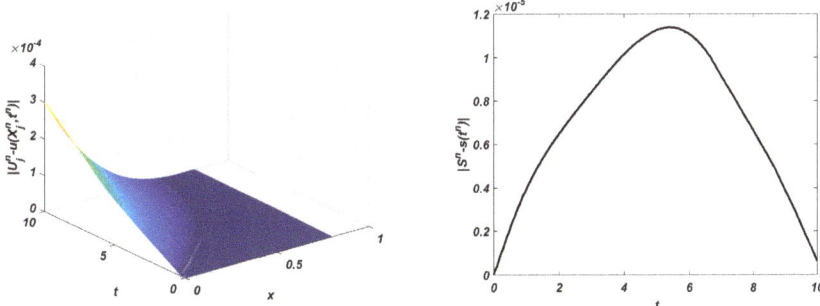

Figure 1. $k = 1.25 \times 10^{-2}$, $h = 1.25 \times 10^{-1}$. Plot on the left: evolution of the error in the consumer population; plot on the right: evolution of the error in the resource population.

Table 1. T = 10. For each value of h and k, the corresponding upper number is $\hat{e}^{\dagger}_{h,k}$; the lower number is $\widehat{order}^{\dagger}_{h,k}$.

h\k	1.25×10^{-2}	6.25×10^{-3}	3.125×10^{-3}	1.5625×10^{-3}	7.8125×10^{-4}
1.25×10^{-1}	3.003180×10^{-4}	5.059363×10^{-5}	9.929120×10^{-6}	2.615432×10^{-5}	2.965525×10^{-5}
6.25×10^{-2}	3.260311×10^{-4}	7.522357×10^{-5}	1.271462×10^{-5}	2.473141×10^{-6}	6.535886×10^{-6}
		2.00	1.99	2.01	2.00
3.125×10^{-2}	3.334010×10^{-4}	8.200264×10^{-5}	1.880835×10^{-5}	3.186199×10^{-6}	6.163402×10^{-7}
		1.99	2.00	2.00	2.00
1.5625×10^{-2}	3.359480×10^{-4}	8.344165×10^{-5}	2.046955×10^{-5}	4.704060×10^{-6}	7.979058×10^{-7}
		2.00	2.00	2.00	2.00
7.8125×10^{-3}	3.363417×10^{-4}	8.380967×10^{-5}	2.093071×10^{-5}	5.118703×10^{-6}	1.176429×10^{-6}
		2.00	2.00	2.00	2.00

5. Conclusions

We proposed a numerical method specially adapted to integrate a model that describes the evolution of a size-structured consumer population feeding on a dynamical resource. The model couples a first-order hyperbolic partial differential equation with nonlocal and nonlinear boundary condition and a differential equation driving the dynamics of the resource. The novelty of the model is the appearance of a delay in the resource evolution rate. Numerical methods are the only feasible approach, except in exceptional cases, for the approximation of the solution of the problem and eventually studying the dynamical behavior of the model. The authors have in preparation [15], using this model, a numerical study on the effect of the delay in the resource evolution rate on the dynamics of a population of the *Daphnia magna* (water flea).

The analyzed numerical method integrated the problem along its characteristic curves. We proved optimal second-order of convergence to the solution; this was confirmed numerically by means of an academic test problem. This is the first convergence analysis to our knowledge for the discretization of this kind of problem.

Author Contributions: All authors contributed equally to this article. All authors have read and agreed to the published version of the manuscript.

Funding: This research was funded in part by project MTM2017-85476-C2-1-P of the Spanish Ministerio de Economía y Competitividad and European FEDER Funds. VA138G18 of the Consejería de Educación, Junta de Castilla y León, Spain.

Conflicts of Interest: The authors declare no conflict of interest. The funders had no role in the design of the study; in the collection, analyses or interpretation of data; in the writing of the manuscript, or in the decision to publish the results.

References

1. Ruan, S. Delay differential equations in single species dynamics. In *Delay Differential Equations and Applications*; Arino, O., Hbid, M.L., Ait Dads, E., Eds.; Springer: Berlin, Germany, 2006; pp. 477–517.
2. Kooijman, S.A.L.M.; Metz, J.A.J. On the dynamics of chemically stressed populations: The deduction of population consequences from effects on individuals. *Ecotoxicol. Environ. Saf.* **1984**, *8*, 254–274. [CrossRef]
3. Thieme, H.R. Well-posedness of physiologically structured population models for daphnia magna. *J. Math. Biol.* **1988**, *26*, 299–317. [CrossRef]
4. de Roos, A.M.; Metz, J.A.J.; Evers, E.; Leipoldt, A. A size dependent predator-prey interaction: Who pursues whom? *J. Math. Biol.* **1990**, *28*, 609–643. [CrossRef]
5. Diekmann, O.; Gyllenberg, M.; Metz, J.A.J.; Nakaoka, S.; de Roos, A.M. Daphnia revisited: Local stability and bifurcation theory for physiologically structured population models explained by way of an example. *J. Math. Biol.* **2010**, *61*, 277–318. [CrossRef] [PubMed]
6. de Roos, A.M.; Diekmann, O.; Getto, P.; Kirkilionis, M.A. Numerical equilibrium analysis for structured consumer resource models. *Bull. Math. Biol.* **2010**, *72*, 259–297. [CrossRef] [PubMed]
7. Diekmann, O.; Getto, P.; Gyllenberg, M. Stability and bifurcation analysis of volterra functional equations in the light of suns and stars. *SIAM J. Math. Anal.* **2007**, *39*, 1023–1069. [CrossRef]
8. de Roos, A.M. Numerical methods for structured population models: The escalator boxcar train. *Numer. Methods Partial. Differ. Equations* **1988**, *4*, 173–195. [CrossRef]
9. Angulo, O.; López-Marcos, J.C.; López-Marcos, M.A. Analysis of an efficient integrator for a size-structured population model with a dynamical resource. *Comput. Math. Appl.* **2014**, *68*, 941–961. [CrossRef]
10. Brännström, Å.; Carlsson, L.; Simpson, D. On the convergence of the escalator boxcar train. *SIAM J. Numer. Anal.* **2013**, *51*, 3213–3231. [CrossRef]
11. Carrillo, J.A.; Gwiazda, P.; Kropielnicka, K.; Marciniak-Czochra, A.K. The escalator boxcar train method for a system of age-structured equations in the space of measures. *SIAM J. Numer. Anal.* **2019**, *57*, 1842–1874. [CrossRef]
12. Angulo, O.; López-Marcos, J.C. Numerical integration of fully nonlinear size-structured population models. *Appl. Numer. Math.* **2004**, *50*, 291–327. [CrossRef]
13. Angulo, O.; López-Marcos, J.C. Numerical schemes for size-structured population equations. *Math. Biosci.* **1999**, *157*, 169–188. [CrossRef]
14. López-Marcos, J.C; Sanz-Serna. J.M. Stability and convergence in numerical analysis III: Linear investigation of nonlinear stability. *IMA J. Numer. Anal.* **1988**, *8*, 71–84.
15. Abia, L.; Angulo, O.; López-Marcos, J.C.; López-Marcos, M.A. A numerical study on the effect of a delay in the resource evolution rate in the dynamics of a population of *Daphnia magna*. in preparation.

© 2020 by the authors. Licensee MDPI, Basel, Switzerland. This article is an open access article distributed under the terms and conditions of the Creative Commons Attribution (CC BY) license (http://creativecommons.org/licenses/by/4.0/).

Article

Mean Square Convergent Non-Standard Numerical Schemes for Linear Random Differential Equations with Delay

Julia Calatayud [1], Juan Carlos Cortés [1,*], Marc Jornet [1] and Francisco Rodríguez [2]

[1] Instituto Universitario de Matemática Multidisciplinar, Building 8G, Access C, 2nd Floor, Universitat Politècnica de València, Camino de Vera s/n, 46022 Valencia, Spain; jucagre@doctor.upv.es (J.C.); marjorsa@doctor.upv.es (M.J.)

[2] Department of Applied Mathematics, University of Alicante, Apdo. 99, 03080 Alicante, Spain; f.rodriguez@ua.es

* Correspondence: jccortes@imm.upv.es

Received: 28 July 2020; Accepted: 21 August 2020; Published: 24 August 2020

Abstract: In this paper, we are concerned with the construction of numerical schemes for linear random differential equations with discrete delay. For the linear deterministic differential equation with discrete delay, a recent contribution proposed a family of non-standard finite difference (NSFD) methods from an exact numerical scheme on the whole domain. The family of NSFD schemes had increasing order of accuracy, was dynamically consistent, and possessed simple computational properties compared to the exact scheme. In the random setting, when the two equation coefficients are bounded random variables and the initial condition is a regular stochastic process, we prove that the randomized NSFD schemes converge in the mean square (m.s.) sense. M.s. convergence allows for approximating the expectation and the variance of the solution stochastic process. In practice, the NSFD scheme is applied with symbolic inputs, and afterward the statistics are explicitly computed by using the linearity of the expectation. This procedure permits retaining the increasing order of accuracy of the deterministic counterpart. Some numerical examples illustrate the approach. The theoretical m.s. convergence rate is supported numerically, even when the two equation coefficients are unbounded random variables. M.s. dynamic consistency is assessed numerically. A comparison with Euler's method is performed. Finally, an example dealing with the time evolution of a photosynthetic bacterial population is presented.

Keywords: delay random differential equation; non-standard finite difference method; mean square convergence

1. Introduction

Modeling physical systems for which the future state depends on history due to hereditary characteristics, such as aftereffects or time lags, usually requires the use of delay differential models. The delay may be discrete or continuous, depending on whether a specific or complete past information is used. The inclusion of a delay requires specific techniques for the theoretical study of the differential model [1–4]. In practice, delay differential models play a key role in different scientific and technical fields [5–10].

In the context of delay differential equations, the construction of non-standard finite difference (NSFD) numerical schemes has not been much explored. Historically, NSFD schemes were developed by Mickens in the years 1994 and 2000 [11,12], together with a later edited book in 2005 [13]. Mickens observed that traditional standard finite difference schemes may be modified, on the basis of exact numerical schemes for basic ordinary differential equations, so that the essential properties of the governing continuous model are mimicked [14]. Until relatively recently, NSFD schemes were successfully designed and applied for ordinary, partial and fractional differential equations [15]. However, delay differential equations have not been addressed in detail.

Recently, [16] proposed a NSFD scheme for the general linear delay problem

$$\begin{cases} x'(t) = \alpha x(t) + \beta x(t-\tau), & t > 0, \\ x(t) = f(t), & -\tau \leq t \leq 0, \end{cases} \quad (1)$$

($\tau > 0$) from an exact scheme on the whole domain, providing high order of accuracy and consistent dynamical behavior with simple computational properties. Such approach was extended to the non-scalar case in [17].

In modeling, the variability of data, due to limited knowledge and fluctuation of the process under study, lack of information, bad calibration machines, etc., gives rise to variability in the model coefficients. Therefore, for a more realistic description of the process, coefficients should be regarded as random quantities on an abstract probability space. When the coefficients are random variables and regular stochastic processes, the solution to the model becomes a differentiable stochastic process, whose realizable trajectories solve the deterministic version of the model. A common treatment of random differential models uses mean square (m.s.) calculus [18–24]. Of special importance is the computation of the mean and the variance of the solution stochastic process, or even its probability density function.

We are interested on delay random differential equations. Specifically, the randomization of (1) as

$$\begin{cases} x'(t,\omega) = \alpha(\omega)x(t,\omega) + \beta(\omega)x(t-\tau,\omega), & t > 0, \omega \in \Omega, \\ x(t,\omega) = f(t,\omega), & -\tau \leq t \leq 0, \omega \in \Omega. \end{cases} \quad (2)$$

Here α and β are random variables and f is a stochastic process on a complete probability space $(\Omega, \mathcal{F}, \mathbb{P})$, where Ω is the sample space formed by the outcomes $\omega \in \Omega$, $\mathcal{F}$ is the σ-algebra of events, and $\mathbb{P} : \mathcal{F} \to [0,1]$ is the probability measure. The solution x is a differentiable stochastic process.

Only recently, a theoretical study on delay random differential equations was started. General delay random differential equations were analyzed in the m.s. sense in [25], with the goal of extending some of the existing results on random differential equations with no delay from the book [18]. Problem (2) was solved in the m.s. sense in [26], and later generalized to equations with a random forcing term in [27]. On the other hand, in [28] the authors studied (2), but considered the solution in the sample-path sense and computed its probability density function via the random variable transformation technique, for certain forms of the initial condition process.

In this paper, we are concerned with computational aspects of delay random differential equations. Standard finite difference methods have already been applied to random ordinary, partial and fractional differential equations, by establishing the m.s. convergence, and even the convergence of densities, of the numerical discretizations towards the stochastic process solution [29–34]. Here we aim at extending the NSFD method from [16] to (2), by assessing the m.s. convergence of the discretizations. This permits approximating the expectation and the variance of the solution with high accuracy, whenever computationally feasible.

The organization of this paper is the following. In Section 2, the main results on m.s. calculus are exposed. The material for this section is essentially taken from [18]. In Section 3, the NSFD numerical scheme from [16] is presented. The randomization of the scheme, its m.s. convergence and its usefulness for approximating moments are discussed in Section 4. Illustration of the theory with numerical examples is conducted in Section 5. Finally, Section 6 draws the main conclusions.

2. M.s. Calculus

We are interested in second order real random variables $y : \Omega \to \mathbb{R}$, satisfying

$$\mathbb{E}[y^2] = \int_\Omega y(\omega)^2 \, d\mathbb{P}(\omega) < \infty. \tag{3}$$

We refer the reader to ([18], Ch. 4, [35]). The set of these random variables is a Hilbert space, denoted as $L^2(\Omega)$ and endowed with the inner product $\langle y_1, y_2 \rangle = \mathbb{E}[y_1 y_2]$. This inner product gives rise to the norm $\|y\|_2 = (\mathbb{E}[y^2])^{1/2}$. By Cauchy–Schwarz inequality ([18], p. 19) $\mathbb{E}[|y_1 y_2|] \leq \|y_1\|_2 \|y_2\|_2$. Random variables in $L^2(\Omega)$ are characterized by having finite variance:

$$\mathbb{V}[y] = \mathbb{E}[(y - \mathbb{E}[y])^2] = \mathbb{E}[y^2] - (\mathbb{E}[y])^2 < \infty. \tag{4}$$

This is one of the principal reasons for working with second order random variables, since the main statistical information for uncertainty quantification, namely the average value and the dispersion, are well-defined.

Given a stochastic process $\{z(t) : t \in I \subseteq \mathbb{R}\}$, it is of second order if the random variable $z(t)$ is of second order, for all $t \in I$. By Cauchy–Schwarz inequality, it is straightforward to check that a second order stochastic process possesses a correlation function, $\mathbb{E}[z(t_1)z(t_2)]$.

Convergence in $L^2(\Omega)$ is defined through its norm $\|\cdot\|_2$: a sequence of random variables $\{y_n\}_{n=1}^\infty$ converges to y in $L^2(\Omega)$ if $\lim_{n\to\infty} \|y_n - y\|_2 = 0$. This is referred to as m.s. convergence.

M.s. convergence preserves the convergence of the expectation and the variance. This is a key fact. In general, if $\{x_n\}_{n=1}^\infty$ and $\{y_n\}_{n=1}^\infty$ are two sequences of second order random variables such that $x_n \to x$ and $y_n \to y$ as $n \to \infty$ in the m.s. sense, then $\mathbb{E}[x_n y_n] \to \mathbb{E}[xy]$ ([18], p. 88).

In the particular case that $\{x_n\}_{n=1}^\infty$ is a sequence of second order random variables such that its mean and its variance tend to zero, i.e., $\mathbb{E}[x_n] \to 0$ and $\mathbb{V}[x_n] \to 0$ as $n \to \infty$, then $\{x_n\}_{n=1}^\infty$ is m.s. convergent to zero, since $(\|x_n\|_2)^2 = \mathbb{E}[(x_n)^2] = \mathbb{V}[x_n] + (\mathbb{E}[x_n])^2 \to 0$ as $n \to \infty$. The converse is also true.

M.s. convergence gives rise to m.s. calculus, where continuity, differentiability and Riemann integrability of a stochastic process are naturally defined by taking m.s. limits in the classical definitions. A stochastic process $\{z(t) : t \in I \subseteq \mathbb{R}\}$ is m.s. continuous at $t_0 \in I$ if $z(t) \to z(t_0)$ as $t \to t_0$ in the m.s. sense. It is m.s. differentiable at $t_0 \in I$ if $\lim_{h\to 0} \frac{z(t_0+h)-z(t_0)}{h}$ exists in the m.s. sense, which is denoted as $z'(t_0)$. Finally, $z(t)$ is m.s. Riemann integrable on an interval $[a,b] \subseteq I$ if there exists a sequence of partitions $\{P_n\}_{n=1}^\infty$ with mesh tending to 0, $P_n = \{a = t_0^n < t_1^n < \ldots < t_{r_n}^n = b\}$, such that for any choice of points $s_i^n \in [t_{i-1}^n, t_i^n]$, $i = 1, \ldots, r_n$, the limit $\lim_{n\to\infty} \sum_{i=1}^{r_n} z(s_i^n)(t_i^n - t_{i-1}^n)$ exists in the m.s. sense, and it is denoted as $\int_a^b z(t) \, dt$.

The following important properties will be used: m.s. continuity on an interval implies m.s. Riemann integrability ([18] Section 4.5.1 (1)), $\|\int_a^b z(t) \, dt\|_2 \leq \int_a^b \|z(t)\|_2 \, dt$ for any m.s. Riemann integrable process $z(t)$ ([18] Section 4.5.1 (3)), and the fundamental theorem of m.s. calculus ([18] Section 4.5.1 (5), (6)).

Finally, we mention that the essential supremum norm is defined as

$$\|y\|_\infty = \inf\{C \geq 0 : |y| \leq C \text{ almost surely}\}. \tag{5}$$

The set of random variables satisfying $\|y\|_\infty < \infty$ gives rise to the Banach space $L^\infty(\Omega)$. Obviously, for two random variables $y_1 \in L^\infty(\Omega)$ and $y_2 \in L^2(\Omega)$, it holds $\|y_1 y_2\|_2 \leq \|y_1\|_\infty \|y_2\|_2 < \infty$.

3. NSFD Methods for Linear Deterministic Differential Equations with Delay

Based on the explicit solution to (1),

$$x(t) = f(0) \sum_{k=0}^{m-1} \frac{\beta^k (t-k\tau)^k}{k!} e^{\alpha(t-k\tau)}$$

$$+ \sum_{k=0}^{m-2} \frac{\beta^{k+1}}{k!} \int_{-\tau}^{0} (t-(k+1)\tau - s)^k e^{\alpha(t-(k+1)\tau - s)} f(s)\, ds$$

$$+ \frac{\beta^m}{(m-1)!} \int_{-\tau}^{t-m\tau} (t - m\tau - s)^{m-1} e^{\alpha(t-m\tau-s)} f(s)\, ds, \tag{6}$$

for $(m-1)\tau < t \leq m\tau$, $m \geq 1$, an exact numerical difference scheme for (1) is obtained in [16,17]. It is detailed in the following theorem.

Theorem 1. ([16], Theorem 2), ([17], Theorem 2) *Consider a size mesh $h > 0$ such that $Nh = \tau$, for some integer $N \geq 1$. Write $t_n = nh$ and $x_n = x(t_n)$, for $n \geq -N$. Then the numerical solution given by $x_n = f(t_n)$, for $-N \leq n \leq 0$, and by*

$$x_{n+1} = e^{\alpha h} \sum_{k=0}^{m-1} \frac{\beta^k h^k}{k!} x_{n-kN} + \frac{\beta^m}{(m-1)!} \int_{t_n - m\tau}^{t_n - m\tau + h} (t_n - m\tau + h - s)^{m-1} e^{\alpha(t_n - m\tau + h - s)} f(s)\, ds, \tag{7}$$

where $(m-1)\tau \leq nh < m\tau$ and $m \geq 1$, defines an exact numerical scheme for (1).

Having an exact numerical scheme is ideal, since it reproduces the exact values of the solution at the points of the mesh. However, a drawback of (7) is that definite integrals need to be numerically computed. The number of definite integrals increases with increasing times. Thus, a NSFD method is proposed to maintain sufficient accuracy and adequate dynamical properties, but reduce the complexity by avoiding definite integrals. In the first M intervals $[0, \tau], \ldots, [(M-1)\tau, M\tau]$, the exact solution (7) is used (or any other numerical method with sufficiently high accuracy), but afterward the integral part from (7) is discarded. The precision of the method increases with M.

Theorem 2. ([16], Theorem 3), ([17], Theorem 3) *Fix $M \geq 1$, and compute the numerical solution to (1) in the intervals $(m-1)\tau \leq nh \leq m\tau$, for $0 \leq m \leq M$ with the exact method (7) or with any other numerical method of global error at most $\mathcal{O}(h^M)$. Then, for $m \geq M+1$ and $(m-1)\tau \leq nh < m\tau$, the expression*

$$x_{n+1} = e^{\alpha h} \sum_{k=0}^{M} \frac{\beta^k h^k}{k!} x_{n-kN} \tag{8}$$

defines a NSFD scheme of global error $\mathcal{O}(h^M)$.

As detailed in ([16], Remark 1), the method (8) has the characteristics of a NSFD method:

$$\frac{x_{n+1} - x_n}{(e^{\alpha h} - 1)/\alpha} = \alpha x_n + \frac{\alpha e^{\alpha h}}{e^{\alpha h} - 1} \sum_{k=1}^{M} \frac{\beta^k h^k}{k!} x_{n-kN}. \tag{9}$$

Furthermore, in the rest of [16], it is proved and illustrated that the method from Theorem 2 is dynamically consistent with (1), for asymptotic stability, positive preserving properties, and oscillation behavior.

4. NSFD Methods for Linear Random Differential Equations with Delay: Approximations of Moments

When α and β are random variables and f is a stochastic process, problem (1) is randomized. These inputs depend on each outcome $\omega \in \Omega$ and (2) is obtained. The numerical schemes from Section 3 are also randomized. The exact scheme (7) becomes

$$x_{n+1}(\omega) = e^{\alpha(\omega)h} \sum_{k=0}^{m-1} \frac{\beta(\omega)^k h^k}{k!} x_{n-kN}(\omega)$$
$$+ \frac{\beta(\omega)^m}{(m-1)!} \int_{t_n-m\tau}^{t_n-m\tau+h} (t_n - m\tau + h - s)^{m-1} e^{\alpha(\omega)(t_n - m\tau + h - s)} f(s,\omega)\, ds, \tag{10}$$

where the integral is considered in the m.s. sense (it is assumed that f is a m.s. integrable stochastic process), while (8) becomes

$$x_{n+1}(\omega) = e^{\alpha(\omega)h} \sum_{k=0}^{M} \frac{\beta(\omega)^k h^k}{k!} x_{n-kN}(\omega). \tag{11}$$

We translate Theorem 2 into m.s. convergence.

Theorem 3. *Suppose that α and β are bounded random variables, and that f is a m.s. continuous stochastic process on $[-\tau, 0]$. Fix $M \geq 1$, and compute the numerical stochastic solution to (2) in the intervals $(m-1)\tau \leq nh \leq m\tau$, for $0 \leq m \leq M$ with the exact method (10). Then, for $m \geq M+1$ and $(m-1)\tau \leq nh < m\tau$, the expression (11) defines a random NSFD scheme of m.s. global error $\mathcal{O}(h^M)$.*

Proof. For $n = MN$, we have $t_n = nh = M\tau$ and $t_{n+1} = (n+1)h > M\tau$. For t_k, $k \leq n$, the exact scheme (10) is used, so that $\|x(t_k) - x_k\|_2 = 0$. By (10) and (11), one gets

$$\|x(t_{n+j+1}) - x_{n+j+1}\|_2 \leq \|e^{\alpha h}\|_\infty \sum_{k=0}^{M} \frac{\|\beta\|_\infty^k h^k}{k!} \|x(t_{n+j-kN}) - x_{n+j-kN}\|_2$$
$$+ \frac{\|\beta\|_\infty^m}{(m-1)!} \int_{t_{n+j}-m\tau}^{t_{n+j}-m\tau+h} (t_{n+j} - m\tau + h - s)^{m-1} \|e^{\alpha(t_{n+j}-m\tau+h-s)}\|_\infty \|f(s)\|_2\, ds, \tag{12}$$

for $(m-1)\tau \leq t_{n+j} < m\tau$, $m - 1 \geq M$, $j \geq 0$.

Let $M_0 = \|\beta\|_\infty$, and $M_1, M_2 > 0$ such that $\|e^{\alpha s}\|_\infty < M_1$ and $\|f(s)\|_2 < M_2$ for $s \in [0, h]$. We first consider $j = 0, \ldots, N-1$ and $m - 1 = M$. By (12),

$$\|x(t_{n+1}) - x_{n+1}\|_2 \leq \frac{\|\beta\|_\infty^m}{(m-1)!} \int_{t_n-m\tau}^{t_n-m\tau+h} (t_n - m\tau + h - s)^{m-1} \|e^{\alpha(t_n-m\tau+h-s)}\|_\infty \|f(s)\|_2\, ds$$
$$\leq \frac{M_0^m M_1 M_2}{(m-1)!} \int_{t_n-m\tau}^{t_n-m\tau+h} (t_n - m\tau + h - s)^{m-1}\, ds$$
$$= \frac{M_0^m M_1 M_2}{m!} h^m \leq C_1 h^m = C_1 h^{M+1}, \tag{13}$$

where C_1 is a constant independent of m and h,

$$\|x(t_{n+2}) - x_{n+2}\|_2 \leq \|e^{\alpha h}\|_\infty \|x(t_{n+1}) - x_{n+1}\|_2 + C_1 h^{M+1} \leq C_1 \left(e^{\|\alpha\|_\infty h} + 1\right) h^{M+1}, \tag{14}$$

$$\|x(t_{n+3}) - x_{n+3}\|_2 \leq \|e^{\alpha h}\|_\infty \|x(t_{n+2}) - x_{n+2}\|_2 + C_1 h^{M+1} \leq C_1 \left(e^{2\|\alpha\|_\infty h} + e^{\|\alpha\|_\infty h} + 1\right) h^{M+1}, \tag{15}$$

...

$$\|x(t_{n+N}) - x_{n+N}\|_2 \leq C_1 \left(\sum_{j=0}^{N-1} e^{j\|\alpha\|_\infty h}\right) h^{M+1}$$

$$= C_1 \frac{e^{N\|\alpha\|_\infty h} - 1}{e^{\|\alpha\|_\infty h} - 1} h^{M+1} = C_1 \frac{e^{\|\alpha\|_\infty \tau} - 1}{e^{\|\alpha\|_\infty h} - 1} h^{M+1}$$

$$= \left(C_1 \left(e^{\|\alpha\|_\infty \tau} - 1\right) \frac{\|\alpha\|_\infty h}{e^{\|\alpha\|_\infty h} - 1} \frac{1}{\|\alpha\|_\infty}\right) h^M \leq C_2 h^M, \quad (16)$$

where C_2 is a constant that only depends on τ, $\|\alpha\|_\infty$ and $\|\beta\|_\infty$. Thus,

$$\max_{n+1 \leq j \leq n+N} \|x(t_j) - x_j\|_2 = \mathcal{O}(h^M). \quad (17)$$

We continue by evaluating $\|x(t_{n+N+j}) - x_{n+N+j}\|_2$, $1 \leq j \leq N$, by starting from (12) and by employing the bounds already obtained:

$$\|x(t_{n+N+j}) - x_{n+N+j}\|_2 \leq e^{\|\alpha\|_\infty h}\|x(t_{n+N+j-1}) - x_{n+N+j-1}\|_2 + e^{\|\alpha\|_\infty h}\|\beta\|_\infty h C_2 h^M + C_1 h^{M+1}. \quad (18)$$

By solving this first order recursive inequality, we derive

$$\|x(t_{n+N+j}) - x_{n+N+j}\|_2 \leq e^{j\|\alpha\|_\infty h}\|x(t_{n+N}) - x_{n+N}\|_2 + \sum_{k=1}^{j} h^{M+1}\left(C_1 + C_2\|\beta\|_\infty e^{\|\alpha\|_\infty h}\right) e^{\|\alpha\|_\infty h(j-k)}$$

$$\leq C_2 h^M e^{N\|\alpha\|_\infty h} + h^{M+1}\left(C_1 + C_2\|\beta\|_\infty e^{\|\alpha\|_\infty h}\right) \frac{e^{\|\alpha\|_\infty h j} - 1}{e^{\|\alpha\|_\infty h} - 1}$$

$$\leq C_2 h^M e^{\|\alpha\|_\infty \tau} + h^M \left(C_1 + C_2\|\beta\|_\infty e^{\|\alpha\|_\infty h}\right) \left(e^{\|\alpha\|_\infty \tau} - 1\right) \left(\frac{\|\alpha\|_\infty h}{e^{\|\alpha\|_\infty h} - 1}\right) \frac{1}{\|\alpha\|_\infty}$$

$$\leq C_3 h^M, \quad (19)$$

where C_3 is a constant that only depends on τ, $\|\alpha\|_\infty$ and $\|\beta\|_\infty$. Therefore,

$$\max_{n+N+1 \leq j \leq n+2N} \|x(t_j) - x_j\|_2 = \mathcal{O}(h^M). \quad (20)$$

In general, we proceed by induction. Suppose that $\|x(t_j) - x_j\|_2 \leq C h^M$ for $n+1 \leq j \leq n+lN$, where $C = C(\tau, \|\alpha\|_\infty, \|\beta\|_\infty) > 0$ is constant and $l \geq 1$. We prove that $\max_{n+lN+1 \leq j \leq n+(l+1)N} \|x(t_j) - x_j\|_2 = \mathcal{O}(h^M)$. Fix $1 \leq j \leq N$. From (12) and by induction hypothesis,

$$\|x(t_{n+lN+j}) - x_{n+lN+j}\|_2 \leq e^{\|\alpha\|_\infty h}\|x(t_{n+lN+j-1}) - x_{n+lN+j-1}\|_2$$

$$+ e^{\|\alpha\|_\infty h} \sum_{k=1}^{M} \frac{\|\beta\|_\infty^k h^k}{k!} C h^M + C h^{M+1}. \quad (21)$$

By solving this first order recursive inequality, we derive (for $h < 1$)

$$\|x(t_{n+lN+j}) - x_{n+lN+j}\|_2 \leq e^{j\|\alpha\|_\infty h}\|x(t_{n+lN}) - x_{n+lN}\|_2$$

$$+ \sum_{k=1}^{j} C h^{M+1} \left(e^{\|\alpha\|_\infty h} \sum_{r=1}^{M} \frac{\|\beta\|_\infty^r h^{r-1}}{r!} + 1\right) e^{\|\alpha\|_\infty h(j-k)}$$

$$\leq C h^M e^{\|\alpha\|_\infty \tau} + C h^{M+1} \left(e^{\|\alpha\|_\infty + \|\beta\|_\infty} + 1\right) \frac{e^{\|\alpha\|_\infty h j} - 1}{e^{\|\alpha\|_\infty h} - 1} \quad (22)$$

$$\leq C h^M \left(e^{\|\alpha\|_\infty \tau} + \left(e^{\|\alpha\|_\infty + \|\beta\|_\infty} + 1\right)\left(e^{\|\alpha\|_\infty \tau} - 1\right) \frac{h}{e^{\|\alpha\|_\infty h} - 1}\right)$$

$$\leq \tilde{C} h^M,$$

where $\tilde{C} = \tilde{C}(\tau, \|\alpha\|_\infty, \|\beta\|_\infty) > 0$ is constant. This concludes the proof by induction. □

Remark 1. *As shall be seen in the numerical computations from Section 5, the boundedness of α and β from Theorem 3 is sufficient, but not necessary. Nonetheless, for unbounded α and/or β, if one wants to ensure the m.s. convergence of the NSFD scheme a priori, it is possible to properly truncate the support of α and β. Indeed, since $\lim_{m\to\infty} \mathbb{P}[\alpha \in (-m,m)] = 1$, one may take a sufficiently big interval $(-m^*, m^*)$ in such a way that $\mathbb{P}[\alpha \in (-m^*, m^*)] \approx 1$, and truncate the support of α to $(-m^*, m^*)$ (analogously for β). In fact, by applying the generalized Markov's inequality, it may be demonstrated that any second order random variable can be truncated to the interval $[\text{mean} \pm 10 \times \text{deviation}]$, so that this interval contains 99% of the probability mass irrespective of the probability distribution. In the theory of m.s. calculus, the boundedness of the random input coefficient must be usually imposed: as proved in ([36], Example p. 541), in order for an autonomous and homogeneous first order linear random differential equation (i.e., $x'(t) = ax(t)$, where a is a random variable) to possess a m.s. solution for every m.s. integrable initial condition $x(0) = x_0$, the coefficient a must be bounded.*

M.s. convergence guarantees convergence of the expectation and the variance. In the computer, $x_n(\omega)$ is explicitly and symbolically expressed in terms of $\alpha(\omega)$, $\beta(\omega)$ and $f(\cdot, \omega)$, by employing either the exact scheme (10) along all the integration region or, at cheaper cost, the NSFD scheme from Theorem 3. By using the linearity of the expectation $\mathbb{E}$, one can explicitly compute $\mathbb{E}[x_n]$. If the NSFD scheme is being used, one is approximating the true expectation of the solution to (2). Complexity is severely increased for large times t, large N (small h), and moderate or high dimension of the random space. The variance may also be approximated by symbolically expressing $x_n(\omega)^2$ and explicitly computing $\mathbb{V}[x_n] = \mathbb{E}[x_n^2] - (\mathbb{E}[x_n])^2$, although the complexity becomes significantly affected because the symbolic expression handled is larger.

Notice that working with these symbolic expressions for $x_n(\omega)$ seems to be necessary. Indeed, if one applies the expectation operator directly in (11), for instance, then $\mathbb{E}[x_{n+1}] = \sum_{k=0}^{M} \mathbb{E}[e^{\alpha h} \frac{\beta^k h^k}{k!} x_{n-kN}]$. Since each x_{n-kN} depends on α and β, the expectation $\mathbb{E}[e^{\alpha h} \frac{\beta^k h^k}{k!} x_{n-kN}]$ cannot be split as $\mathbb{E}[e^{\alpha h}] \frac{\mathbb{E}[\beta^k]h^k}{k!} \mathbb{E}[x_{n-kN}]$, unless both α and β are nonrandom. So there does not seem to exist a recursive relation formula for $\{\mathbb{E}[x_n]\}_n$.

Notice that by Jensen's and Cauchy–Schwarz inequalities,

$$|\mathbb{E}[x_n] - \mathbb{E}[x(t_n)]| = |\mathbb{E}[x_n - x(t_n)]| \leq \mathbb{E}[|x_n - x(t_n)|] \leq \|x_n - x(t_n)\|_2. \tag{23}$$

By triangular, Jensen's and Cauchy–Schwarz inequalities,

$$\begin{aligned}
|\mathbb{V}[x_n] - \mathbb{V}[x(t_n)]| &= |\mathbb{E}[(x_n)^2] - (\mathbb{E}[x_n])^2 - \mathbb{E}[(x(t_n))^2] + (\mathbb{E}[x(t_n)])^2| \\
&\leq \mathbb{E}[|(x_n)^2 - (x(t_n))^2|] + |(\mathbb{E}[x_n])^2 - (\mathbb{E}[x(t_n)])^2| \\
&= \mathbb{E}[|x_n - x(t_n)||x_n + x(t_n)|] + |\mathbb{E}[x_n] - \mathbb{E}[x(t_n)]||\mathbb{E}[x_n] + \mathbb{E}[x(t_n)]| \\
&\leq \|x_n - x(t_n)\|_2 \|x_n + x(t_n)\|_2 + |\mathbb{E}[x_n] - \mathbb{E}[x(t_n)]|(|\mathbb{E}[x_n]| + |\mathbb{E}[x(t_n)]|) \\
&\leq \|x_n - x(t_n)\|_2 (\|x_n\|_2 + \|x(t_n)\|_2) + |\mathbb{E}[x_n] - \mathbb{E}[x(t_n)]|(|\mathbb{E}[x_n]| + |\mathbb{E}[x(t_n)]|).
\end{aligned} \tag{24}$$

So the approximations of the expectation and the variance inherit the rate of convergence corresponding to the m.s. norm, which is $\mathcal{O}(h^M)$ when the exact numerical scheme (10) is used for the first M intervals of length τ and (11) is used for the subsequent intervals (Theorem 3).

In the following section, the m.s. convergence of the NSFD scheme is illustrated with some numerical computations. We point out that the boundedness of α and β from Theorem 3 is sufficient, but not necessary. It may be possible that a random coefficient is unbounded and the NSFD scheme converges in the m.s. sense.

5. Numerical Examples

The theoretical discussion is illustrated with some numerical computations. We consider specific probability distributions for α, β and/or f, and a fixed delay $\tau > 0$. We denote by x_n the discretization of the random NSFD method from Theorem 3. The exact solution $x(t_n)$ is computed with the exact scheme (10). Both random variables are explicitly and symbolically expressed in terms of $\alpha(\omega)$, $\beta(\omega)$ and $f(\cdot, \omega)$. These expansions are employed to compute the expectation and the variance, by using the linearity of the expectation. To check the accuracy, the absolute errors in the approximations of the mean value, $\epsilon_{N,M} = |\mathbb{E}[x(t_n)] - \mathbb{E}[x_n]|$, and the variance, $\delta_{N,M} = |\mathbb{V}[x(t_n)] - \mathbb{V}[x_n]|$, are calculated for different values of N and M. According to the theoretical discussion, the errors should decay as $\mathcal{O}(h^M)$ when $h \to 0$, which entails accuracy up to a significant number of digits. We remark that such level of accuracy cannot be achieved by Monte Carlo simulation, since its error decreases as the reciprocal of the square root of the number of realizations.

The implementations and computations are performed with Mathematica® (Wolfram Research, Inc, Mathematica, Version 12.0, Champaign, IL, USA, 2019), owing to its capability to handle both symbolic and numeric computations.

Example 1. *Let $\tau = 0.35$. Consider $f(t) = 1$ and $\alpha = -1$, while β is a random variable, uniformly distributed on the interval $[0.1, 0.2]$.*

Figure 1 plots absolute errors $\epsilon_{N,M}$ of the approximation of the expectation. First, $N = 10$ is fixed and $M \in \{1, 2, 3, 4\}$ varies. Second, $M = 1$ is fixed and $N \in \{5, 7, 10\}$ varies. In addition, third, these errors are divided by h to show, because of the overlapping, the decrease $\mathcal{O}(h^M)$ as $h \to 0$. Observe that the error is exactly 0 on the first M intervals of length τ. In the fourth panel, $\mathbb{E}[x_n]$ is plotted, where x_n is the output of the NSFD scheme of Theorem 3; the estimated expected values are validated by Monte Carlo simulation. Figure 2 is analogous with the variance and the absolute error of its approximation, $\delta_{N,M}$. According to ([16], Lemma 4, Theorem 7), since $\alpha + \beta < 0$ and $\alpha \leq \beta$ almost surely, the NSFD scheme converges to 0 almost surely as $t \to \infty$ (it is asymptotically stable almost surely). In both figures, observe that $\mathbb{E}[x_n]$ and $\mathbb{V}[x_n]$ tend to 0 as $t \to \infty$, which means that the NSFD scheme is asymptotically stable in the m.s. sense. Finally, the numerical solution is always positive because $\beta > 0$ almost surely and $f(t) > 0$ ([16], Theorem 8).

Example 2. *Let $\tau = 0.35$. Consider $f(t) = 1$, $\alpha = 0$, and β random with Gaussian distribution, of zero mean and 0.3 standard deviation. Notice that the support of β is unbounded; however, we will see that m.s. convergence of the NSFD scheme described in Theorem 3 holds.*

Figure 3 reports absolute errors $\epsilon_{N,M}$ of the approximation of the expectation, where N and M take on the same values as in Example 1. The decay $\mathcal{O}(h^M)$ as $h \to 0$ is captured again. The last panel of the figure plots the expectation of the numerical solution from the NSFD scheme of Theorem 3, $\mathbb{E}[x_n]$, together with Monte Carlo simulation. Figure 4 is analogous with the variance and the absolute error of its approximation, $\delta_{N,M}$. This example is interesting from the dynamics viewpoint. By ([16], Lemma 4), the probability that the zero solution to the realizations of (2) is asymptotically stable is the probability that $\beta < 0$ and $\tau < \tau^* = 1/|\beta|$, i.e., $-1/\tau < \beta < 0$. Taking into account the Gaussian distribution of β, this probability is ≈ 0.5 up to 12 decimals, i.e., approximately half of the time a realizable NSFD scheme tends to 0 as $t \to \infty$, and half of the time it does not. The m.s. treatment mixes these two behaviors. In the figures, both $\mathbb{E}[x_n]$ and $\mathbb{V}[x_n]$ seem to increase as t advances, which means that the NSFD scheme is unstable in the m.s. sense. Finally, notice that β has one half of probability of being negative and the mean of the solution is positive ([16], Theorem 8).

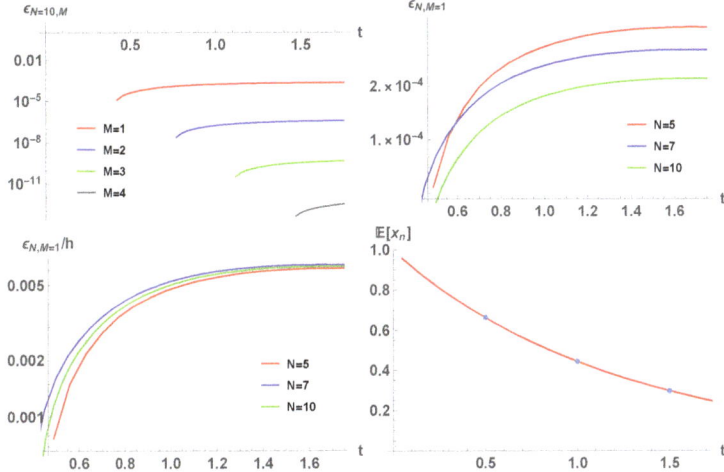

Figure 1. Upper left panel: Absolute errors (log-scale) in the approximation of the mean value with the NSFD scheme, with $M = 1, 2, 3, 4$ and $N = 10$ ($h = \tau/N$). Upper right panel: Absolute errors (log-scale) in the approximation of the mean value with the NSFD scheme, with $N = 5, 7, 10$ ($h = \tau/N$) and $M = 1$. Lower left panel: Errors from the upper right panel divided by h. Lower right panel: Approximation of the expectation with the NSFD scheme, with $M = 1$ and $N = 7$, and comparison with Monte Carlo simulation (circles) using 10,000 realizations. This figure corresponds to Example 1.

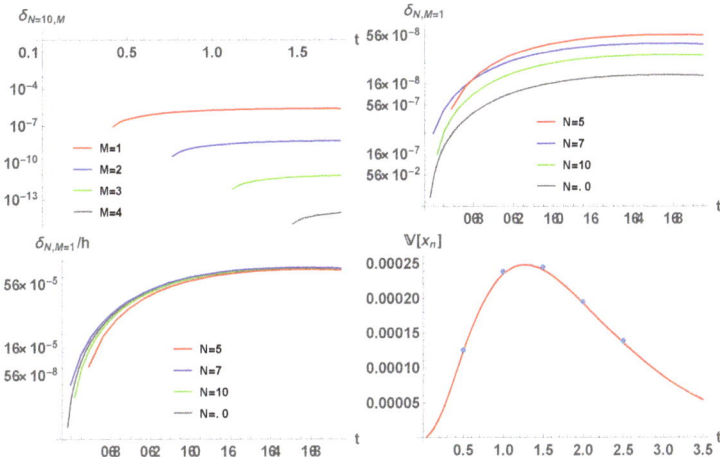

Figure 2. Upper left panel: Absolute errors (log-scale) in the approximation of the variance with the NSFD scheme, with $M = 1, 2, 3, 4$ and $N = 10$ ($h = \tau/N$). Upper right panel: Absolute errors (log-scale) in the approximation of the variance with the NSFD scheme, with $N = 5, 7, 10, 20$ ($h = \tau/N$) and $M = 1$. Lower left panel: Errors from the upper right panel divided by h. Lower right panel: Approximation of the variance with the NSFD scheme, with $M = 1$ and $N = 7$, and comparison with Monte Carlo simulation (circles) using 10,000 realizations. This figure corresponds to Example 1.

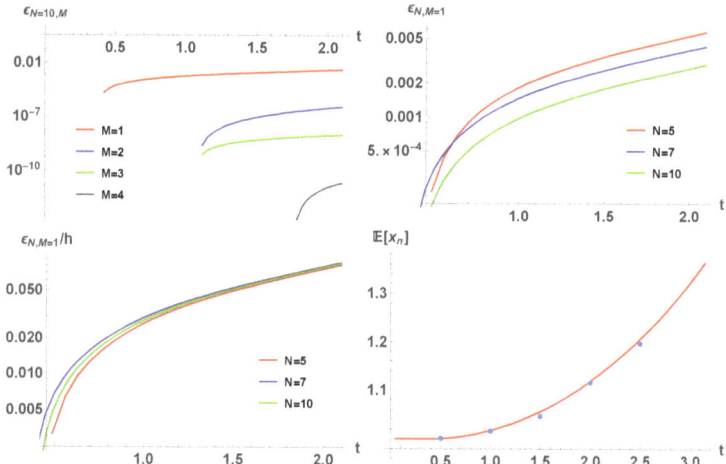

Figure 3. Upper left panel: Absolute errors (log-scale) in the approximation of the mean value with the NSFD scheme, with $M = 1, 2, 3, 4$ and $N = 10$ ($h = \tau/N$). Upper right panel: Absolute errors (log-scale) in the approximation of the mean value with the NSFD scheme, with $N = 5, 7, 10$ ($h = \tau/N$) and $M = 1$. Lower left panel: Errors from the upper right panel divided by h. Lower right panel: Approximation of the expectation with the NSFD scheme, with $M = 1$ and $N = 7$, and comparison with Monte Carlo simulation (circles) using 10,000 realizations. This figure corresponds to Example 2.

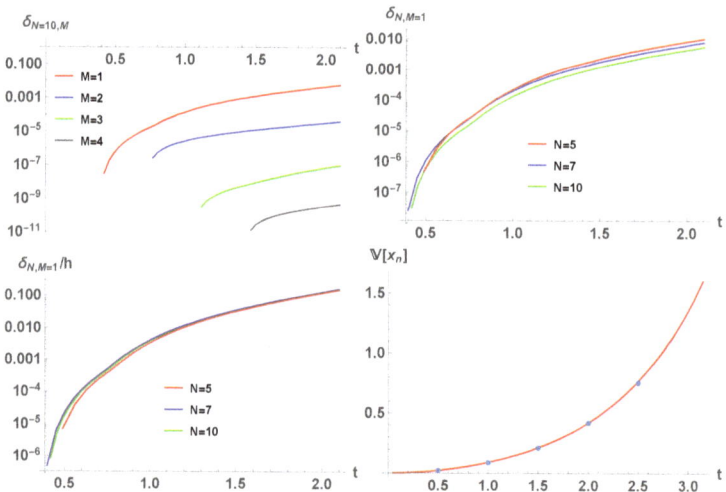

Figure 4. Upper left panel: Absolute errors (log-scale) in the approximation of the variance with the NSFD scheme, with $M = 1, 2, 3, 4$ and $N = 10$ ($h = \tau/N$). Upper right panel: Absolute errors (log-scale) in the approximation of the variance with the NSFD scheme, with $N = 5, 7, 10$ ($h = \tau/N$) and $M = 1$. Lower left panel: Errors from the upper right panel divided by h. Lower right panel: Approximation of the variance with the NSFD scheme, with $M = 1$ and $N = 7$, and comparison with Monte Carlo simulation (circles) using 10,000 realizations. This figure corresponds to Example 2.

Example 3. *Let $\tau = 0.35$. Consider $f(t) = \gamma$, where γ is a random variable. It is assumed that α, β and γ are independent random quantities, uniformly distributed on the interval $[0.1, 0.2]$.*

As Example 1, Figure 5 reports absolute errors $\epsilon_{N,M}$ of the approximation of the expectation. First, for fixed $N = 10$ and $M \in \{1, 2, 3, 4\}$. Second, for fixed $M = 2$ and $N \in \{5, 7, 10\}$. In addition, third, these errors are divided by h^2 to highlight, because of the overlapping, the decrease $\mathcal{O}(h^M)$ as $h \to 0$. The fourth panel plots $\mathbb{E}[x_n]$ with the discretization x_n computed via the NSFD scheme of Theorem 3, which is validated by Monte Carlo simulation. For the variance, computations become more expensive, due to the dimension three of the random space. In particular, the symbolic expression of the exact scheme (10) becomes unfeasible, so the exact error $\delta_{N,M}$ of the variance approximation cannot be reported. In Figure 6, we plot $\mathbb{V}[x_n]$ with the discretization x_n from the NSFD scheme of Theorem 3. Comparison is performed with Monte Carlo simulation, showing agreement of the estimates. Based on ([16], Lemma 4, Theorem 7), the condition $\alpha + \beta > 0$ almost surely entails that the NSFD scheme does not approach 0 as $t \to \infty$ (almost sure instability). This fact agrees with the plots of $\mathbb{E}[x_n]$ and $\mathbb{V}[x_n]$, which seem to increase as t grows; this behavior entails that the NSFD scheme is unstable in the m.s. sense. Finally, $\beta > 0$ and $\gamma > 0$ almost surely implies the positivity of the numerical solution ([16], Theorem 8).

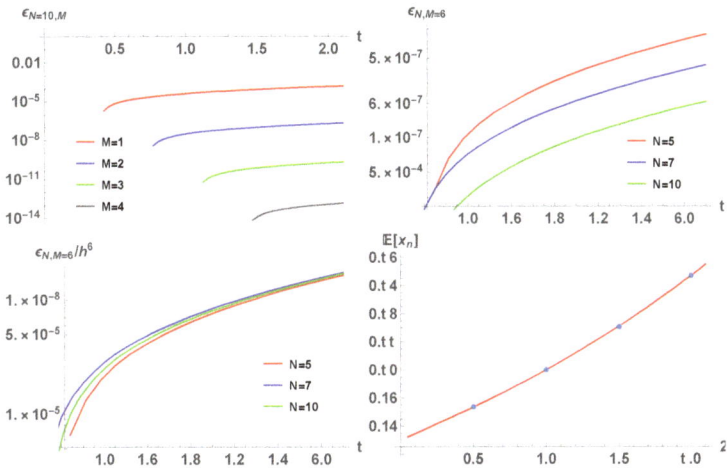

Figure 5. Upper left panel: Absolute errors (log-scale) in the approximation of the mean value with the NSFD scheme, with $M = 1, 2, 3, 4$ and $N = 10$ ($h = \tau/N$). Upper right: Absolute errors (log-scale) in the approximation of the mean value with the NSFD scheme, with $N = 5, 7, 10$ ($h = \tau/N$) and $M = 2$. Lower left panel: Errors from the upper right panel divided by h^2. Lower right panel: Approximation of the expectation with the NSFD scheme, with $M = 2$ and $N = 7$, and comparison with Monte Carlo simulation (circles) using 10,000 realizations. This figure corresponds to Example 3.

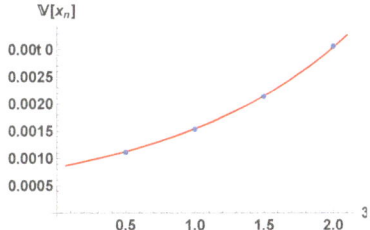

Figure 6. Approximation of the variance with the NSFD scheme, with $M = 1$ and $N = 5$ ($h = \tau/N$), and comparison with Monte Carlo simulation (circles) using 10,000 realizations. This figure corresponds to Example 3.

We would like to remark that even when $M = 1$ and the global error of the NSFD scheme is $\mathcal{O}(h)$, its error is lower than Euler's method, given by $x_{n+1} = (1 + \alpha h)x_n + \beta h x_{n-N}$. Euler's method has already been employed for random differential equations (ordinary and fractional) in the m.s. sense [29,30,34]. In this Example 3, Figure 7 plots errors ϵ_N for approximations of the mean, for comparing Euler's method and the NSFD scheme with $M = 1$ fixed. Observe that in log scale, both errors are located in parallel, but the error corresponding to the NSFD scheme is lower; this may be due to the non-standard nature of the method and being error-free on $[0, \tau]$. Although the proposed NSFD method is restricted to the linear random differential equation with delay, it may provide the foundation for designing new non-standard numerical methods for delay nonlinear equations.

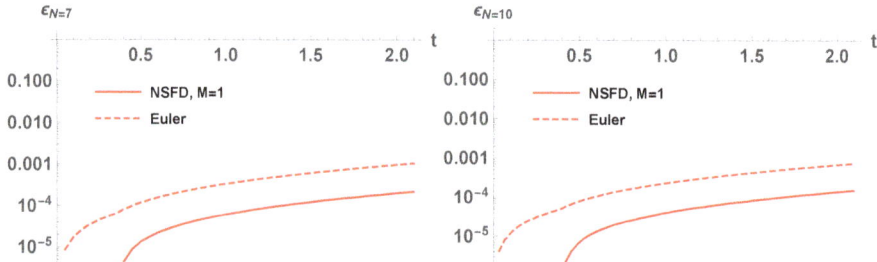

Figure 7. Absolute errors (log-scale) in the approximation of the mean value, with Euler's method (dashed lines) and with the NSFD scheme $M = 1$ (solid lines). For step size $h = \tau/N$, the left panel corresponds to $N = 7$ and the right panel to $N = 10$. This figure corresponds to Example 3.

Remark 2. *In the recent literature, the m.s. convergence of Euler's method has not been formally proved for delay random differential equations. If randomness is not incorporated into the system by the coefficients, but by a Wiener noise instead (Itô stochastic delay differential equation, which gives rise to non-differentiable solutions), then Euler's method (Euler-Maruyama's method, which considers discrete increments of the driving Wiener process) was rigorously studied and its m.s. convergence was proved in [37]. In the context of delay random differential equations (those with randomness manifested in coefficients and no Wiener noise), we focus on the linear case (2) studied in this paper assuming, as in Theorem 3, that α and β are bounded random variables, so that $\|\alpha\|_\infty$ and $\|\beta\|_\infty$ are finite. The exact m.s. solution to (2) satisfies $x(t_{n+1}) = x(t_n) + \alpha \int_{t_n}^{t_n+h} x(s)\,ds + \beta \int_{t_n}^{t_n+h} x(s - \tau)\,ds$, where the integrals are m.s. Riemann. If $e_n = x(t_n) - x_n$ denotes the difference between the exact solution and the discretization at the mesh point t_n, then*

$$e_{n+1} = e_n + \alpha \int_{t_n}^{t_n+h} (x(s) - x_n)\,ds + \beta \int_{t_n}^{t_n+h} (x(s - \tau) - x_{n-N})\,ds, \tag{25}$$

by a simple subtraction. Given any interval $[-\tau, T]$, $T > 0$, the m.s. Lipschitz condition $\|x(s_1) - x(s_2)\|_2 = \|\int_{s_2}^{s_1} x'(s)\,ds\|_2 \leq |\int_{s_2}^{s_1} \|x'(s)\|_2\,ds| \leq \lambda|s_1 - s_2|$ holds, $\lambda = \lambda(T) = \max_{[-\tau,T]} \|x'(s)\|_2 > 0$. Then $\|x(s) - x_m\|_2 \leq \|x(s) - x(t_m)\|_2 + \|e_m\|_2 \leq \lambda h + \|e_m\|_2$, $m \geq -N$, $s \in [t_m, t_m + h]$, $t_m + h \leq T$. Consequently,

$$\|e_{n+1}\|_2 \leq \|e_n\|_2 + \|\alpha\|_\infty h (\lambda h + \|e_n\|_2) + \|\beta\|_\infty h (\lambda h + \|e_{n-N}\|_2)$$
$$= (1 + \|\alpha\|_\infty h) \|e_n\|_2 + \|\beta\|_\infty h \|e_{n-N}\|_2 + (\|\alpha\|_\infty + \|\beta\|_\infty) h^2 \lambda. \tag{26}$$

For $0 \leq n \leq N$ (notice that $e_{n-N} = 0$), by solving the first order recursive inequality for $\|e_n\|_2$, we derive the following bounds:

$$\begin{aligned}
\|e_n\|_2 &\leq \sum_{i=1}^{n} \left(\|\alpha\|_\infty + \|\beta\|_\infty\right) h^2 \lambda \left(1 + \|\alpha\|_\infty h\right)^{n-i} \\
&= h^2 \lambda \left(\|\alpha\|_\infty + \|\beta\|_\infty\right) \frac{(1 + \|\alpha\|_\infty h)^n - 1}{\|\alpha\|_\infty h} \\
&\leq h\lambda \frac{\|\alpha\|_\infty + \|\beta\|_\infty}{\|\alpha\|_\infty} \left[(1 + \|\alpha\|_\infty \tau/N)^N - 1\right] \\
&\leq h\lambda \frac{\|\alpha\|_\infty + \|\beta\|_\infty}{\|\alpha\|_\infty} \left(e^{\|\alpha\|_\infty \tau} - 1\right) \\
&= C_1 h,
\end{aligned} \quad (27)$$

where $C_1 = C_1(\lambda, \tau, \|\alpha\|_\infty, \|\beta\|_\infty)$ is constant. Then

$$\max_{0 \leq n \leq N} \|e_n\|_2 = \mathcal{O}(h). \quad (28)$$

For $N+1 \leq n \leq 2N$, based on similar calculations,

$$\begin{aligned}
\|e_n\|_2 &\leq (1 + \|\alpha\|_\infty h)^{n-N} \|e_N\|_2 + \sum_{i=N+1}^{n} \left[\|\beta\|_\infty \|e_{i-N}\|_2 + (\|\alpha\|_\infty + \|\beta\|_\infty) \lambda h\right] h \left(1 + \|\alpha\|_\infty h\right)^{n-i} \\
&\leq C_1 h e^{\|\alpha\|_\infty \tau} + [C_1 \|\beta\|_\infty + (\|\alpha\|_\infty + \|\beta\|_\infty) \lambda] h^2 \frac{(1+\|\alpha\|_\infty h)^{n-N}-1}{\|\alpha\|_\infty h} \\
&\leq C_2 h,
\end{aligned} \quad (29)$$

where $C_2 = C_2(\lambda, \tau, \|\alpha\|_\infty, \|\beta\|_\infty)$ is constant. Then

$$\max_{N+1 \leq n \leq 2N} \|e_n\|_2 = \mathcal{O}(h). \quad (30)$$

For $2N+1 \leq n \leq 3N$, $3N+1 \leq n \leq 4N$, etc. one proceeds similarly. This proves that the random Euler's method has m.s. global error $\mathcal{O}(h)$. This proof corresponds to the linear case, although it may be extendible to delay random differential equations satisfying a m.s. Lipschitz condition.

Example 4. *In this example, the linear model (2) is considered for fitting the time evolution of a photosynthetic bacterial population, Rhodobacter capsulatus (R. capsulatus) [38], under infrared lighting conditions. Direct cell counts were made for the first 7 days, every two to three days, during which the population grew with no effect of competition for resources (light and/or CO_2) that would yield logistic nonlinearities. For days 0, 2, 4 and 7, the population sizes, measured in cells/mL scaled by one million, were 0.583, 0.635, 1.08 and 3.20, respectively. For delay $\tau = 1$ and initial function $f(t) = 0.583$ on $[-\tau, 0]$, the least-squares estimates for α and β are 1.20426 and -1.18024, respectively. The effect of small random displacements on the coefficients is studied here. Let us suppose 0.5% displacements of α and β with respect to their least-squares estimates, with zero mean values. According to the maximum entropy principle [39], α and $-\beta$ follow truncated exponential distributions, with rates $1/1.20426$ and $1/1.18024$ respectively. The expectation and the variance of the output are approximated with the random NSFD scheme. Figure 8 plots the results for $M = 2$ and distinct N (mean values in solid line, and mean $\pm 2 \times$ standard deviation in dashed lines), together with the least-squares fitting. Observe that a small uncertainty of 0.5% for parameters may cause significant changes in the final solution, up to 30% variation for the seventh day compared to an idealized situation containing no uncertainty. Observe also that as N increases, the approximations from the NSFD scheme tend to overlap, thus indicating convergence.*

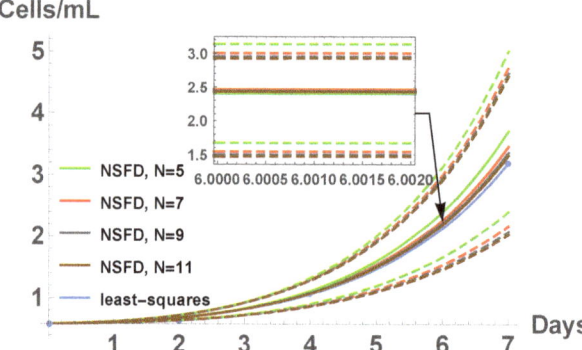

Figure 8. Application of the random NSFD scheme for the model of the *R. capsulatus* bacterial population, for $M = 2$ and different values of N (mean values in solid line, and mean $\pm 2 \times$ standard deviation in dashed lines). The least-squares fitting is also plotted. A zoom for a particular region is included for a better appreciation of convergence as N grows. This figure corresponds to Example 4.

6. Conclusions

In this paper, we have extended a NSFD numerical scheme recently proposed for deterministic linear differential equations with delay to the random framework. Incorporating randomness into models is important to account for measurement errors in data. M.s. convergence of the numerical discretizations has been established when the two equation coefficients are bounded random variables and the initial condition is a regular stochastic process, with rate of convergence given by $\mathcal{O}(h^M)$, where h is the step size and M is the number of intervals of length τ where the exact scheme is applied. M.s. convergence allows for approximating the expectation and the variance of the solution at inherited rate $\mathcal{O}(h^M)$, by symbolically expanding the discretizations in terms of the random inputs. The numerical examples have illustrated and assessed the proposed approach. The convergence rate $\mathcal{O}(h^M)$ has been supported numerically, even when the two equation coefficients are unbounded random variables. The asymptotic behavior of the expectation and the variance as the time t grows has been evaluated, graphically and taking into account theoretical results on deterministic stability and instability of the zero solution. A comparison with Euler's method has been performed when $M = 1$; although both methods have global errors $\mathcal{O}(h)$, the error of the NSFD scheme is lower, possibly due to the non-standard nature of the scheme and being error-free on $[0, \tau]$. Also, we have considered an example dealing with actual experimental data for a bacterial specie growing under infrared lighting conditions, and have calculated numerical solutions after randomizing the input parameters according to the maximum entropy principle.

The advantage of the random NSFD scheme is the high accuracy to approximate some statistics, which cannot be achieved with Monte Carlo methods. In addition, the procedure is simple: one only symbolically expands the discretizations in terms of the random inputs and afterward applies the corresponding statistical operator. However, this strategy possesses some limitations. Obviously, the necessity of symbolically expressing the discretizations restricts the applicability of the NSFD scheme to moderate step size h and time variable t, as well as small dimension of the random space. Although the calculation of the expectation of the discretization seems to be quite feasible in the computer, the calculation of the variance may become a big issue with this approach, let alone other statistics of order greater than two. We ask ourselves about the possibility of accurately calculating statistics with the random NSFD scheme without relying on symbolic expansions. We admit that for the moment, Monte Carlo simulation seems the best option for large time variable t, small step size h or large dimension of the random space, where each realization of the governing delay model is numerically solved by employing a NSFD scheme. For estimating densities, the symbolic expression is too complex and kernel methods are the preferable.

Mickens' methodology on NSFD schemes has shown fruitful applications along the years on ordinary, partial and fractional deterministic differential equations. However, only very recently, the application of NSFD numerical schemes to delay deterministic differential equations has been explored, in the context of linear models. Thus, this is the first contribution that proposes the use of NSFD schemes for quantifying uncertainty for delay random differential equations. Further study of NSFD methods for delay deterministic and random differential equations needs to be conducted, especially for nonlinear equations, for applications to modeling of real-life systems with aftereffects or time lags.

We propose specific lines of research for possible future developments:

- In the deterministic setting, the extension of the NSFD method to nonlinear delay differential equations. We believe that this extension may be done by linearization or by applying the empirical rules proposed by Mickens.
- Randomization of the NSFD method for delay random differential equations and applications without relying on symbolic expansions. Symbolic computations are the main drawback of the method proposed in the present paper.
- A theoretical analysis of m.s. dynamic consistency.

Author Contributions: Investigation, J.C. and M.J.; Methodology, J.C. and M.J.; Software, M.J.; Supervision, J.C.C. and F.R.; Validation, all authors; Visualization, all authors; Writing—original draft, J.C. and M.J.; Writing—review & editing, all authors. All authors have read and agreed to the published version of the manuscript.

Funding: This work has been supported by the Spanish Ministerio de Economía, Industria y Competitividad (MINECO), the Agencia Estatal de Investigación (AEI) and Fondo Europeo de Desarrollo Regional (FEDER UE) grant MTM2017–89664–P.

Conflicts of Interest: The authors declare that there is no conflict of interests regarding the publication of this article.

References

1. Smith, H. *An Introduction to Delay Differential Equations with Applications to the Life Sciences*; Texts in Applied Mathematics; Springer: New York, NY, USA, 2011.
2. Driver, Y. *Ordinary and Delay Differential Equations*; Applied Mathematical Science Series; Springer: New York, NY, USA, 1977.
3. Diekmann, O.; van Cils, S.A.; Verduyn Lunel, S.M.; Walther, H.-O. *Delay Equations*; Springer: New York, NY, USA, 1995.
4. Saaty, T.L. *Modern Nonlinear Equations*; Dover Publications: New York, NY, USA, 1981.
5. Bocharov, G.A.; Rihan, F.A. Numerical modelling in biosciences using delay differential equations. *J. Comput. Appl. Math.* **2000**, *125*, 183–199. [CrossRef]
6. Jackson, M.; Chen-Charpentier, B.M. Modeling plant virus propagation with delays. *J. Comput. Appl. Math.* **2017**, *309*, 611–621. [CrossRef]
7. Chen-Charpentier, B.M.; Diakite, I. A mathematical model of bone remodeling with delays. *J. Comput. Appl. Math.* **2016**, *291*, 76–84. [CrossRef]
8. Erneux, T. *Applied Delay Differential Equations*; Surveys and Tutorials in the Applied Mathematical Sciences Series; Springer: New York, NY, USA, 2009.
9. Kyrychko, Y.N.; Hogan, S.J. On the use of delay equations in engineering applications. *J. Vib. Control* **2017**, *16*, 943–960. [CrossRef]
10. Harding, L.; Neamtu, M. A dynamic model of unemployment with migration and delayed policy intervention. *Comput. Econ.* **2018**, *51*, 427–462. [CrossRef]
11. Mickens, R.E. *Nonstandard Finite Difference Models of Differential Equations*; World Scientific: Singapore, 1994.
12. Mickens, R.E. *Applications of Nonstandard Finite Difference Schemes*; World Scientific: Singapore, 2000.
13. Mickens, R.E. *Advances on Applications of Nonstandard Finite Difference Schemes*; World Scientific: Singapore, 2005.

14. Mickens, R.E. Dynamic consistency: A fundamental principle for constructing nonstandard finite difference schemes for differential equations. *J. Differ. Equ. Appl.* **2005**, *11*, 645–653. [CrossRef]
15. Patidar, K.C. Nonstandard finite difference methods: Recent trends and further developments. *J. Differ. Equ. Appl.* **2016**, *22*, 817–849. [CrossRef]
16. García, M.A.; Castro, M.A.; Martín, J.A.; Rodríguez, F. Exact and nonstandard numerical schemes for linear delay differential models. *Appl. Math. Comput.* **2018**, *338*, 337–345. [CrossRef]
17. Castro, M.Á.; García, M.A.; Martín, J.A.; Rodríguez, F. Exact and Nonstandard Finite Difference Schemes for Coupled Linear Delay Differential Systems. *Mathematics* **2019**, *7*, 1038. [CrossRef]
18. Soong, T.T. *Random Differential Equations in Science and Engineering*; Academic Press: New York, NY, USA, 1973.
19. Villafuerte, L.; Braumann, C.A.; Cortés, J.-C.; Jódar, L. Random differential operational calculus: Theory and applications. *Comput. Math. Appl.* **2010**, *59*, 115–125. [CrossRef]
20. Cortés, J.-C.; Jódar, L.; Roselló, M.-D.; Villafuerte, L. Solving initial and two-point boundary value linear random differential equations: A mean square approach. *Appl. Math. Comput.* **2012**, *219*, 2204–2211. [CrossRef]
21. Calatayud, J.; Cortés, J.-C.; Jornet, M.; Villafuerte, L. Random non-autonomous second order linear differential equations: Mean square analytic solutions and their statistical properties. *Adv. Differ. Equ.* **2018**, *392*, 1–29. [CrossRef]
22. Calatayud, J.; Cortés, J.-C.; Jornet, M. Improving the approximation of the first- and second-order statistics of the response stochastic process to the random Legendre differential equation. *Mediterr. J. Math.* **2019**, *16*, 68. [CrossRef]
23. Licea, J.A.; Villafuerte, L.; Chen-Charpentier, B.M. Analytic and numerical solutions of a Riccati differential equation with random coefficients. *J. Comput. Appl. Math.* **2013**, *239*, 208–219. [CrossRef]
24. Burgos, C.; Calatayud, J.; Cortés, J.-C.; Villafuerte, L. Solving a class of random non-autonomous linear fractional differential equations by means of a generalized mean square convergent power series. *Appl. Math. Lett.* **2018**, *78*, 95–104. [CrossRef]
25. Calatayud, J.; Cortés, J.-C.; Jornet, M. Random differential equations with discrete delay. *Stoch. Anal. Appl.* **2019**, *37*, 699–707. [CrossRef]
26. Calatayud, J.; Cortés, J.-C.; Jornet, M. L^p-calculus approach to the random autonomous linear differential equation with discrete delay. *Mediterr. J. Math.* **2019**, *16*, 85. [CrossRef]
27. Cortés, J.-C.; Jornet, M. L^p-solution to the random linear delay differential equation with a stochastic forcing term. *Mathematics* **2020**, *8*, 1013. [CrossRef]
28. Caraballo, T.; Cortés, J.-C.; Navarro-Quiles, A. Applying the Random Variable Transformation method to solve a class of random linear differential equation with discrete delay. *Appl. Math. Comput.* **2019**, *356*, 198–218. [CrossRef]
29. Cortés, J.-C.; Jódar, L.; Villafuerte, L. Numerical solution of random differential equations: A mean square approach. *Math. Comput. Model.* **2007**, *45*, 757–765. [CrossRef]
30. Cortés, J.-C.; Jódar, L.; Villafuerte, L. Mean square numerical solution of random differential equations: Facts and possibilities. *Comput. Math. Appl.* **2007**, *53*, 1098–1106. [CrossRef]
31. El-Tawil, M.A. The approximate solutions of some stochastic differential equations using transformations. *Appl. Math. Comput.* **2005**, *164*, 167–178. [CrossRef]
32. Calatayud, J.; Cortés, J.-C.; Díaz, J.A.; Jornet, M. Density function of random differential equations via finite difference schemes: A theoretical analysis of a random diffusion-reaction Poisson-type problem. *Stochastics* **2020**, *92*, 627–641. [CrossRef]
33. Calatayud, J.; Cortés, J.-C.; Díaz, J.A.; Jornet, M. Constructing reliable approximations of the probability density function to the random heat PDE via a finite difference scheme. *Appl. Numer. Math.* **2020**, *151*, 413–424. [CrossRef]
34. Burgos, C.; Cortés, J.-C.; Villafuerte, L.; Villanueva, R.-J. Mean square convergent numerical solutions of random fractional differential equations: Approximations of moments and density. *J. Comput. Appl. Math.* **2020**, *378*, 112925. [CrossRef]
35. Loève, M. *Probability Theory*; Springer: New York, NY, USA, 1977; Volume I–II.
36. Strand, J.L. Random ordinary differential equations. *J. Differ. Equ.* **1970**, *7*, 538–553. [CrossRef]

37. Buckwar, E. Introduction to the numerical analysis of stochastic delay differential equations. *J. Comp. Appl. Math.* **2000**, *125*, 297–307. [CrossRef]
38. Stanescu, D.; Chen-Charpentier, B.; Jensen, B.J.; Colberg, P.J. Random coefficient differential models of growth of anaerobic photosynthetic bacteria. *Electron. Trans. Numer. Anal.* **2009**, *34*, 44–58.
39. Dorini, F.A.; Sampaio, R. Some results on the random wear coefficient of the Archard model. *J. Appl. Mech.* **2012**, *79*, 051008–051014. [CrossRef]

© 2020 by the authors. Licensee MDPI, Basel, Switzerland. This article is an open access article distributed under the terms and conditions of the Creative Commons Attribution (CC BY) license (http://creativecommons.org/licenses/by/4.0/).

Article

L^p-solution to the Random Linear Delay Differential Equation with a Stochastic Forcing Term

Juan Carlos Cortés * and Marc Jornet

Instituto Universitario de Matemática Multidisciplinar, Building 8G, access C, 2nd floor, Universitat Politècnica de València, Camino de Vera s/n, 46022 Valencia, Spain; marjorsa@doctor.upv.es
* Correspondence: jccortes@imm.upv.es

Received: 25 May 2020; Accepted: 18 June 2020; Published: 20 June 2020

Abstract: This paper aims at extending a previous contribution dealing with the random autonomous-homogeneous linear differential equation with discrete delay $\tau > 0$, by adding a random forcing term $f(t)$ that varies with time: $x'(t) = ax(t) + bx(t - \tau) + f(t), t \geq 0$, with initial condition $x(t) = g(t)$, $-\tau \leq t \leq 0$. The coefficients a and b are assumed to be random variables, while the forcing term $f(t)$ and the initial condition $g(t)$ are stochastic processes on their respective time domains. The equation is regarded in the Lebesgue space L^p of random variables with finite p-th moment. The deterministic solution constructed with the method of steps and the method of variation of constants, which involves the delayed exponential function, is proved to be an L^p-solution, under certain assumptions on the random data. This proof requires the extension of the deterministic Leibniz's integral rule for differentiation to the random scenario. Finally, we also prove that, when the delay τ tends to 0, the random delay equation tends in L^p to a random equation with no delay. Numerical experiments illustrate how our methodology permits determining the main statistics of the solution process, thereby allowing for uncertainty quantification.

Keywords: random linear delay differential equation; stochastic forcing term; random L^p-calculus; uncertainty quantification

1. Introduction

In this paper, we are concerned with random delay differential equations, defined as classical delay differential equations whose inputs (coefficients, forcing term, initial condition, ...) are considered as random variables or regular stochastic processes on an underlying complete probability space $(\Omega, \mathcal{F}, \mathbb{P})$, which may take a wide variety of probability distributions, such as Binomial, Poisson, Gamma, Gaussian, etc.

Equations of this kind should not be confused with stochastic differential equations of Itô type, forced by an irregular error term called White noise process (formal derivative of Brownian motion). In contrast to random differential equations, the solutions to stochastic differential equations exhibit nondifferentiable sample-paths. See [1] (pp. 96–98) for a detailed explanation of the difference between random and stochastic differential equations. See [1–6], for instance, for applications of random differential equations in engineering, physics, biology, etc. Thus, random differential equations require their own treatment and study: they model smooth random phenomena, with any type of input probability distributions.

From a theoretical viewpoint, random differential equations may be studied in two senses: the sample-path sense or the L^p-sense. The former case considers the trajectories of the stochastic processes involved, so that the realizations of the random system correspond to deterministic versions of the problem. The latter case works with the topology of the Lebesgue space $(L^p, \|\cdot\|_p)$ of random variables with finite absolute p-th moment, where the norm $\|\cdot\|_p$ is defined as: $\|U\|_p = \mathbb{E}[|U|^p]^{1/p}$

for $1 \leq p < \infty$ ($\mathbb{E}$ denotes the expectation operator), and $\|U\|_\infty = \inf\{C \geq 0 : |U| \leq C \text{ almost surely}\}$ (essential supremum), $U : \Omega \to \mathbb{R}$ being any random variable. The Lebesgue space ($L^p, \|\cdot\|_p$) has the structure of a Banach space. Continuity, differentiability, Riemann integrability, etc., can be considered in the aforementioned space L^p, which gives rise to the random L^p-calculus.

In order to fix concepts, given a stochastic process $x(t) \equiv x(t, \omega)$ on $I \times \Omega$, where $I \subseteq \mathbb{R}$ is an interval (notice that as usual the ω-sample notation might be hidden), we say that x is L^p-continuous at $t_0 \in I$ if $\lim_{h \to 0} \|x(t_0 + h) - x(t_0)\|_p = 0$. We say that x is L^p-differentiable at $t_0 \in I$ if $\lim_{h \to 0} \|\frac{x(t_0+h)-x(t_0)}{h} - x'(t_0)\|_p = 0$, for certain random variable $x'(t_0)$ (called the derivative of x at t_0). Finally, if $I = [a, b]$, we say that x is L^p-Riemann integrable on $[a, b]$ if there exists a sequence of partitions $\{P_n\}_{n=1}^\infty$ with mesh tending to 0, $P_n = \{a = t_0^n < t_1^n < \ldots < t_{r_n}^n = b\}$, such that, for any choice of points $s_i^n \in [t_{i-1}^n, t_i^n]$, $i = 1, \ldots, r_n$, the limit $\lim_{n \to \infty} \sum_{i=1}^{r_n} x(s_i^n)(t_i^n - t_{i-1}^n)$ exists in L^p. In this case, these Riemann sums have the same limit, which is a random variable and is denoted by $\int_a^b x(t)\,dt$.

This L^p-approach has been widely used in the context of random differential equations with no delay, especially the case $p = 2$ which corresponds to the Hilbert space L^2 and yields the so-called mean square calculus; see [5,7–15]. Only recently, a theoretical probabilistic analysis of random differential equations with discrete constant delay has been addressed in [16–18]. In [16], general random delay differential equations in L^p were analyzed, with the goal of extending some of the existing results on random differential equations with no delay from the celebrated book [5]. In [17], we started our study on random delay differential equations with the basic autonomous-homogeneous linear equation, by proving the existence and uniqueness of L^p-solution under certain conditions. In [18], the authors studied the same autonomous-homogeneous random linear differential equation with discrete delay as [17], but considered the solution in the sample-path sense and computed its probability density function via the random variable transformation technique, for certain forms of the initial condition process. Other recent contributions for random delay differential equations, but focusing on numerical methods instead, are [19–21].

There is still a lack of theoretical analysis for important random delay differential equations. Motivated by this issue, the aim of this contribution is to advance further in the theoretical analysis of relevant random differential equations with discrete delay. In particular, in this paper we extend the recent study performed in [17] for the basic linear equation by adding a stochastic forcing term:

$$\begin{cases} x'(t, \omega) = a(\omega)x(t, \omega) + b(\omega)x(t - \tau, \omega) + f(t, \omega), & t \geq 0, \omega \in \Omega, \\ x(t, \omega) = g(t, \omega), & -\tau \leq t \leq 0, \omega \in \Omega. \end{cases} \quad (1)$$

The delay $\tau > 0$ is constant. The coefficients a and b are random variables. The forcing term $f(t)$ and the initial condition $g(t)$ are stochastic processes on $[0, \infty)$ and $[-\tau, 0]$ respectively, which depend on the outcome $\omega \in \Omega$ of a random experiment which might be sometimes omitted in notation. The term $x(t)$ represents the differentiable solution stochastic process in a certain probabilistic sense. Formally, according to the deterministic theory [22], we may express the solution process as

$$\begin{aligned} x(t, \omega) &= e^{a(\omega)(t+\tau)} e_\tau^{b_1(\omega),t} g(-\tau, \omega) \\ &+ \int_{-\tau}^0 e^{a(\omega)(t-s)} e_\tau^{b_1(\omega),t-\tau-s}(g'(s, \omega) - a(\omega)g(s, \omega))\,ds \\ &+ \int_0^t e^{a(\omega)(t-s)} e_\tau^{b_1(\omega),t-\tau-s} f(s, \omega)\,ds, \end{aligned} \quad (2)$$

where $b_1 = e^{-a\tau}b$ and

$$e_\tau^{c,t} = \begin{cases} 0, & -\infty < t < -\tau, \\ 1, & -\tau \leq t < 0, \\ 1 + c\dfrac{t}{1!}, & 0 \leq t < \tau, \\ 1 + c\dfrac{t}{1!} + c^2\dfrac{(t-\tau)^2}{2!}, & \tau \leq t < 2\tau, \\ \vdots & \vdots \\ \sum_{k=0}^n c^k \dfrac{(t-(k-1)\tau)^k}{k!}, & (n-1)\tau \leq t < n\tau, \end{cases}$$

is the delayed exponential function [22] (Definition 1), $c, t \in \mathbb{R}$, and $n = \lfloor t/\tau \rfloor + 1$ (here $\lfloor \cdot \rfloor$ denotes the integer part defined by the so-called floor function). This formal solution is obtained with the method of steps and the method of variation of constants.

The primary objective of this paper is to set probabilistic conditions under which $x(t)$ is an L^p-solution to (1). We decompose the original problem (1) as

$$\begin{cases} y'(t,\omega) = a(\omega)y(t,\omega) + b(\omega)y(t-\tau,\omega), & t \geq 0, \\ y(t,\omega) = g(t,\omega), & -\tau \leq t \leq 0, \end{cases} \quad (3)$$

and

$$\begin{cases} z'(t,\omega) = a(\omega)z(t,\omega) + b(\omega)z(t-\tau,\omega) + f(t,\omega), & t \geq 0, \\ z(t,\omega) = 0, & -\tau \leq t \leq 0. \end{cases} \quad (4)$$

System (3) does not possess a stochastic forcing term, and it was deeply studied in the recent contribution [17]. Under certain assumptions, its L^p-solution is expressed as

$$y(t,\omega) = e^{a(\omega)(t+\tau)} e_\tau^{b_1(\omega),t} g(-\tau,\omega) + \int_{-\tau}^0 e^{a(\omega)(t-s)} e_\tau^{b_1(\omega),t-\tau-s}(g'(s,\omega) - a(\omega)g(s,\omega))\,ds, \quad (5)$$

as a generalization of the deterministic solution to (3) obtained via the method of steps [22] (Theorem 1). Problem (4) is new and requires an analysis in the L^p-sense, in order to solve the initial problem (1). Its formal solution is given by

$$z(t,\omega) = \int_0^t e^{a(\omega)(t-s)} e_\tau^{b_1(\omega),t-\tau-s} f(s,\omega)\,ds, \quad (6)$$

see [22] (Theorem 2). In order to differentiate (6) in the L^p-sense, one requires the extension of the deterministic Leibniz's integral rule for differentiation to the random scenario. This extension is an important piece of this paper.

In Section 2, we show preliminary results on L^p-calculus that are used through the exposition, which correspond to those preliminary results from [17] and the new random Leibniz's rule for L^p-Riemann integration. Auxiliary but novel results to demonstrate the random Leibniz's integral rule are Fubini's theorem and a chain rule theorem. In Section 3, we prove in detail that $x(t)$ defined by (2) is the unique L^p-solution to (1), by analyzing problem (4). We also find closed-form expressions for some statistics (expectation and variance) of $x(t)$ related to its moments. Section 4 deals with the L^p-convergence of $x(t)$ as the delay τ tends to 0. We then show a numerical example that illustrates the theoretical findings. Finally, Section 5 draws the main conclusions.

In order to complete a fair overview of the existing literature, it must be pointed out that, apart for random delay differential equations (which is the context of this paper), other complementary approaches are stochastic delay differential equations and fuzzy delay differential equations. Stochastic delay differential equations are those in which uncertainty appears due to stochastic processes with irregular sample-paths: the Brownian motion process, Wiener process, Poisson process, etc. A new tool is required to tackle equations of this type, called Itô calculus [23]. Studies on stochastic delay differential equations can be read in [24–28], for example. On the other hand, in fuzzy delay differential equations, uncertainty is driven by fuzzy processes; see [29] for instance. In any of these approaches, the delay might even be considered random; see [30,31].

2. Results on L^p-calculus

In this section, we state the preliminary results on L^p-calculus needed for the following sections. Proposition 1 is the chain rule theorem in L^p-calculus, which was first proved in [8] (Theorem 3.19) in the setting of mean square calculus ($p = 2$). Both Lemma 1 and Lemma 2 provide conditions under which the product of three stochastic processes is L^p-continuous or L^p-differentiable. Proposition 2 is a result concerning L^p-differentiation under the L^p-Riemann integral sign, when the interval of integration is fixed. These four results have been already used and stated in the recent contribution [17], and will be required through our forthcoming exposition.

For the sake of completeness, we demonstrate Proposition 2 with an alternative proof to [17], based on Fubini's theorem for L^p-Riemann integration. In the random framework, Fubini's theorem has not been tackled yet in the recent literature. It states that, if a stochastic process depending on two variables is L^p-continuous, then the two iterated L^p-Riemann integrals can be interchanged.

We present a new result, Proposition 3, in which we put conditions in order to L^p-differentiate an L^p-Riemann integral whose interval of integration depends on t. This proposition supposes the extension of the so-called Leibniz's rule for integration to the random scenario. The proof relies on a new chain rule theorem.

Proposition 1 (Chain rule theorem ([17] Proposition 2.1))**.** *Let $\{X(t) : t \in [a,b]\}$ be a stochastic process, where $[a,b]$ is any interval in $\mathbb{R}$. Let f be a deterministic C^1 function on an open set that contains $X([a,b])$. Fix $1 \leq p < \infty$. Let $t \in [a,b]$ be any point such that:*

(i) *X is L^{2p}-differentiable at t;*
(ii) *X is path continuous on $[a,b]$;*
(iii) *there exist $r > 2p$ and $\delta > 0$ such that $\sup_{s \in [-\delta, \delta]} \mathbb{E}[|f'(X(t+s))|^r] < \infty$.*

Then $f \circ X$ is L^p-differentiable at t and $(f \circ X)'(t) = f'(X(t))X'(t)$.

Lemma 1 ([17] Lemma 2.1)**.** *Let $Y_1(t,s)$, $Y_2(t,s)$ and $Y_3(t,s)$ be three stochastic processes and fix $1 \leq p < \infty$. If Y_1 and Y_2 are L^q-continuous for all $1 \leq q < \infty$, and Y_3 is $L^{p+\eta}$-continuous for certain $\eta > 0$, then the product process $Y_1 Y_2 Y_3$ is L^p-continuous.*

On the other hand, if Y_1 and Y_2 are L^∞-continuous, and Y_3 is L^p-continuous, then the product process $Y_1 Y_2 Y_3$ is L^p-continuous.

Lemma 2 ([17] Lemma 2.2)**.** *Let $Y_1(t)$, $Y_2(t)$ and $Y_3(t)$ be three stochastic processes, and $1 \leq p < \infty$. If Y_1 and Y_2 are L^q-differentiable for all $1 \leq q < \infty$, and Y_3 is $L^{p+\eta}$-differentiable for certain $\eta > 0$, then the product process $Y_1 Y_2 Y_3$ is L^p-differentiable and $\frac{d}{dt}(Y_1(t)Y_2(t)Y_3(t)) = Y_1'(t)Y_2(t)Y_3(t) + Y_1(t)Y_2'(t)Y_3(t) + Y_1(t)Y_2(t)Y_3'(t)$.*

Additionally, if Y_1 and Y_2 are assumed to be L^∞-differentiable, and Y_3 is L^p-differentiable, then $Y_1 Y_2 Y_3$ is L^p-differentiable, with $\frac{d}{dt}(Y_1(t)Y_2(t)Y_3(t)) = Y_1'(t)Y_2(t)Y_3(t) + Y_1(t)Y_2'(t)Y_3(t) + Y_1(t)Y_2(t)Y_3'(t)$.

Lemma 3 (Fubini's theorem for iterated L^p-Riemann integrals)**.** *Let $H(t,s)$ be a process on $[a,b] \times [c,d]$. If H is L^p-continuous, then $\int_a^b \int_c^d H(t,s) \, ds \, dt = \int_c^d \int_a^b H(t,s) \, dt \, ds$, where the integrals are regarded as L^p-Riemann integrals.*

Proof. The proof is a variation of Fubini's theorem for Itô stochastic integration with respect to the standard Brownian motion ([32] Theorem 2.10.1). The stochastic processes $H(t,s)$, $\int_c^d H(t,s) \, ds$ and $\int_a^b H(t,s) \, dt$ are L^p-continuous, so the iterated integrals exist. Let $\{P_n\}_{n=1}^\infty$ be a sequence of partitions of $[a,b]$ with mesh tending to 0. Write $P_n = \{a = t_0^n < t_1^n < \cdots < t_n^n = b\}$, and let $r_i^n \in [t_{i-1}^n, t_i^n]$, $1 \leq i \leq n$, $n \geq 1$. Consider the processes $G_n(t,s) = \sum_{i=1}^n H(r_i^n, s) \mathbb{1}_{[t_{i-1}^n, t_i^n]}(t)$ (here $\mathbb{1}$ denotes the characteristic function of a set) and $F_n(s) = \int_a^b G_n(t,s)\,dt = \sum_{i=1}^n H(r_i^n, s)(t_i^n - t_{i-1}^n)$. Notice that, by definition of L^p-Riemann integral, $\lim_{n\to\infty} F_n(s) = \int_a^b H(t,s)\,dt$ in L^p.

By definition of L^p-Riemann integral,

$$\int_c^d F_n(s)\,ds = \sum_{i=1}^n \left(\int_c^d H(r_i^n, s)\,ds\right)(t_i^n - t_{i-1}^n) \xrightarrow{n\to\infty} \int_a^b \int_c^d H(t,s)\,ds\,dt$$

in L^p. On the other hand,

$$\left\| \int_c^d \int_a^b H(t,s)\,dt\,ds - \int_c^d F_n(s)\,ds \right\|_p = \left\| \int_c^d \left(\int_a^b H(t,s)\,dt - F_n(s) \right) ds \right\|_p$$

$$\leq \int_c^d \left\| \int_a^b H(t,s)\,dt - F_n(s) \right\|_p ds,$$

where the last inequality is due to a property of L^p-integration ([5] p. 102). As $H(t,s)$ and $F_n(s)$ are L^p-bounded on $[a,b] \times [c,d]$ and $[c,d]$, respectively (uniformly on $n \geq 1$), the dominated convergence theorem allows concluding that $\lim_{n\to\infty} \int_c^d F_n(s)\,ds = \int_c^d \int_a^b H(t,s)\,dt\,ds$ in L^p. □

Proposition 2 (L^p-differentiation under the L^p-Riemann integral sign). *Let $F(t,s)$ be a stochastic process on $[a,b] \times [c,d]$. Fix $1 \leq p < \infty$. Suppose that $F(t,\cdot)$ is L^p-continuous on $[c,d]$, for each $t \in [a,b]$, and that there exists the L^p-partial derivative $\frac{\partial F}{\partial t}(t,s)$ for all $(t,s) \in [a,b] \times [c,d]$, which is L^p-continuous on $[a,b] \times [c,d]$. Let $G(t) = \int_c^d F(t,s)\,ds$ (the integral is understood as an L^p-Riemann integral). Then G is L^p-differentiable on $[a,b]$ and $G'(t) = \int_c^d \frac{\partial F}{\partial t}(t,s)\,ds$.*

Proof. We present an alternative and simpler proof to ([17] Proposition 2.2), based on Fubini's theorem (Lemma 3). Since $\frac{\partial F}{\partial t}$ is L^p-continuous, by Barrow's rule ([5] p. 104) we can write $G(t) = \int_c^d F(a,s)\,ds + \int_c^d \int_a^t \frac{\partial F}{\partial t}(\tau,s)\,d\tau\,ds = \int_c^d F(a,s)\,ds + \int_a^t \int_c^d \frac{\partial F}{\partial t}(\tau,s)\,ds\,d\tau$. The stochastic process $\tau \in [a,b] \mapsto \int_c^d \frac{\partial F}{\partial t}(\tau,s)\,ds$ is L^p-continuous; therefore, $G'(t) = \int_c^d \frac{\partial F}{\partial t}(t,s)\,ds$ in L^p, as a consequence of the fundamental theorem for L^p-calculus; see ([5] p. 103). □

Lemma 4 (Version of the chain rule theorem). *Let $G(t,s)$ be a stochastic process on $[a,b] \times [c,d]$. Let $u: [a,b] \to [c,d]$ be a differentiable deterministic function. Suppose that $G(t,s)$ has L^p-partial derivatives, with $\frac{\partial G}{\partial t}(t,s)$ being L^p-continuous on $[a,b] \times [c,d]$, and $\frac{\partial G}{\partial s}(t,\cdot)$ being L^p-continuous on $[c,d]$, for each $t \in [a,b]$. Then $\frac{d}{dt} G(t, u(t)) = \frac{\partial G}{\partial t}(t, u(t)) + u'(t) \frac{\partial G}{\partial s}(t, u(t))$ in L^p.*

Proof. For $h \neq 0$, by the triangular inequality,

$$\left\| \frac{G(t+h, u(t+h)) - G(t, u(t))}{h} - \frac{\partial G}{\partial t}(t, u(t)) - u'(t) \frac{\partial G}{\partial s}(t, u(t)) \right\|_p$$

$$\leq \underbrace{\left\| \frac{G(t+h, u(t+h)) - G(t, u(t+h))}{h} - \frac{\partial G}{\partial t}(t, u(t)) \right\|_p}_{I_1(t,h)}$$

$$+ \underbrace{\left\| \frac{G(t, u(t+h)) - G(t, u(t))}{h} - u'(t) \frac{\partial G}{\partial s}(t, u(t)) \right\|_p}_{I_2(t,h)}.$$

By Barrow's rule ([5] p. 104) and an inequality from ([5] p. 102),

$$I_1(t,h) = \left\| \frac{1}{h} \int_t^{t+h} \frac{\partial G}{\partial t}(\tau, u(t+h)) \, d\tau - \frac{\partial G}{\partial t}(t, u(t)) \right\|_p$$

$$= \left\| \frac{1}{h} \int_t^{t+h} \left(\frac{\partial G}{\partial t}(\tau, u(t+h)) - \frac{\partial G}{\partial t}(t, u(t)) \right) d\tau \right\|_p$$

$$\leq \frac{1}{|h|} \left| \int_t^{t+h} \left\| \frac{\partial G}{\partial t}(\tau, u(t+h)) - \frac{\partial G}{\partial t}(t, u(t)) \right\|_p d\tau \right|.$$

The process $\frac{\partial G}{\partial t}(t, u(r))$ is L^p-uniform continuous on $[a,b] \times [a,b]$; therefore,

$$I_1(t,h) \leq \sup_{\tau \in [t, t+h] \cup [t+h, t]} \left\| \frac{\partial G}{\partial t}(\tau, u(t+h)) - \frac{\partial G}{\partial t}(t, u(t)) \right\|_p \xrightarrow{h \to 0} 0.$$

On the other hand, let $Y(r) = G(t,r)$, for $t \in [a,b]$ fixed. To conclude that $\lim_{h \to 0} I_2(t,h) = 0$, we need $(Y \circ u)'(t) = Y'(u(t))u'(t)$. We have that Y is L^p-$C^1([c,d])$ and that u is differentiable on $[a,b]$, so the following existing version of the chain rule theorem applies: ([33] Theorem 2.1). □

Remark 1. *Although not needed in the subsequent development, Lemma 4 gives in fact a general multidimensional chain rule theorem for L^p-calculus, for the composition of a stochastic process $G(t,s)$ and two deterministic functions $(v(r), u(r))$. This is the generalization of ([33] Theorem 2.1) to several variables. Indeed, let $G(t,s)$ be a stochastic process on an open set $\Lambda \subseteq \mathbb{R}^2$, with L^p-partial derivatives, $\frac{\partial G}{\partial t}(t,s)$ and $\frac{\partial G}{\partial s}(t,s)$, being L^p-continuous on Λ. Let $v,u : [a,b] \to \mathbb{R}$ be two C^1 deterministic functions with $(v(r), u(r)) \in \Lambda$. Then $\frac{d}{dr} G(v(r), u(r)) = v'(r) \frac{\partial G}{\partial t}(v(r), u(r)) + u'(r) \frac{\partial G}{\partial s}(v(r), u(r))$. For the proof, just define $\overline{G}(t,r) = G(v(t), r)$. By ([33] Theorem 2.1), $\frac{\partial \overline{G}}{\partial t}(t,r) = v'(t) \frac{\partial G}{\partial t}(v(t), r)$, which is L^p-continuous on (t,r). Additionally, $\frac{\partial \overline{G}}{\partial r}(t,r) = \frac{\partial G}{\partial s}(v(t), r)$ is L^p-continuous. Then $G(v(r), u(r)) = \overline{G}(r, u(r))$ can be L^p-differentiated at each r, by our Lemma 4: $\frac{d}{dr} G(v(r), u(r)) = \frac{\partial \overline{G}}{\partial t}(r, u(r)) + u'(r) \frac{\partial \overline{G}}{\partial r}(r, u(r)) = v'(r) \frac{\partial G}{\partial t}(v(r), u(r)) + u'(r) \frac{\partial G}{\partial s}(v(r), u(r))$.*

Proposition 3 (Random Leibniz's rule for L^p-calculus). *Let $F(t,s)$ be a stochastic process on $[a,b] \times [c,d]$. Let $u,v : [a,b] \to [c,d]$ be two differentiable deterministic functions. Suppose that $F(t, \cdot)$ is L^p-continuous on $[c,d]$, for each $t \in [a,b]$, and that $\frac{\partial F}{\partial t}(t,s)$ exists in the L^p-sense and is L^p-continuous on $[a,b] \times [c,d]$. Then $H(t) = \int_{u(t)}^{v(t)} F(t,s) \, ds$ is L^p-differentiable and*

$$H'(t) = v'(t) F(t, v(t)) - u'(t) F(t, u(t)) + \int_{u(t)}^{v(t)} \frac{\partial F}{\partial t}(t,s) \, ds$$

(the integral is considered as an L^p-Riemann integral).

Proof. First, notice that $H(t)$ is well-defined, since $F(t, \cdot)$ is L^p-continuous. Decompose $H(t)$ as $H(t) = \int_a^{v(t)} F(t,s) \, ds - \int_a^{u(t)} F(t,s) \, ds$. Let $G(t,r) = \int_a^r F(t,s) \, ds$, $t \in [a,b]$, $r \in [c,d]$. We have $H(t) = G(t, v(t)) - G(t, u(t))$.

Let us check the conditions of Lemma 4. By Lemma 2, $\frac{\partial G}{\partial t}(t,r) = \int_a^r \frac{\partial F}{\partial t}(t,s) \, ds$, which is L^p-continuous on $[a,b] \times [c,d]$ as a consequence of the L^p-continuity of $\frac{\partial F}{\partial t}(t,s)$. On the other hand, $\frac{\partial G}{\partial r}(t,r) = F(t,r)$, by the fundamental theorem of L^p-calculus ([5] p. 103), with $\frac{\partial G}{\partial r}(t, \cdot) = F(t, \cdot)$ being L^p-continuous. Thus, by Lemma 4,

$$H'(t) = \frac{\partial G}{\partial t}(t, v(t)) + v'(t) \frac{\partial G}{\partial r}(t, v(t)) - \frac{\partial G}{\partial t}(t, u(t)) - u'(t) \frac{\partial G}{\partial r}(t, u(t))$$

$$= v'(t) F(t, v(t)) - u'(t) F(t, u(t)) + \int_{u(t)}^{v(t)} \frac{\partial F}{\partial t}(t,s) \, ds.$$

□

Remark 2 (Proposition 3 against another proof of the random Leibniz's rule). *In [10, Proposition 6], a result pointing towards the conclusion of Proposition 3 was stated (in the mean square case $p = 2$, with $v(t) = t$, $u(t) = 0$ and $[c,d] = [a,b]$). However, the proof presented therein is not correct. In the notation therein, the authors proved an inequality of the form*

$$\|K(t, \Delta t)\|_2 \leq (t-a) \max_{x \in [a,t]} \|K_1(x, t, \Delta t)\|_2 + \max_{x \in [t, t+\Delta t]} \|K_2(x, t, \Delta t)\|_2.$$

The authors justified correctly that $\|K_1(x, t, \Delta t)\|_2 \to 0$ and $\|K_2(x, t, \Delta t)\|_2 \to 0$ as $\Delta t \to 0$, for each $x \in [a, b]$. However, this fact does not imply

$$\max_{x \in [a,t]} \|K_1(x, t, \Delta t)\|_2 \xrightarrow{\Delta t \to 0} 0, \quad \max_{x \in [t, t+\Delta t]} \|K_2(x, t, \Delta t)\|_2 \xrightarrow{\Delta t \to 0} 0,$$

as they stated at the end of their proof. For K_1, one has to utilize the dominated convergence theorem. For K_2, one should use uniform continuity.

Remark 3 (Random Leibniz's rule cannot be proved with a mean value theorem). *In the deterministic setting, both Proposition 2 and Proposition 3 can be proven with the mean value theorem. However, such proofs do not work in the random scenario, as there is no version of the stochastic mean value theorem. In previous contributions (see [15] Lemma 2.4, Corollary 2.5; [34] Lemma 3.1, Theorem 3.2), there is an incorrect version of it. For instance, if $U \sim \text{Uniform}(0,1)$ and $Y(t) = \mathbb{1}_{\{t > U\}}(t)$, $t \in [0,1]$, then Y is mean square continuous on $[0,1]$ (notice that $\|Y(t) - Y(s)\|_2^2 = |t - s|$). Suppose that there exists $\eta \in [0,1]$ such that $\int_0^1 Y(s)\,ds = Y(\eta)$ almost surely. Then $Y(\eta) = 1 - U$ almost surely. But this is not possible, since $1 - U \in (0,1)$ and $Y(\eta) \in \{0,1\}$. Thus, Y does not satisfy any mean square mean value theorem.*

3. L^p-solution to the Random Linear Delay Differential Equation with a Stochastic Forcing Term

In this section we solve (1) in the L^p-sense. To do so, we will demonstrate that $x(t)$ defined by (2) is the unique L^p-solution to (1). We will take advantage of the decomposition of problem (1) into its homogeneous part, (3), and its complete part, (4). The formal solution to (3) is given by $y(t)$ defined as (5), while the formal solution to (4) is given by $z(t)$ expressed as (6). The previous contribution [17] provides conditions under which $y(t)$ defined by (5) solves (3) in the L^p-sense. Thus, our primary goal will be to find conditions under which $z(t)$ given by (6) is a true solution to (4) in the L^p-sense.

Again, recall that the integrals that appear in the expressions (2), (5) and (6) are L^p-Riemann integrals.

The uniqueness (not existence for now) of (1) is proved analogously to ([17] Theorem 3.1), by invoking results from [7] that connect L^p-solutions with sample-path solutions, which satisfy analogous properties to deterministic solutions. The precise uniqueness statement is as follows.

Theorem 1 (Uniqueness). *The random differential equation problem with delay (1) has at most one L^p-solution, for $1 \leq p < \infty$.*

Proof. Assume that (1) has an L^p-solution. We will prove it is unique. Let $x_1(t)$ and $x_2(t)$ be two L^p-solutions to (1). Let $u(t) = x_1(t) - x_2(t)$, which satisfies the random differential equation problem with delay

$$\begin{cases} u'(t, \omega) = a(\omega)u(t, \omega) + b(\omega)u(t - \tau, \omega), \ t \geq 0, \\ u(t, \omega) = 0, \ -\tau \leq t \leq 0. \end{cases}$$

If $t \in [0, \tau]$, then $t - \tau \in [-\tau, 0]$; therefore, $u(t - \tau) = 0$. Thus, $u(t)$ satisfies a random differential equation problem with no delay on $[0, \tau]$:

$$\begin{cases} u'(t, \omega) = a(\omega)u(t, \omega), \ t \in [0, \tau], \\ u(0, \omega) = 0. \end{cases} \tag{7}$$

In [7], it was proved that any L^p-solution to a random initial value problem has a product measurable representative which is an absolutely continuous solution in the sample-path sense. Since the sample-path solution to (7) must be 0 (from the deterministic theory), we conclude that $u(t) = 0$ on $[0, \tau]$, as desired. For the subsequent intervals $[\tau, 2\tau]$, $[2\tau, 3\tau]$, etc., the same reasoning applies. □

Now we move on to existence results. First, recall that the random delayed exponential function is the solution to the random linear homogeneous differential equation with pure delay that satisfies the unit initial condition.

Proposition 4 (L^p-derivative of the random delayed exponential function ([17] Prop 3.1)). *Consider the random system with discrete delay*

$$\begin{cases} x'(t, \omega) = c(\omega)x(t - \tau, \omega), \ t \geq 0, \\ x(t, \omega) = 1, \ -\tau \leq t \leq 0, \end{cases} \tag{8}$$

where $c(\omega)$ is a random variable.
If c has absolute moments of any order, then $e_\tau^{c,t}$ is the unique L^p-solution to (8), for all $1 \leq p < \infty$. On the other hand, if c is bounded, then $e_\tau^{c,t}$ is the unique L^∞-solution to (8).

In [17], two results on the existence of solution to (3) were stated and proved. In terms of notation, the moment-generating function of a random variable a is denoted as $\phi_a(\zeta) = \mathbb{E}[e^{a\zeta}]$, $\zeta \in \mathbb{R}$.

Theorem 2 (Existence and uniqueness for (3), first version ([17] Theorem 3.2)). *Fix $1 \leq p < \infty$. Suppose that $\phi_a(\zeta) < \infty$ for all $\zeta \in \mathbb{R}$, b has absolute moments of any order, and g belongs to $C^1([-\tau, 0])$ in the $L^{p+\eta}$-sense, for certain $\eta > 0$. Then the stochastic process $y(t)$ defined by (5) is the unique L^p-solution to (3).*

Theorem 3 (Existence and uniqueness for (3), second version ([17] Theorem 3.4)). *Fix $1 \leq p < \infty$. Suppose that a and b are bounded random variables, and g belongs to $C^1([-\tau, 0])$ in the L^p-sense. Then the stochastic process $y(t)$ defined by (5) is the unique L^p-solution to (3).*

In what follows, we establish two theorems on the existence of a solution to (4); see Theorem 4 and Theorem 5. As a corollary, we will derive the solution to (1); see Theorem 6 and Theorem 7.

Theorem 4 (Existence and uniqueness for (4), first version). *Fix $1 \leq p < \infty$. Suppose that $\phi_a(\zeta) < \infty$ for all $\zeta \in \mathbb{R}$, b has absolute moments of any order, and f is continuous on $[0, \infty)$ in the $L^{p+\eta}$-sense, for certain $\eta > 0$. Then the stochastic process $z(t)$ defined by (6) is the unique L^p-solution to (4).*

Proof. At the beginning of the proof of ([17] Theorem 3.2), it was proved that $b_1 = e^{-a\tau}b$ has absolute moments of any order, as a consequence of Cauchy-Schwarz inequality, therefore Proposition 4 tells us that the process $e_\tau^{b_1,t}$ is L^q-differentiable, for each $1 \leq q < \infty$, and $\frac{d}{dt} e_\tau^{b_1,t} = b_1 e_\tau^{b_1,t-\tau}$. It was also proved that, by the chain rule theorem (Proposition 1), the process e^{at} is L^q-differentiable, for each $1 \leq q < \infty$, and $\frac{d}{dt} e^{at} = ae^{at}$. To justify these two assertions on $e_\tau^{b_1,t}$ and e^{at}, the hypotheses $\phi_a(\zeta) < \infty$ and b having absolute moments of any order are required.

Fix $0 \leq s \leq t$. Let $Y_1(t, s) = e^{a(t-s)}$, $Y_2(t, s) = e_\tau^{b_1, t-\tau-s}$ and $Y_3(s) = f(s)$, according to the notation of Lemma 1. Set the product of the three processes $F(t, s) = Y_1(t, s)Y_2(t, s)Y_3(s)$, so that our candidate solution process becomes $z(t) = \int_0^t F(t, s) \, ds$. We check the conditions of the random

Leibniz's rule, see Proposition 3, to differentiate $z(t)$. By the first paragraph of this proof, in which we stated that both $e_\tau^{b_1,t}$ and e^{at} are L^q-differentiable, for each $1 \leq q < \infty$, we derive that Y_1 and Y_2 are L^q-continuous on both variables, for all $1 \leq q < \infty$. Since Y_3 is $L^{p+\eta}$-continuous, for certain $\eta > 0$ by assumption, we deduce that F is L^p-continuous on both variables, as a consequence of Lemma 1.

Fixed s, let $Y_1(t) = e^{a(t-s)}$, $Y_2(t) = e_\tau^{b_1,t-\tau-s}$ and $Y_3 = f(s)$. We have that Y_1 and Y_2 are L^q-differentiable, for each $1 \leq q < \infty$. The random variable Y_3 belongs to $L^{p+\eta}$. By Lemma 2, $F(\cdot, s)$ is L^p-differentiable at each t, with

$$\frac{\partial F}{\partial t}(t,s) = \left\{ a e^{a(t-s)} e_\tau^{b_1,t-\tau-s} + e^{a(t-s)} b_1 e_\tau^{b_1,t-2\tau-s} \right\} f(s).$$

Let us see that $\frac{\partial F}{\partial t}(t,s)$ is L^p-continuous at (t,s). Since a has absolute moments of any order (by finiteness of its moment-generating function) and $e^{a(t-s)}$ is L^q-continuous at (t,s), for each $1 \leq q < \infty$, we derive that $a e^{a(t-s)}$ is L^q-continuous at each (t,s), for every $1 \leq q < \infty$, by Hölder's inequality. Thus, we have that $Y_1(t,s) = a e^{a(t-s)}$ and $Y_2(t,s) = e_\tau^{b_1,t-\tau-s}$ are L^q-continuous at (t,s), for each $1 \leq q < \infty$, while $Y_3(s) = f(s)$ is $L^{p+\eta}$-continuous. By Lemma 1, $a e^{a(t-s)} e_\tau^{b_1,t-\tau-s} f(s)$ is L^p-continuous at each (t,s). Analogously, $e^{a(t-s)} b_1 e_\tau^{b_1,t-2\tau-s} f(s)$ is L^p-continuous at (t,s). Therefore, $\frac{\partial F}{\partial t}(t,s)$ is L^p-continuous at (t,s). By Proposition 3, the process $z(t)$ is L^p-differentiable and $z'(t) = F(t,t) + \int_0^t \frac{\partial F}{\partial t}(t,s)\,ds = f(t) + az(t) + bz(t-\tau)$ (by definition of $F(t,s)$ in the proof, $F(t,t) = e^{a(t-t)} e_\tau^{b_1,t-\tau-t} f(t) = e_\tau^{b_1,-\tau} f(t) = f(t)$, where $e_\tau^{b_1,-\tau} = 1$ by definition of delayed exponential function), and we are done.

Once the existence of L^p-solution has been proved, uniqueness follows from Theorem 1. □

Theorem 5 (Existence and uniqueness for (4), second version). *Fix $1 \leq p < \infty$. Suppose that a and b are bounded random variables, and f is continuous on $[0, \infty)$ in the L^p-sense. Then the stochastic process $z(t)$ defined by (6) is the unique L^p-solution to (4).*

Proof. As was shown in ([17] Theorem 3.4), the process $e_\tau^{b_1,t}$ is L^∞-differentiable and $\frac{d}{dt} e_\tau^{b_1,t} = b_1 e_\tau^{b_1,t-\tau}$, because $b_1 = e^{-a\tau} b$ is bounded. Additionally, the process e^{at} is L^∞-differentiable and $\frac{d}{dt} e^{at} = a e^{at}$, as a consequence of the deterministic mean value theorem and the boundedness of a.

The rest of the proof is completely analogous to the previous Theorem 4, by applying the second part of both Lemma 1 and Lemma 2. □

Theorem 6 (Existence and uniqueness for (1), first version). *Fix $1 \leq p < \infty$. Suppose that $\phi_a(\zeta) < \infty$ for all $\zeta \in \mathbb{R}$, b has absolute moments of any order, g belongs to $C^1([-\tau, 0])$ in the $L^{p+\eta}$-sense and f is continuous on $[0, \infty)$ in the $L^{p+\eta}$-sense, for certain $\eta > 0$. Then the stochastic process $x(t)$ defined by (2) is the unique L^p-solution to (1).*

Proof. This is a consequence of Theorem 2 and Theorem 4, with $x(t) = y(t) + z(t)$. Uniqueness follows from Theorem 1. □

Theorem 7 (Existence and uniqueness for (1), second version). *Fix $1 \leq p < \infty$. Suppose that a and b are bounded random variables, g belongs to $C^1([-\tau, 0])$ in the L^p-sense and f is continuous on $[0, \infty)$ in the L^p-sense. Then the stochastic process $x(t)$ defined by (2) is the unique L^p-solution to (1).*

Proof. This is a consequence of Theorem 3 and Theorem 5, with $x(t) = y(t) + z(t)$. Uniqueness follows from Theorem 1. □

Remark 4. *As emphasized in ([17] Remark 3.6), the condition of boundedness for a and b in Theorem 7 is necessary if we only assume that $g \in C^1([-\tau, 0])$ in the L^p-sense. See ([7] Example p. 541), where it is proved that, in order for a random autonomous and homogeneous linear differential equation of first-order to have an L^p-solution for every initial condition in L^p, one needs the random coefficient to be bounded.*

Assume the conditions from Theorem 6 or Theorem 7. From expression (2), it is possible to approximate the statistical moments of $x(t)$. We focus on its expectation, $\mathbb{E}[x(t)]$, and on its variance, $\mathbb{V}[x(t)] = \mathbb{E}[x(t)^2] - (\mathbb{E}[x(t)])^2$. These statistics provide information on the average and the dispersion of $x(t)$, and they are very useful for uncertainty quantification for $x(t)$. For ease of notation, denote the stochastic processes

$$F_1(t,\omega) = e^{a(\omega)(t+\tau)} e_\tau^{b_1(\omega),t} g(-\tau,\omega),$$

$$F_2(t,s,\omega) = e^{a(\omega)(t-s)} e_\tau^{b_1(\omega),t-\tau-s}(g'(s,\omega) - a(\omega)g(s,\omega)),$$

$$F_3(t,s,\omega) = e^{a(\omega)(t-s)} e_\tau^{b_1(\omega),t-\tau-s} f(s,\omega).$$

Due to the linearity of the expectation and its interchangeability with the L^1-Riemann integral ([5] p. 104), if $p \geq 1$,

$$\mathbb{E}[x(t)] = \mathbb{E}[F_1(t)] + \int_{-\tau}^{0} \mathbb{E}[F_2(t,s)]\,ds + \int_{0}^{t} \mathbb{E}[F_3(t,s)]\,ds. \tag{9}$$

To compute $\mathbb{V}[x(t)]$ when $p \geq 2$, we start by

$$x(t)^2 = F_1(t)^2 + \int_{-\tau}^{0}\int_{-\tau}^{0} F_2(t,s_1)F_2(t,s_2)\,ds_2\,ds_1$$
$$+ \int_{0}^{t}\int_{0}^{t} F_3(t,s_1)F_3(t,s_2)\,ds_2\,ds_1 + 2\int_{-\tau}^{0} F_1(t)F_2(t,s)\,ds$$
$$+ 2\int_{0}^{t} F_1(t)F_3(t,s)\,ds + 2\int_{-\tau}^{0}\int_{0}^{t} F_2(t,s_1)F_3(t,s_2)\,ds_2\,ds_1.$$

Each of these integrals has to be considered in $L^{p/2}$; see ([35] Remark 2). This is due to the loss of integrability of the product, by Hölder's inequality. By applying expectations,

$$\mathbb{E}[x(t)^2] = \mathbb{E}[F_1(t)^2] + \int_{-\tau}^{0}\int_{-\tau}^{0} \mathbb{E}[F_2(t,s_1)F_2(t,s_2)]\,ds_2\,ds_1$$
$$+ \int_{0}^{t}\int_{0}^{t} \mathbb{E}[F_3(t,s_1)F_3(t,s_2)]\,ds_2\,ds_1 + 2\int_{-\tau}^{0} \mathbb{E}[F_1(t)F_2(t,s)]\,ds$$
$$+ 2\int_{0}^{t} \mathbb{E}[F_1(t)F_3(t,s)]\,ds + 2\int_{-\tau}^{0}\int_{0}^{t} \mathbb{E}[F_2(t,s_1)F_3(t,s_2)]\,ds_2\,ds_1. \tag{10}$$

As a consequence, one derives an expression for $\mathbb{V}[x(t)]$, by utilizing the relation $\mathbb{V}[x(t)] = \mathbb{E}[x(t)^2] - (\mathbb{E}[x(t)])^2$. Other statistics related to moments could be derived in a similar fashion.

In Example 1, we will show how useful these expressions are to determine $\mathbb{E}[x(t)]$ and $\mathbb{V}[x(t)]$ in practice. Our procedure is an alternative to the usual techniques for uncertainty quantification: Monte Carlo simulation, generalized polynomial chaos (gPC) expansions, etc. [1,2].

4. L^p-convergence to a Random Complete Linear Differential Equation When the Delay Tends to 0

Given a discrete delay $\tau > 0$, we denote the L^p-solution (2) to (1) by $x_\tau(t)$. We denote the L^p-solutions (5) and (6) to (3) and (4) by $y_\tau(t)$ and $z_\tau(t)$, respectively, so that $x_\tau(t) = y_\tau(t) + z_\tau(t)$. Thus, we are making the dependence on the delay τ explicit. If we put $\tau = 0$ into (1), (3) and (4), we obtain random linear differential equations with no delay:

$$\begin{cases} x_0'(t,\omega) = (a(\omega) + b(\omega))x_0(t,\omega) + f(t,\omega), & t \geq 0, \\ x_0(0,\omega) = g(0,\omega), \end{cases} \tag{11}$$

$$\begin{cases} y_0'(t,\omega) = (a(\omega) + b(\omega))y_0(t,\omega), & t \geq 0, \\ y_0(0,\omega) = g(0,\omega), \end{cases} \tag{12}$$

$$\begin{cases} z_0'(t,\omega) = (a(\omega) + b(\omega))z_0(t,\omega) + f(t,\omega), \ t \geq 0, \\ z_0(0,\omega) = 0, \end{cases} \quad (13)$$

respectively. The following results establish conditions under which (11), (12) and (13) have L^p-solutions.

Theorem 8 ([17] Corollary 4.1). *Fix $1 \leq p < \infty$. If $\phi_a(\zeta) < \infty$ and $\phi_b(\zeta) < \infty$ for all $\zeta \in \mathbb{R}$, and $g(0) \in L^{p+\eta}$ for certain $\eta > 0$, then the stochastic process $y_0(t) = g(0)e^{(a+b)t}$ is the unique L^p-solution to (12).*

On the other hand, if a and b are bounded random variables and $g(0) \in L^p$, then the stochastic process $y_0(t) = g(0)e^{(a+b)t}$ is the unique L^p-solution to (12).

Theorem 9. *Fix $1 \leq p < \infty$. If $\phi_a(\zeta) < \infty$ and $\phi_b(\zeta) < \infty$ for all $\zeta \in \mathbb{R}$, and f is continuous on $[0,\infty)$ in the $L^{p+\eta}$-sense for certain $\eta > 0$, then the stochastic process $z_0(t) = \int_0^t e^{(a+b)(t-s)} f(s)\,ds$ is the unique L^p-solution to (13).*

On the other hand, if a and b are bounded random variables and f is continuous on $[0,\infty)$ in the L^p-sense, then the stochastic process $z_0(t) = \int_0^t e^{(a+b)(t-s)} f(s)\,ds$ is the unique L^p-solution to (13).

Proof. Take the first set of assumptions. Let $F(t,s) = e^{(a+b)(t-s)} f(s)$ be the process inside the integral sign. Since $\phi_a < \infty$ and $\phi_b < \infty$, the chain rule theorem (Proposition 1) allows differentiating $e^{(a+b)t}$ in L^q, for each $1 \leq q < \infty$. In particular, $e^{(a+b)(t-s)}$ is L^q-continuous at (t,s), for $1 \leq q < \infty$. As f is continuous on $[0,\infty)$ in the $L^{p+\eta}$-sense, we derive that F is L^p-continuous at (t,s). It also exists $\frac{\partial F}{\partial t}(t,s) = (a+b)e^{(a+b)(t-s)} f(s)$ in L^p. Since $a+b$ has absolute moments of any order, $(a+b)e^{(a+b)(t-s)}$ is L^q-continuous at (t,s), for $1 \leq q < \infty$. Then $\frac{\partial F}{\partial t}$ is L^p-continuous at (t,s). By Proposition 3, z_0 is L^p-differentiable and $z_0'(t) = F(t,t) + \int_0^t \frac{\partial F}{\partial t}(t,s)\,ds = f(t) + (a+b)z_0(t)$, and we are done.

Suppose that a and b are bounded random variables and f is continuous on $[0,\infty)$ in the L^p-sense. If a and b are bounded, then $e^{(a+b)t}$ is L^∞-differentiable (this is because of an application of the deterministic mean value theorem; see ([17] Theorem 3.4)). Then an analogous proof to the previous paragraph works in this case, by only assuming that f is continuous on $[0,\infty)$ in the L^p-sense. □

Theorem 10. *Fix $1 \leq p < \infty$. If $\phi_a(\zeta) < \infty$ and $\phi_b(\zeta) < \infty$ for all $\zeta \in \mathbb{R}$, $g(0) \in L^{p+\eta}$, and f is continuous on $[0,\infty)$ in the $L^{p+\eta}$-sense for certain $\eta > 0$, then the stochastic process $x_0(t) = g(0)e^{(a+b)t} + \int_0^t e^{(a+b)(t-s)} f(s)\,ds$ is the unique L^p-solution to (11).*

On the other hand, if a and b are bounded random variables, $g(0) \in L^p$, and f is continuous on $[0,\infty)$ in the L^p-sense, then the stochastic process $x_0(t) = g(0)e^{(a+b)t} + \int_0^t e^{(a+b)(t-s)} f(s)\,ds$ is the unique L^p-solution to (11).

Proof. It is a consequence of Theorem 8 and Theorem 9 with $x_0(t) = y_0(t) + z_0(t)$. □

Our goal is to establish conditions under which $\lim_{\tau \to 0} x_\tau(t) = x_0(t)$ in L^p, for each $t \geq 0$. To do so, we will utilize $\lim_{\tau \to 0} y_\tau(t) = y_0(t)$ and $\lim_{\tau \to 0} z_\tau(t) = z_0(t)$.

The first limit, $\lim_{\tau \to 0} y_\tau(t) = y_0(t)$, was demonstrated in ([17] Theorem 4.5), by using inequalities for the deterministic and random delayed exponential function ([36] Theorem A.3), ([17] Lemma 4.2, Lemma 4.3, Lemma 4.4).

Theorem 11 ([17] Theorem 4.5). *Fix $1 \leq p < \infty$. Let a and b be bounded random variables and let g be a stochastic process that belongs to $C^1([-\tau, 0])$ in the L^p-sense. Then, $\lim_{\tau \to 0} y_\tau(t) = y_0(t)$ in L^p, uniformly on $[0,T]$, for each $T > 0$.*

Next we prove the convergence $\lim_{\tau \to 0} z_\tau(t) = z_0(t)$. As a corollary, we will be able to derive $\lim_{\tau \to 0} x_\tau(t) = x_0(t)$.

Theorem 12. Fix $1 \leq p < \infty$. Let a and b be bounded random variables and let f be a continuous stochastic process on $[0, \infty)$ in the L^p-sense. Then, $\lim_{\tau \to 0} z_\tau(t) = z_0(t)$ in L^p, uniformly on $[0, T]$, for each $T > 0$.

Proof. Notice that $z_\tau(t)$ defined by (6) (see the first paragraph of this section) exists by Theorem 5, which used the boundedness of a and b and the L^p-continuity of f on $[0, \infty)$. Analogously, $z_0(t)$ exists by Theorem 9.

Fix $t \in [0, T]$. We bound

$$\|z_\tau(t) - z_0(t)\|_p \leq \int_0^t \left\| e^{a(t-s)} f(s) \left(e_\tau^{b_1, t-\tau-s} - e^{b(t-s)} \right) \right\|_p ds$$

$$\leq \int_0^t \left\| e^{a(t-s)} \right\|_\infty \|f(s)\|_p \left\| e_\tau^{b_1, t-\tau-s} - e^{b(t-s)} \right\|_\infty ds.$$

We have $\|e^{a(t-s)}\|_\infty \leq e^{\|a\|_\infty T}$ and $\|f(s)\|_p \leq C_f = \max_{s \in [0,T]} \|f(s)\|_p$. These bounds yield

$$\|z_\tau(t) - z_0(t)\|_p \leq C_f e^{\|a\|_\infty T} \int_0^t \left\| e_\tau^{b_1, t-\tau-s} - e^{b(t-s)} \right\|_\infty ds. \tag{14}$$

Let k be a number such that $k \geq \|b_1\|_\infty = \|e^{a\tau} b\|_\infty$, for all $\tau \in (0, 1]$. By ([17] Lemma 4.3),

$$\left\| e_\tau^{b_1, t-\tau-s} - e^{b_1(t-s)} \right\|_\infty \leq C_{T,k} \cdot \tau, \tag{15}$$

for $t \in [0, T]$, $0 \leq s \leq t$ and $\tau \in (0, 1]$. On the other hand, by the deterministic mean value theorem (applied for each outcome ω),

$$e^{b_1(t-s)} - e^{b(t-s)} = e^{e^{-a\tau} b(t-s)} - e^{b(t-s)}$$

$$= b(t-s) e^{\zeta_{\tau,\omega} b(t-s)} (e^{-a\tau} - 1),$$

where $\zeta_{\tau,\omega} \in (1, e^{-a\tau}) \cup (e^{-a\tau}, 1)$. In particular, $|\zeta_{\tau,\omega}| \leq 1 + e^{\|a\|_\infty}$. We apply again the deterministic mean value theorem to $e^{-a\tau} - 1$:

$$e^{-a\tau} - 1 = e^{\bar{\zeta}_{\tau,\omega}} (-a\tau),$$

where $\bar{\zeta}_{\tau,\omega} \in (-a\tau, 0) \cup (0, -a\tau)$. In particular,

$$\|e^{-a\tau} - 1\|_\infty \leq e^{\|a\|_\infty} \|a\|_\infty \tau.$$

As a consequence,

$$\|e^{b_1(t-s)} - e^{b(t-s)}\|_\infty \leq \underbrace{\|b\|_\infty T e^{(1+e^{\|a\|_\infty}) \|b\|_\infty T} e^{\|a\|_\infty} \|a\|_\infty}_{\overline{C}_{T, \|a\|_\infty, \|b\|_\infty}} \tau. \tag{16}$$

By combining (15) and (16) and by the triangular inequality,

$$\left\| e_\tau^{b_1, t-\tau-s} - e^{b(t-s)} \right\|_\infty \leq \left\| e_\tau^{b_1, t-\tau-s} - e^{b_1(t-s)} \right\|_\infty$$
$$+ \left\| e^{b_1(t-s)} - e^{b(t-s)} \right\|_\infty \leq \left(C_{T,k} + \overline{C}_{T, \|a\|_\infty, \|b\|_\infty} \right) \tau.$$

Substituting this inequality into (14),

$$\|z_\tau(t) - z_0(t)\|_p \leq C_f e^{\|a\|_\infty T} \left(C_{T,k} + \overline{C}_{T, \|a\|_\infty, \|b\|_\infty} \right) \tau \xrightarrow{\tau \to 0} 0,$$

uniformly on $[0, T]$. □

Theorem 13. *Fix $1 \leq p < \infty$. Let a and b be bounded random variables, let g be a stochastic process that belongs to $C^1([-\tau, 0])$ in the L^p-sense, and let f be a continuous stochastic process on $[0, \infty)$ in the L^p-sense. Then, $\lim_{\tau \to 0} x_\tau(t) = x_0(t)$ in L^p, uniformly on $[0, T]$, for each $T > 0$.*

Proof. This is a consequence of Theorem 11 and Theorem 12, with $x_\tau(t) = y_\tau(t) + z_\tau(t)$ and $x_0(t) = y_0(t) + z_0(t)$. □

Example 1. This is a test example, with arbitrary distributions, to show how (9) and (10) may be employed to compute the expectation and the variance of the stochastic solution. Theoretical results are also illustrated. Let $a \sim \text{Beta}(2,3)$ and $b \sim \text{Uniform}(0.2, 1)$. Define $g(t, \omega) = \sin(\sin(d(\omega)t^2))$ and $f(t, \omega) = \cos(te(\omega)^2)$, where d and e are random variables with $d \sim \text{Triangular}(1, 1.15, 1.3)$ and $e \sim \text{Uniform}(0.1, 0.2)$. By using the chain rule theorem, Proposition 1, it is easy to prove that both g and f are C^∞ in the L^p-sense, $1 \leq p < \infty$. The random variables a, b, d and e are assumed to be independent. Consider the solution stochastic process $x_\tau(t)$ defined by (2). It is an L^p-solution for all $1 \leq p < \infty$, by Theorem 7. With expressions (9) and (10), we can compute $\mathbb{E}[x_\tau(t)]$ and $\mathbb{V}[x_\tau(t)]$; see Figure 1. The results agree with Monte Carlo simulation on (1). Observe that, as τ approaches 0, the solution stochastic process tends to the solution with no delay defined in Theorem 10, as predicted by Theorem 13.

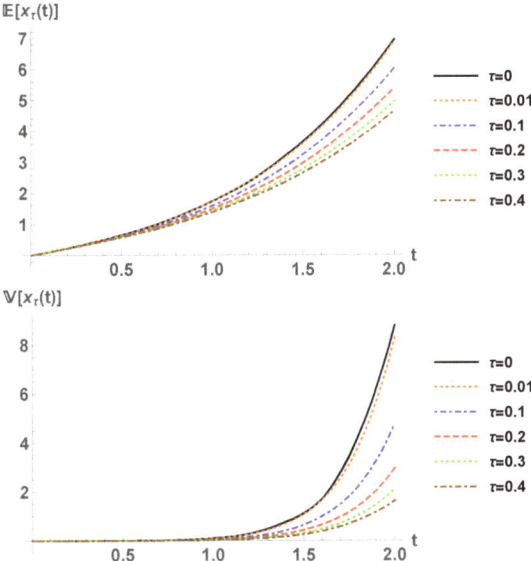

Figure 1. Expectation (up) and variance (down) of $x_\tau(t)$, Example 1.

Example 2. In this example, we specify new probability distributions for the input coefficients. Let $a \sim \text{Uniform}(0.2, 1)$, $b \sim \text{Uniform}(-1, 0)$, $d \sim \text{Beta}(1, 1.3)$ and $e \sim \text{Uniform}(-0.2, -0.1)$, all of them independent. The stochastic process $x_\tau(t)$ given by (2) is an L^p-solution for all $1 \leq p < \infty$, by Theorem 7. We compute $\mathbb{E}[x_\tau(t)]$ and $\mathbb{V}[x_\tau(t)]$ with (9) and (10), see Figure 2. Observe that the convergence when $\tau \to 0$ agrees with Theorem 13.

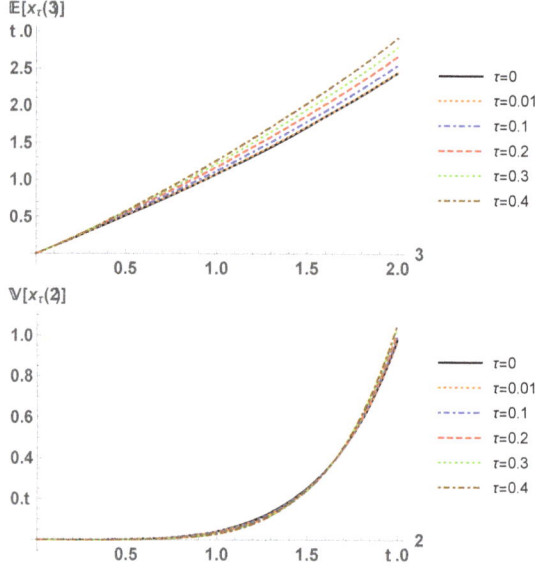

Figure 2. Expectation (up) and variance (down) of $x_\tau(t)$, Example 2.

We now comment on some computational aspects. We have used the software Mathematica®, version 11.2 [37]. The integrals and expectations from (9) and (10) have been computed as multidimensional integrals with the built-in function NIntegrate (recall that the expectation is an integral with respect to the corresponding probability density function). Expression (9) does not pose serious numerical challenges, and one can use a standard NIntegrate routine with no specified options. However, for expression (10), we have set the option quasi-Monte Carlo with 10^5 sampling points (otherwise the computational time would increase dramatically). We have checked numerically that the following factors increase the computational time: large ratio t/τ; probability distributions with unbounded support for the input data; and moderate or large dimensions of the random space.

5. Conclusions

In this paper, we have performed a comprehensive stochastic analysis of the random linear delay differential equation with stochastic forcing term. The equation considered has one discrete delay $\tau > 0$, two random coefficients a and b (corresponding to the non-delay and the delay term, respectively) and two stochastic processes $f(t)$ and $g(t)$ (corresponding to the forcing term on $[0,\infty)$ and the initial condition on $[-\tau, 0]$, respectively). Our setting supposes a step further than the previous contribution [17], in which no forcing term was considered (i.e., $f(t) = 0$). We have rigorously addressed the problem of extending the deterministic theory to the random scenario, by proving that the deterministic solution constructed via the method of steps and the method of variation of constants is an L^p-solution, under certain assumptions on the random data. A new result, the random Leibniz's rule for L^p-Riemann integration has been necessary to derive our conclusions. We have also studied the behavior in L^p of the random delay equation when the delay tends to zero.

Our approach has been shown to be useful to approximate the statistical moments of the solution stochastic process, in particular its expectation and its variance. Thus, it is possible to perform uncertainty quantification. Our procedure is an alternative to the usual techniques for uncertainty quantification: Monte Carlo simulation, generalized polynomial chaos (gPC) expansions, etc.

Our approach could be extendable to other random differential equations with or without delay. As usual, one could prove that the deterministic solution also works in the random framework. To do so, a rigorous and careful analysis of the probabilistic properties of the solution based on L^p-calculus should be conducted.

Finally, we humbly think that advancing in theoretical aspects of random differential equations with delay will permit rigorously applying this class of equations to modeling phenomena involving memory and aftereffects together with uncertainties. In particular, they may be crucial to capture uncertainties inherent to some complex modeling problems, since input parameters of this type of equations may belong to a wider range of probability distributions than the ones considered in Itô differential equations.

Author Contributions: Investigation, M.J.; methodology, M.J.; software, M.J.; supervision, J.C.C.; validation, J.C.C.; visualization, J.C.C. and M.J.; writing—original draft, M.J.; writing—review and editing, J.C.C. and M.J. All authors have read and agreed to the published version of the manuscript.

Funding: This work has been supported by the Spanish Ministerio de Economía, Industria y Competitividad (MINECO), the Agencia Estatal de Investigación (AEI) and Fondo Europeo de Desarrollo Regional (FEDER UE) grant MTM2017–89664–P.

Conflicts of Interest: The authors declare that there is no conflict of interests regarding the publication of this article.

References

1. Smith, R.C. *Uncertainty Quantification: Theory, Implementation, and Applications*; SIAM: Philadelphia, PA, USA, 2013; Volume 12.
2. Xiu, D. *Numerical Methods for Stochastic Computations: A Spectral Method Approach*; Princeton University Press: Princeton, NJ, USA, 2010.
3. Le Maître, O.P.; Knio, O.M. *Spectral Methods for Uncertainty Quantification: With Applications to Computational Fluid Dynamics*; Springer Science & Business Media: Berlin, Germany, 2010.
4. Xiu, D.; Karniadakis, G.E. Supersensitivity due to uncertain boundary conditions. *Int. J. Numer. Methods Eng.* **2004**, *61*, 2114–2138. [CrossRef]
5. Soong, T.T. *Random Differential Equations in Science and Engineering*; Academic Press: New York, NY, USA, 1973.
6. Casabán, M.C.; Cortés, J.C.; Navarro-Quiles, A.; Romero, J.V.; Roselló, M.D.; Villanueva, R.J. A comprehensive probabilistic solution of random SIS-type epidemiological models using the random variable transformation technique. *Commun. Nonlinear Sci. Numer. Simul.* **2016**, *32*, 199–210. [CrossRef]
7. Strand, J.L. Random ordinary differential equations. *J. Differ. Equ.* **1970**, *7*, 538–553. [CrossRef]
8. Villafuerte, L.; Braumann, C.A.; Cortés, J.C.; Jódar, L. Random differential operational calculus: Theory and applications. *Comput. Math. Appl.* **2010**, *59*, 115–125. [CrossRef]
9. Saaty, T.L. *Modern Nonlinear Equations*; Dover Publications: New York, NY, USA, 1981.
10. Cortés, J.C.; Jódar, L.; Roselló, M.D.; Villafuerte, L. Solving initial and two-point boundary value linear random differential equations: A mean square approach. *Appl. Math. Comput.* **2012**, *219*, 2204–2211. [CrossRef]
11. Calatayud, J.; Cortés, J.C.; Jornet, M.; Villafuerte, L. Random non-autonomous second order linear differential equations: mean square analytic solutions and their statistical properties. *Adv. Differ. Equ.* **2018**, *2018*, 392. [CrossRef]
12. Calatayud, J.; Cortés, J.C.; Jornet, M. Improving the approximation of the first-and second-order statistics of the response stochastic process to the random Legendre differential equation. *Mediterr. J. Math.* **2019**, *16*, 68. [CrossRef]
13. Licea, J.A.; Villafuerte, L.; Chen-Charpentier, B.M. Analytic and numerical solutions of a Riccati differential equation with random coefficients. *J. Comput. Appl. Math.* **2013**, *239*, 208–219. [CrossRef]
14. Burgos, C.; Calatayud, J.; Cortés, J.C.; Villafuerte, L. Solving a class of random non-autonomous linear fractional differential equations by means of a generalized mean square convergent power series. *Appl. Math. Lett.* **2018**, *78*, 95–104. [CrossRef]
15. Nouri, K.; Ranjbar, H. Mean square convergence of the numerical solution of random differential equations. *Mediterr. J. Math.* **2015**, *12*, 1123–1140. [CrossRef]

16. Calatayud, J.; Cortés, J.C.; Jornet, M. Random differential equations with discrete delay. *Stoch. Anal. Appl.* **2019**, *37*, 699–707. [CrossRef]
17. Calatayud, J.; Cortés, J.C.; Jornet, M. L^p-calculus approach to the random autonomous linear differential equation with discrete delay. *Mediterr. J. Math.* **2019**, *16*, 85. [CrossRef]
18. Caraballo, T.; Cortés, J.C.; Navarro-Quiles, A. Applying the Random Variable Transformation method to solve a class of random linear differential equation with discrete delay. *Appl. Math. Comput.* **2019**, *356*, 198–218. [CrossRef]
19. Zhou, T. A stochastic collocation method for delay differential equations with random input. *Adv. Appl. Math. Mech.* **2014**, *6*, 403–418. [CrossRef]
20. Shi, W.; Zhang, C. Generalized polynomial chaos for nonlinear random delay differential equations. *Appl. Numer. Math.* **2017**, *115*, 16–31. [CrossRef]
21. Licea-Salazar, J.A. The Polynomial Chaos Method With Applications To Random Differential Equations. Ph.D. Thesis, University of Texas at Arlington, Arlington, TX, USA, 2013.
22. Khusainov, D.Y.; Ivanov, A.; Kovarzh, I.V. Solution of one heat equation with delay. *Nonlinear Oscil.* **2009**, *12*, 260–282. [CrossRef]
23. Øksendal, B. *Stochastic Differential Equations: An Introduction with Applications*; Springer Science & Business Media: New York, NY, USA, 1998.
24. Shaikhet, L. *Lyapunov Functionals and Stability of Stochastic Functional Differential Equations*; Springer Science & Business Media: New York, NY, USA, 2013.
25. Shaikhet, L. Stability of equilibrium states of a nonlinear delay differential equation with stochastic perturbations. *Int. J. Robust Nonlinear Control.* **2017**, *27*, 915–924. [CrossRef]
26. Benhadri, M.; Zeghdoudi, H. Mean square asymptotic stability in nonlinear stochastic neutral Volterra-Levin equations with Poisson jumps and variable delays. *Funct. Approx. Comment. Math.* **2018**, *58*, 157–176. [CrossRef]
27. Santonja, F.J.; Shaikhet, L. Analysing social epidemics by delayed stochastic models. *Discret. Dyn. Nat. Soc.* **2012**, *2012*, [CrossRef]
28. Liu, L.; Caraballo, T. Analysis of a stochastic 2D-Navier–Stokes model with infinite delay. *J. Dyn. Differ. Equ.* **2019**, *31*, 2249–2274. [CrossRef]
29. Lupulescu, V.; Abbas, U. Fuzzy delay differential equations. *Fuzzy Optim. Decis. Mak.* **2012**, *11*, 99–111. [CrossRef]
30. Krapivsky, P.L.; Luck, J.M.; Mallick, K. On stochastic differential equations with random delay. *J. Stat. Mech. Theory Exp.* **2011**, *2011*, P10008. [CrossRef]
31. Garrido-Atienza, M.J.; Ogrowsky, A.; Schmalfuß, B. Random differential equations with random delays. *Stochastics Dyn.* **2011**, *11*, 369–388. [CrossRef]
32. Calatayud, J. A Theoretical Study of a Short Rate Model. Master's Thesis, Universitat de Barcelona, Barcelona, Spain, 2016.
33. Cortés, J.C.; Villafuerte, L.; Burgos, C. A mean square chain rule and its application in solving the random Chebyshev differential equation. *Mediterr. J. Math.* **2017**, *14*, 35. [CrossRef]
34. Cortés, J.C.; Jódar, L.; Villafuerte, L. Numerical solution of random differential equations: a mean square approach. *Math. Comput. Model.* **2007**, *45*, 757–765. [CrossRef]
35. Braumann, C.A.; Cortés, J.C.; Jódar, L.; Villafuerte, L. On the random gamma function: Theory and computing. *J. Comput. Appl. Math.* **2018**, *335*, 142–155. [CrossRef]
36. Khusainov, D.Y.; Pokojovy, M. Solving the linear 1D thermoelasticity equations with pure delay. *Int. J. Math. Math. Sci.* **2015**, *2015*, [CrossRef]
37. *Wolfram Mathematica, V.11.2*; Wolfram Research Inc.: Champaign, IL, USA, 2017.

© 2020 by the authors. Licensee MDPI, Basel, Switzerland. This article is an open access article distributed under the terms and conditions of the Creative Commons Attribution (CC BY) license (http://creativecommons.org/licenses/by/4.0/).

Article

Second-Order Dual Phase Lag Equation. Modeling of Melting and Resolidification of Thin Metal Film Subjected to a Laser Pulse

Ewa Majchrzak [1],* and Bohdan Mochnacki [2]

[1] Faculty of Mechanical Engineering, Silesian University of Technology, 44-100 Gliwice, Poland
[2] Department of Technical Sciences, University of Occupational Safety Management, 40-007 Katowice, Poland; bmochnacki@wszop.edu.pl
* Correspondence: ewa.majchrzak@polsl.pl

Received: 2 June 2020; Accepted: 16 June 2020; Published: 18 June 2020

Abstract: The process of partial melting and resolidification of a thin metal film subjected to a high-power laser beam is considered. The mathematical model of the process is based on the second-order dual phase lag equation (DPLE). Until now, this equation has not been used for the modeling of phase changes associated with heating and cooling of thin metal films and the considerations regarding this issue are the most important part of the article. In the basic energy equation, the internal heat sources associated with the laser action and the evolution of phase change latent heat are taken into account. Thermal processes in the domain of pure metal (chromium) are analyzed and it is assumed that the evolution of latent heat occurs at a certain interval of temperature to which the solidification point was conventionally extended. This approach allows one to introduce the continuous function corresponding to the volumetric fraction of solid or liquid state at the neighborhood of the point considered, which significantly simplifies the phase changes modeling. At the stage of numerical computations, the authorial program based on the implicit scheme of the finite difference method (FDM) was used. In the final part of the paper, the examples of numerical computations (including the results of simulations for different laser intensities and different characteristic times of laser pulse) are presented and the conclusions are formulated.

Keywords: second-order dual phase lag equation; laser heating; thin metal films; melting and resolidification; finite difference method

1. Introduction

Heat transfer through thin films subjected to an ultrafast laser pulse is of vital importance in microtechnology applications and is a reason that the problems related to the fast heating of solids have become a very active research area. The problems of melting/resolidification processes modeling, which may be the result of heating with a laser beam, are also important from the technical point of view. So far, the method using the equation based on the second order model with two delay times for the phase changes modeling was not presented in literature. This was the most important motivation for the authors to undertake research in this area. In Section 5, the comparison of the results obtained with the similar solution on the basis of the first-order dual phase lag equation (DPLE) is presented.

The mathematical model of macroscale heat conduction is based on the parabolic Fourier equation. This equation was formulated under the assumption of instantaneous propagation of the thermal wave in the domain considered. It is obvious that this assumption is not correct, but for the problems concerning the analysis of macroscale heat conduction processes, the obtained results are fully satisfactory. Despite this, the attempts have been made to modify the Fourier equation to a form that better reproduces the real conditions of heat conduction in solids. Thus, about seventy years ago,

Cattaneo proposed a modification of the Fourier equation now called the Cattaneo–Vernotte equation. This is the hyperbolic partial differential equation (PDE) and contains the parameter τ_q called the relaxation time (the lag time of the heat flux in relation to the temperature gradient) [1,2]. Especially important differences between the Fourier model and the real course of thermal processes appear in the case of microscale heat transfer problems. For example, the very high heating rates accompanying the heating of thin metal films with a laser beam mean that the inclusion of the finite value of thermal wave velocity must be taken into account. The deviations appear mainly when the mean free path of the heat carriers becomes comparable to the characteristic length of the domain considered and the time scale of interest becomes comparable to or smaller than the relaxation time of the heat carriers [3,4].

For the analysis of this type of process, the model with two delay times called a dual-phase lag model is presently applied. In addition to the relaxation time, the thermalization time is introduced. The relaxation time τ_q takes into account the small-scale response in time, while the thermalization time τ_T takes into account the small-scale response in space [3–6]. The dual phase lag equation (DPLE) results from the generalized form of the Fourier law

$$\mathbf{q}(\mathbf{x}, t + \tau_q) = -\lambda \, \nabla T(\mathbf{x}, t + \tau_T) \qquad (1)$$

where $\mathbf{q}$ is a heat flux vector; ∇T is a temperature gradient; λ is a thermal conductivity; and $\mathbf{x}$ and t denote the geometrical co-ordinates and time, respectively. Both sides of the last dependence are developed into a power series and, finally (depending on the number of components), the first- or second-order DPLE can be obtained.

The literature on equations with two delay times is very extensive (especially in the case of the first order equations), and here we quote only a few important articles. The first publications concerning the model with two delay times appeared in the early nineties of the last century. There may be mentioned, for example, the papers [7–9]. Currently, one can already find books devoted to this type of non-Fourier heat conduction model, for example, [10–13].

In this brief literature review, the selected papers on analytical and numerical solutions of first-order DPLE will be listed. First of all, the works containing the analytical solutions of first-order DPLE (usually 1D tasks) will be mentioned [14–18]. The paper [14] concerns the laser heating of ultra-thin metal film; in the papers [15,16], the bio-heat transfer problems are discussed; while in the paper [18], the multi-layered metal domain is considered.

A much larger number of articles concern the application of numerical methods in the tasks based on the models with two delay times. It should be pointed out that, first of all, different variants of finite difference method (FDM) are applied (see, for example, [19–24]). In the paper [19], the numerical model of heating of the double-layered thin film has been applied for the analysis of the thermal deformation process. In the paper [20], the 3D FDM numerical model of the thin metal film heating has been presented. In [21], the explicit scheme of the FDM has been used and the problem of biological tissue freezing process has been discussed. The stability problem of the algorithm of this type is analyzed in [22]. In [23], the problem based on DPLE has been solved using the alternating direction implicit FDM scheme. In the paper [24], the adaptation of typical boundary conditions for non-Fourier equations has been shown. The FDM numerical solutions of the inverse problems can also be found (e.g., [25]).

The number of works presenting solutions using the other numerical methods is significantly smaller. Here, one can mention the papers [26–32]. In particular, solutions based on the finite element method [26–28], the boundary element method [29], the control volume method [30], or the lattice Boltzmann method [31,32] are discussed in the above-mentioned papers.

Literature on the second-order DPLE is not as extensive as for the first-order equations. As an example, the papers [33–37] can be mentioned. The main subject of these papers (except [37]) is related to the construction of algorithms for numerical modeling of problems described by second-order DPLE (the different variants of FDM are used). Additionally, the transformed second-order equations are shown in [20,34,37] and the changed forms are more convenient at the stage of numerical modeling.

At the stage of melting/resolidification modeling, the concept presented in [38] (for the macroscale problems) is applied. The capacity of the internal heat source related to the phase change is proportional to the melting/solidification rate. To define this function (in particular, the volumetric fraction of liquid state $f_L(T)$) in the form of a continuous one, the melting point corresponding to the temperature T_m is conventionally replaced by a certain interval $[T_m - \Delta T, T_m + \Delta T]$, and then the function discussed can be described by a broken line. For this interval, the substitute thermal capacity is defined and the one domain approach can be used. It should be pointed out that the testing computations show a little impact of the interval ΔT width (within reasonable limits) on the results of numerical simulations.

At the beginning of the part of the article devoted to own research, the assumed form of the dual-phase lag equation and the mathematical formulas determining the laser action and the evolution of phase change latent heat are presented. Both phenomena are taken into account by an introduction to the energy equation of the functions determining the efficiency of internal heat sources. Next, the numerical algorithm based on the implicit scheme of FDM is discussed. In the final part of the paper, the results of numerical computations concerning the heating/cooling process of the thin metal film made of chrome are shown. The conclusions resulting from the performed research are also formulated.

2. Governing Equations

The starting point for the formulation of the energy equation with delays is the generalized Fourier law (1). To obtain the DPLE, the left and right sides of Equation (1) are developed into the Taylor series

$$\mathbf{q}(\mathbf{x},t) + \tau_q \frac{\partial \mathbf{q}(\mathbf{x},t)}{\partial t} + \frac{\tau_q^2}{2} \frac{\partial^2 \mathbf{q}(\mathbf{x},t)}{\partial t^2} + \ldots = \\ -\lambda\left[\nabla T(\mathbf{x},t) + \tau_T \frac{\partial \nabla T(\mathbf{x},t)}{\partial t} + \frac{\tau_T^2}{2} \frac{\partial^2 \nabla T(\mathbf{x},t)}{\partial t^2} + \ldots\right] \tag{2}$$

Let us apply the well-known diffusion equation, namely,

$$c \frac{\partial T(\mathbf{x},t)}{\partial t} = -\nabla \cdot \mathbf{q}(\mathbf{x},t) + Q(\mathbf{x},t) \tag{3}$$

where c is a volumetric specific heat and $Q(\mathbf{x},t)$ is a capacity of volumetric internal heat sources.

When the components containing the second derivatives (Equation (2)) are taken into account, after mathematical manipulations, Equation (3) takes the form

$$\mathbf{x} \in \Omega: \quad c\left[\frac{\partial T(\mathbf{x},t)}{\partial t} + \tau_q \frac{\partial^2 T(\mathbf{x},t)}{\partial t^2} + \frac{\tau_q^2}{2} \frac{\partial^3 T(\mathbf{x},t)}{\partial t^3}\right] = \nabla[\lambda \nabla T(\mathbf{x},t)] + \\ \tau_T \frac{\partial \{\nabla[\lambda \nabla T(\mathbf{x},t)]\}}{\partial t} + \frac{\tau_T^2}{2} \frac{\partial^2 \{\nabla[\lambda \nabla T(\mathbf{x},t)]\}}{\partial t^2} + Q(\mathbf{x},t) + \tau_q \frac{\partial Q(\mathbf{x},t)}{\partial t} + \frac{\tau_q^2}{2} \frac{\partial^2 Q(\mathbf{x},t)}{\partial t^2} \tag{4}$$

When the melting and resolidification problem is considered, the internal heat source in Equation (4) must contain the term controlling the phase change process. This appropriate source function $Q_m(\mathbf{x},t)$ can be defined as

$$Q_m(\mathbf{x},t) = -L \frac{\partial f_L(\mathbf{x},t)}{\partial t} - L\tau_q \frac{\partial^2 f_L(\mathbf{x},t)}{\partial t^2} - L \frac{\tau_q^2}{2} \frac{\partial^3 f_L(\mathbf{x},t)}{\partial t^3} \tag{5}$$

where L is the volumetric latent heat phase change and $f_L(\mathbf{x},t)$ is the volumetric molten state fraction at the neighborhood of the point considered. The last equation is a generalization of what is well known in the thermal theory of foundry processes definition of Q_m (e.g., [38]). The function $f_L(\mathbf{x},t)$ is equal to zero at the beginning of the heating process until $T_1 = T_m - \Delta T$ and $f_L(\mathbf{x},t) = 1$ for $T_2 > T_m + \Delta T$ (T_m is the melting point). In the interval $[T_m - \Delta T, T_m + \Delta T]$, the function $f_L(\mathbf{x},t)$ changes from 0 to 1 in a linear way (such an assumption is fully acceptable). Generally speaking, the volumetric liquid state fraction is given in the form of broken line, this means

$$f_L(\mathbf{x},t) = \begin{cases} 1 & T(\mathbf{x},t) > T_m + \Delta T \\ \frac{T(\mathbf{x},t) - T_m + \Delta T}{2\Delta T} & T_m - \Delta T \leq T(\mathbf{x},t) \leq T_m + \Delta T \\ 0 & T(\mathbf{x},t) < T_m - \Delta T \end{cases} \qquad (6)$$

The derivative of $f_L(\mathbf{x}, t)$ with respect to the temperature is equal to 0 for $T(\mathbf{x}, t) < T_m - \Delta T$ and $T(\mathbf{x}, t) > T_m + \Delta T$, while between the border temperatures $df_L(\mathbf{x}, t)/dT = 1/2\Delta T$. Thus, the source term $Q_m(\mathbf{x}, t)$ acts only for $T(\mathbf{x}, t)$ from the interval $[T_m - \Delta T, T_m + \Delta T]$, and then

$$Q_m(\mathbf{x},t) = -\frac{L}{2\Delta T}\left[\frac{\partial T(\mathbf{x},t)}{\partial t} + \tau_q \frac{\partial T^2(\mathbf{x},t)}{\partial t^2} + \frac{\tau_q^2}{2}\frac{\partial T^3(\mathbf{x},t)}{\partial t^3}\right], \quad T(\mathbf{x},t) \in [T_m - \Delta T, T_m + \Delta T] \qquad (7)$$

Let us introduce the piece-vise constant function $C(T)$

$$C(T) = \begin{cases} c_2 & T(\mathbf{x},t) > T_m + \Delta T \\ 0.5(c_1 + c_2) + \frac{L}{2\Delta T} & T_m - \Delta T \leq T(\mathbf{x},t) \leq T_m + \Delta T \\ c_1 & T(\mathbf{x},t) < T_m - \Delta T \end{cases} \qquad (8)$$

where c_1 and c_2 are the volumetric specific heats of the solid and liquid states, respectively. Then, Equation (4) can be written as follows

$$\mathbf{x} \in \Omega: C(T)\left[\frac{\partial T(\mathbf{x},t)}{\partial t} + \tau_q \frac{\partial^2 T(\mathbf{x},t)}{\partial t^2} + \frac{\tau_q^2}{2}\frac{\partial^3 T(\mathbf{x},t)}{\partial t^3}\right] = \nabla[\lambda \nabla T(\mathbf{x},t)] + \tau_T \frac{\partial \{\nabla[\lambda \nabla T(\mathbf{x},t)]\}}{\partial t} + \frac{\tau_T^2}{2}\frac{\partial^2 \{\nabla[\lambda \nabla T(\mathbf{x},t)]\}}{\partial t^2} + Q_l(\mathbf{x},t) + \tau_q \frac{\partial Q_l(\mathbf{x},t)}{\partial t} + \frac{\tau_q^2}{2}\frac{\partial^2 Q_l(\mathbf{x},t)}{\partial t^2} \qquad (9)$$

Thermal conductivity λ in Equation (9) is defined just like the parameter $C(T)$.

The mathematical formula determining the intensity of the internal heat source $Q_l(x, t)$ resulting from the laser action can be taken in the form [39]

$$Q_l(\mathbf{x}, t) = (1 - R)\frac{I_0}{\delta\, t_p}\exp\left[-\frac{x_1^2 + x_2^2}{r_D^2} - \frac{x_3}{\delta} - 4\ln 2\frac{(t - 2t_p)^2}{t_p^2}\right], \quad \mathbf{x} = \{x_1, x_2, x_3\} \qquad (10)$$

where I_0 [J/m^2] is the laser intensity, t_p [s] is the characteristic time of laser pulse, δ [m] is the optical penetration depth, R is the reflectivity of the irradiated surface, r_D [m] is the laser beam radius, and x_3 is a vertical axis. The derivatives of Q_l with respect to time can be found analytically.

On the outer surface of the system, the adiabatic conditions are assumed (the external heat flux is taken into account in the appropriate source function). The mathematical form of the Neumann boundary condition for the second-order DPLE is as follows [36]

$$\mathbf{x} \in \Gamma: \quad -\lambda\left[\mathbf{n}\cdot\nabla T(\mathbf{x},t) + \tau_T\frac{\partial[\mathbf{n}\cdot\nabla T(\mathbf{x},t)]}{\partial t} + \frac{\tau_T^2}{2}\frac{\partial^2[\mathbf{n}\cdot\nabla T(\mathbf{x},t)]}{\partial t^2}\right] = q_b(\mathbf{x},t) + \tau_q \frac{\partial q_b(\mathbf{x},t)}{\partial t} + \frac{\tau_q^2}{2}\frac{\partial^2 q_b(\mathbf{x},t)}{\partial t^2} \qquad (11)$$

where $\mathbf{n}$ is a normal outward vector and (in the case considered) $q_b(\mathbf{x}, t) = 0$, of course.

The mathematical model is also supplemented by the initial conditions

$$t = 0: \quad T(\mathbf{x},0) = T_p, \quad \left.\frac{\partial T(\mathbf{x},t)}{\partial t}\right|_{t=0} = \frac{Q(\mathbf{x},0)}{c_1}, \quad \left.\frac{\partial^2 T(\mathbf{x},t)}{\partial t^2}\right|_{t=0} = \frac{1}{c_1}\left.\frac{\partial Q(\mathbf{x},t)}{\partial t}\right|_{t=0} \qquad (12)$$

where T_p is an initial temperature.

3. Mathematical Description of 1D Problem

At the stage of numerical modeling, the 1D problem was considered and the basis for the construction of the FDM algorithm is the following system of equations and conditions:

- energy equation for thin metal film domain

$$C(T)\left[\frac{\partial T(x,t)}{\partial t} + \tau_q \frac{\partial^2 T(x,t)}{\partial t^2} + \frac{\tau_q^2}{2}\frac{\partial^3 T(x,t)}{\partial t^3}\right] = \frac{\partial}{\partial x}\left[\lambda\frac{\partial T(x,t)}{\partial x}\right] + \tau_T \frac{\partial^2}{\partial t \partial x}\left[\lambda\frac{\partial T(x,t)}{\partial x}\right] + \frac{\tau_T^2}{2}\frac{\partial^3}{\partial t^2 \partial x}\left[\lambda\frac{\partial T(x,t)}{\partial x}\right] + Z(x,t) \quad (13)$$

where

$$Z(x,t) = Q_l(x,t) + \tau_q \frac{\partial Q_l(x,t)}{\partial t} + \frac{\tau_q^2}{2}\frac{\partial^2 Q_l(x,t)}{\partial t^2} \quad (14)$$

- source function Q_l

$$Q_l(x,t) = (1-R)\frac{I_0}{\delta\, t_p}\exp\left[-\frac{x}{\delta} - 4\ln 2 \frac{(t-2t_p)^2}{t_p^2}\right] \quad (15)$$

adiabatic boundary conditions

$$x = 0 \cup G: \quad \nabla T(x,t) + \tau_T \frac{\partial\left[\frac{\partial T(x,t)}{\partial x}\right]}{\partial t} + \frac{\tau_T^2}{2}\frac{\partial^2\left[\frac{\partial T(x,t)}{\partial x}\right]}{\partial t^2} = 0 \quad (16)$$

- initial conditions

$$t = 0: \quad T(x,0) = T_p, \quad \left.\frac{\partial T(x,t)}{\partial t}\right|_{t=0} = \frac{Q(x,0)}{c_1}, \quad \left.\frac{\partial^2 T(x,t)}{\partial t^2}\right|_{t=0} = \frac{1}{c_1}\left.\frac{\partial Q(x,t)}{\partial t}\right|_{t=0} \quad (17)$$

4. Numerical Model Based on FDM

Let $T_i^f = T(x_i, f\Delta t)$, where $x_i = ih$, $i = 0, 1, \ldots, n$ and $f = 0, 1, \ldots, F$. Taking into account the initial conditions (17), one has

$$\begin{array}{l}T_i^0 = T_p \\ \frac{T_i^1 - T_i^0}{\Delta t} = \frac{1}{c_1} Q_l(x_i, 0) \rightarrow T_i^1 = T_i^0 + \frac{\Delta t}{c_1} Q_l(x_i, 0) \\ \frac{T_i^2 - 2T_i^1 + T_i^0}{(\Delta t)^2} = \frac{1}{c_1}\left.\frac{\partial Q_l(x_i,t)}{\partial t}\right|_{t=0} \rightarrow T_i^2 = 2T_i^1 - T_i^0 + \frac{(\Delta t)^2}{c_1}\left.\frac{\partial Q_j(x_i,t)}{\partial t}\right|_{t=0}\end{array} \quad (18)$$

For the transition $t^{f-1} \rightarrow t^f$ ($f \geq 3$), the approximate form of Equation (13) resulting from the introduction of the assumed differential quotients is as follows

$$C_i^{f-1}\left[\frac{T_i^f - T_i^{f-1}}{\Delta t} + \tau_q \frac{T_i^f - 2T_i^{f-1} + T_i^{f-2}}{(\Delta t)^2} + \frac{\tau_q^2}{2}\frac{T_i^f - 3T_i^{f-1} + 3T_i^{f-2} - T_i^{f-3}}{(\Delta t)^3}\right] = \left[\frac{\partial}{\partial x}\left(\lambda\frac{\partial T}{\partial x}\right)\right]_i^f + \frac{\tau_T}{\Delta t}\left\{\left[\frac{\partial}{\partial x}\left(\lambda\frac{\partial T}{\partial x}\right)\right]_i^f - \left[\frac{\partial}{\partial x}\left(\lambda\frac{\partial T}{\partial x}\right)\right]_i^{f-1}\right\} + \frac{w_T \tau_T^2}{2(\Delta t)^2}\left\{\left[\frac{\partial}{\partial x}\left(\lambda\frac{\partial T}{\partial x}\right)\right]_i^f - 2\left[\frac{\partial}{\partial x}\left(\lambda\frac{\partial T}{\partial x}\right)\right]_i^{f-1} + \left[\frac{\partial}{\partial x}\left(\lambda\frac{\partial T}{\partial x}\right)\right]_i^{f-2}\right\} + Z_i^f \quad (19)$$

where

$$\left[\frac{\partial}{\partial x}\left(\lambda\frac{\partial T}{\partial x}\right)\right]_i^s = \frac{1}{h}\left[\left(\lambda\frac{\partial T}{\partial x}\right)_{i+0.5}^s - \left(\lambda\frac{\partial T}{\partial x}\right)_{i-0.5}^s\right] = \frac{1}{h}\left(\lambda_{i+0.5}^s \frac{T_{i+1}^s - T_i^s}{h} - \lambda_{i-0.5}^s \frac{T_i^s - T_{i-1}^s}{h}\right) = \frac{1}{h}\left(\frac{\lambda_i^s + \lambda_{i+1}^s}{2}\frac{T_{i+1}^s - T_i^s}{h} - \frac{\lambda_{i-1}^s + \lambda_i^s}{2}\frac{T_i^s - T_{i-1}^s}{h}\right) = \frac{1}{2h^2}\left[\left(\lambda_i^s + \lambda_{i+1}^s\right)\left(T_{i+1}^s - T_i^s\right) + \left(\lambda_{i-1}^s + \lambda_i^s\right)\left(T_{i-1}^s - T_i^s\right)\right] \quad (20)$$

while $s = f, s = f - 1$, or $s = f - 2$.

Thus,

$$\frac{T_i^f - T_i^{f-1}}{\Delta t} + \tau_q \frac{T_i^f - 2T_i^{f-1} + T_i^{f-2}}{(\Delta t)^2} + \frac{\tau_q^2}{2} \frac{T_i^f - 3T_i^{f-1} + 3T_i^{f-2} - T_i^{f-3}}{(\Delta t)^3} =$$
$$\frac{2(\Delta t)^2 + 2\tau_T \Delta t + \tau_T^2}{4C_i^{f-1} h^2 (\Delta t)^2} \left[\left(\lambda_i^{f-1} + \lambda_{i+1}^{f-1}\right)\left(T_{i+1}^f - T_i^f\right) + \left(\lambda_{i-1}^{f-1} + \lambda_i^{f-1}\right)\left(T_{i-1}^f - T_i^f\right) \right] -$$
$$\frac{2\tau_T \Delta t + 2\tau_T^2}{4C_i^{f-1} h^2 (\Delta t)^2} \left[\left(\lambda_i^{f-1} + \lambda_{i+1}^{f-1}\right)\left(T_{i+1}^{f-1} - T_i^{f-1}\right) + \left(\lambda_{i-1}^{f-1} + \lambda_i^{f-1}\right)\left(T_{i-1}^{f-1} - T_i^{f-1}\right) \right] +$$
$$\frac{\tau_T^2}{4C_i^{f-1} h^2 (\Delta t)^2} \left[\left(\lambda_i^{f-2} + \lambda_{i+1}^{f-2}\right)\left(T_{i+1}^{f-2} - T_i^{f-2}\right) + \left(\lambda_{i-1}^{f-2} + \lambda_i^{f-2}\right)\left(T_{i-1}^{f-2} - T_i^{f-2}\right) \right] + \frac{Z_i^f}{C_i^{f-1}} \quad (21)$$

and after mathematical manipulations, one has

$$A_i^f T_{i-1}^f + B_i^f T_i^f + C_i^f T_{i+1}^f = D_i^f \quad (22)$$

where

$$A_i^f = -\frac{\left[2(\Delta t)^2 + 2\tau_T \Delta t + \tau_T^2\right]\left(\lambda_{i-1}^{f-1} + \lambda_i^{f-1}\right)}{4C_i^{f-1} h^2 (\Delta t)^2},$$

$$B_i^f = \frac{2(\Delta t)^2 + 2\tau_q \Delta t + \tau_q^2}{2(\Delta t)^3} + \frac{\left[2(\Delta t)^2 + 2\tau_T \Delta t + \tau_T^2\right]\left(\lambda_{i-1}^{f-1} + 2\lambda_i^{f-1} + \lambda_{i+1}^{f-1}\right)}{4C_i^{f-1} h^2 (\Delta t)^2},$$

$$C_i^f = -\frac{\left[2(\Delta t)^2 + 2\tau_T \Delta t + \tau_T^2\right]\left(\lambda_{i+1}^{f-1} + \lambda_i^{f-1}\right)}{4C_i^{f-1} h^2 (\Delta t)^2}, \quad (23)$$

$$D_i^f = \frac{2(\Delta t)^2 + 4\tau_q \Delta t + 3\tau_q^2}{2(\Delta t)^3} T_i^{f-1} - \frac{2\tau_q \Delta t + 3\tau_q^2}{2(\Delta t)^3} T_i^{f-2} + \frac{\tau_q^2}{2(\Delta t)^3} T_i^{f-3} -$$
$$\frac{2\tau_T \Delta t + 2\tau_T^2}{4C_i^{f-1} h^2 (\Delta t)^2} \left[\left(\lambda_i^{f-1} + \lambda_{i+1}^{f-1}\right)\left(T_{i+1}^{f-1} - T_i^{f-1}\right) + \left(\lambda_{i-1}^{f-1} + \lambda_i^{f-1}\right)\left(T_{i-1}^{f-1} - T_i^{f-1}\right) \right] +$$
$$\frac{\tau_T^2}{4C_i^{f-1} h^2 (\Delta t)^2} \left[\left(\lambda_i^{f-2} + \lambda_{i+1}^{f-2}\right)\left(T_{i+1}^{f-2} - T_i^{f-2}\right) + \left(\lambda_{i-1}^{f-2} + \lambda_i^{f-2}\right)\left(T_{i-1}^{f-2} - T_i^{f-2}\right) \right] + \frac{Z_i^f}{C_i^{f-1}}$$

The approximation of boundary conditions (16) is as follows

- for $x = 0$:

$$\frac{T_1^f - T_0^f}{h} + \frac{\tau_T}{\Delta t}\left(\frac{T_1^f - T_0^f}{h} - \frac{T_1^{f-1} - T_0^{f-1}}{h}\right) + \frac{\tau_T^2}{2(\Delta t)^2}\left(\frac{T_1^f - T_0^f}{h} - 2\frac{T_1^{f-1} - T_0^{f-1}}{h} + \frac{T_1^{f-2} - T_0^{f-2}}{h}\right) = 0 \quad (24)$$

- for $x = G$:

$$\frac{T_n^f - T_{n-1}^f}{h} + \frac{\tau_T}{\Delta t}\left(\frac{T_n^f - T_{n-1}^f}{h} - \frac{T_n^{f-1} - T_{n-1}^{f-1}}{h}\right) + \frac{\tau_T^2}{2(\Delta t)^2}\left(\frac{T_n^f - T_{n-1}^f}{h} - 2\frac{T_n^{f-1} - T_{n-1}^{f-1}}{h} + \frac{T_n^{f-2} - T_{n-1}^{f-2}}{h}\right) = 0 \quad (25)$$

or

$$-\left[2(\Delta t)^2 + 2\tau_T \Delta t + \tau_T^2\right]T_0^f + \left[2(\Delta t)^2 + 2\tau_T \Delta t + \tau_T^2\right]T_1^f =$$
$$\left(2\tau_T \Delta t + 2\tau_T^2\right)\left(T_1^{f-1} - T_0^{f-1}\right) - \tau_T^2\left(T_1^{f-2} - T_0^{f-2}\right) \quad (26)$$

and

$$-\left[2(\Delta t)^2 + 2\tau_T \Delta t + \tau_T^2\right]T_{n-1}^f + \left[2(\Delta t)^2 + 2\tau_T \Delta t + \tau_T^2\right]T_n^f =$$
$$\left(2\tau_T \Delta t + 2\tau_T^2\right)\left(T_n^{f-1} - T_{n-1}^{f-1}\right) - \tau_T^2\left(T_n^{f-2} - T_{n-1}^{f-2}\right) \quad (27)$$

The transition from t^{f-1} to t^f ($f \geq 3$) requires the solution of the system of Equations (22), (26), and (27) with a three-band main matrix that can be solved quickest using the Thomas algorithm [40].

5. Results of Computations

The computations were performed for the thin metal film of the thickness $G = 100$ nm made of chromium. The surface $x = 0$ is subjected to the laser pulse with the parameters $R = 0.93$, $I_0 = 3825$ J/m^2, $\delta = 15.3$ nm, and $t_p = 10$ ps—c.f. Formula (15). The following values of chromium parameters are assumed: thermal conductivities $\lambda_1 = 93$ W/(m K), $\lambda_2 = 35$ W/(m K), volumetric specific heats $c_1 = 3.2148$ MJ/(m^3 K), $c_2 = 2.79276$ MJ/(m^3 K) [24], relaxation time $\tau_q = 0.136$ ps, thermalization time $\tau_T = 7.86$ ps, melting temperature $T_m = 2180$ K, and volumetric heat of fusion $L = 2904$ MJ/m^3.

Different temporal-spatial meshes were considered; see Table 1. For each combination of mesh steps, the temperature after 80 ps on the heated surface was recorded. As can be seen, the result does not depend much on the time step, while the number of nodes is important. Analysis of the results obtained showed that the satisfactory accuracy provides the geometrical mesh containing $n = 1000$ nodes ($h = 0.1$ nm) and time step $\Delta t = 0.0005$ ps (see Table 1, column 5).

Table 1. Temperature of the irradiated surface after 80 ps for different time steps and number of nodes.

Δt [ps]	$n = 100$	$n = 200$	$n = 500$	$n = 1000$
0.0001	1484.22	1498.47	1507.06	1511.45
0.00025	1484.19	1498.43	1507.07	1511.43
0.0005	1484.08	1498.44	1507.05	1511.47

Figure 1 shows the temperature courses at the points $x = 0$ (heated surface) and $x = 20$ nm. The results were compared to those published in [24], where the first order DPL equation and slightly different partial melting model were used. As one can see, the differences are small, but visible. For the second-order model (solid lines), the maximum temperature of the heated surface occurs after 28.8 ps and is equal to 2323.28 K, while for the first-order model (dashed lines), the maximum temperature occurs after 28 ps and equals 2307.20 K.

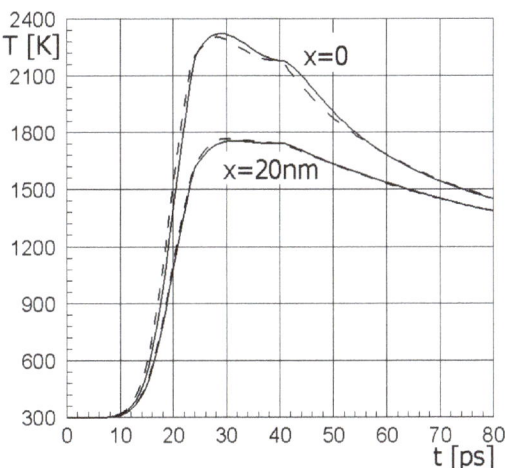

Figure 1. Temperature histories at the points $x = 0$ and $x = 20$ nm; comparison of the results obtained using the first-order (dashed line [24]) and second-order dual phase lag (DPL) (solid line) for $I_0 = 3825$ J/m^2, $t_p = 10$ ps.

In Figure 2, the fragments of heating/cooling curves for different widths of ΔT are shown. One can see that the results are similar. For $\Delta T = 3$ K, the maximum temperature is the highest; for $\Delta T = 7$ K, the maximum temperature is the lowest; while the biggest differences between them are of the order

10 K. After the resolidification process, the cooling curves for all cases are very similar. Further computations were carried out for $\Delta T = 3$ K.

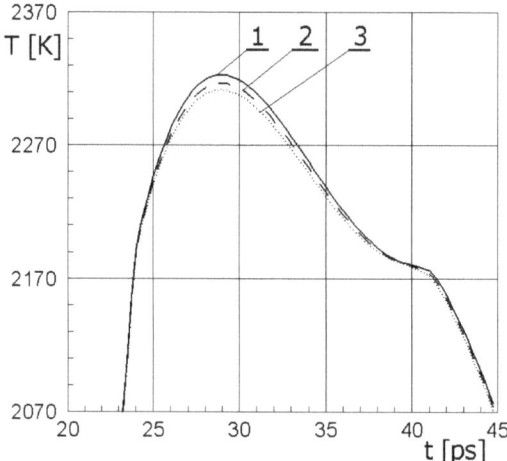

Figure 2. Comparison of obtained temperature courses for different intervals ΔT (on the surface $x = 0$), (1) $\Delta T = 3$ K, (2) $\Delta T = 5$ K, (3) $\Delta T = 7$ K.

In Figure 3, the temperature courses on the irradiated surface for the laser intensity $I_0 = 3825$ J/m^2 and different characteristic times of laser pulse t_p are presented. The impact of changes in this parameter on the course of the heating/cooling curves is clearly visible. It is obvious that an increase in intensity of the laser pulse causes a more rapid heating process. Interesting, however, are the effects of changes of characteristic time t_p. Shortening this time increases the heating rate of the metal film and maxima of the cooling/heating curves shift to the left. At the same time, the maximum values of temperatures increase slightly. A detailed analysis of Equation (10) determining the time-dependent efficiency of the laser-related internal heat source shows that such an effect could be expected.

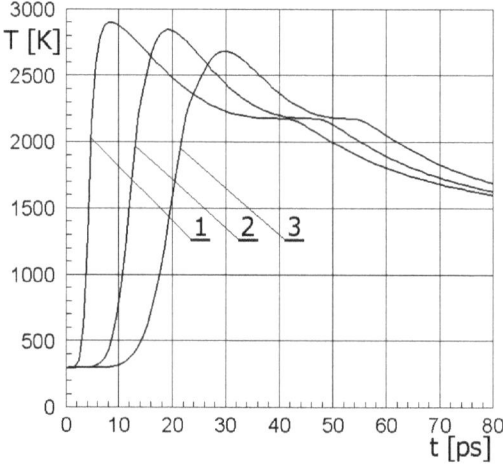

Figure 3. Heating/cooling curves on the surface $x = 0$ for different characteristic times of laser pulse. t_p: (1) $t_p = 2$ ps, (2) $t_p = 6$ ps, (3) $t_p = 10$ ps ($I_0 = 4500$ J/m^2).

Finally, Figure 4 shows the changes of molten layer thicknesses in time. The successive curves correspond to t_p = 2, 6, and 10 ps. For t_p = 2 ps, the maximal depth is equal to g = 17.6 nm; for t_p = 6 ps, it is equal to g = 15.3 nm; while for t_p = 10 ps, it is equal to g = 13 nm.

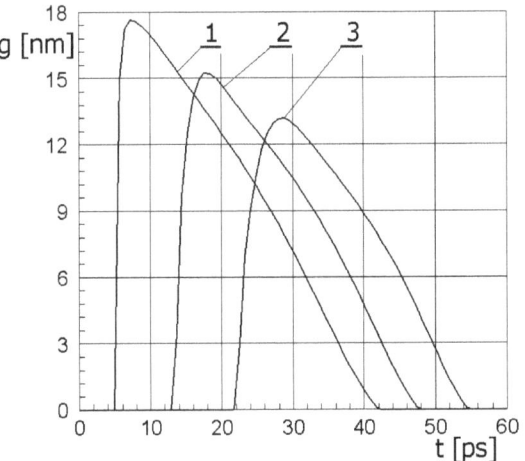

Figure 4. Changes in the thickness of the molten layer for different characteristic times of laser pulse, (1) t_p = 2 ps, (2) t_p = 6 ps, (3) t_p = 10 ps.

Changes in the characteristic time of laser pulse cause a similar effect to that seen in the previous figure. The depth of the molten layer increases with the reduction of time t_p and the process is more dynamic. Such analysis and information obtained on its basis can often be useful in engineering practice.

6. Conclusions

The mathematical model, the numerical algorithm, and exemplary results of the computations concerning the heating of thin metal film subjected to the laser beam are discussed, wherein the problem is described using the second-order DPLE. The laser beam power is so high that, in the domain under consideration, the partial melting and then resolidification of the material take place. To our knowledge, the application of the second-order DPLE to solve this type of the problem has not yet been presented. The task is treated as a non-linear one and the thermophysical parameters of the material are temperature-dependent. The melting point is substituted by a certain interval of temperature. This assumption allows one to introduce the continuous function determining the local and temporary volumetric liquid state fraction of the metal. The influence of the width of this interval on the results of numerical calculations is also examined. The laser action and the evolution of latent heat are taken into account by the introduction of two internal heat sources to the energy equation.

At the stage of numerical modeling, the 1D problem was considered, but the generalization of the algorithm and computer program implementing the numerical computations for the larger number of dimensions is not difficult. The computer program is based on the implicit scheme of the finite difference method. The algorithm is unconditionally stable and the transition from time t to $t + \Delta t$ requires the solution of the system of linear equations, whose main matrix is a three-diagonal one. In this place, the very effective Thomas algorithm was used.

The obtained results are in line with the expectations; their comparison with the solution of a similar problem described by the first-order equation shows visible, but slight differences.

We are planning the following further research in this scope:

- numerical algorithm and computer program for 3D problems;

- numerical algorithm and computer program for axially-symmetrical tasks (this geometry is very convenient because of the typical shape of the function describing the laser action);
- adaptation of the algorithm presented in this work for modeling the ablation process;
- research on other approaches to phase changes modeling.

Author Contributions: Conceptualization, E.M. and B.M.; formal analysis, E.M. and B.M.; software, E.M.; investigation, E.M. and B.M., writing—original draft, B.M. All authors have read and agreed to the published version of the manuscript.

Funding: This research received no external funding.

Conflicts of Interest: The authors declare no conflict of interest.

References

1. Cattaneo, M.C. *Sulla Conduzione de Calore. Atti de Seminario Matematico e Fisico*; Della Universita di Modena: Modena, Italy, 1948; pp. 3–21.
2. Cattaneo, M.C. A form of heat conduction equation which eliminates the paradox of instantaneous propagation. *Comptes Rendus* **1958**, *247*, 431–433.
3. Escobar, R.A.; Ghai, S.S.; Jhon, M.S.; Amon, C.H. Multi-length and time scale thermal transport using the l lattice Boltzmann method with applications to electronics cooling. *Int. J. Heat Mass Transf.* **2006**, *49*, 97–107. [CrossRef]
4. Flik, M.I.; Choi, B.I.; Goodson, K.E. Heat transfer regimes in microstructures. *J. Heat Transf.* **1992**, *114*, 664–674. [CrossRef]
5. Tzou, T. A unified field approach for heat conduction from macro- to micro-scales. *J. Heat Transf.* **1995**, *117*, 8–16. [CrossRef]
6. Cabrera, J.; Castro, M.A.; Rodríquez, F.; Martín, J.A. Difference schemes for numerical solutions of lagging models of heat conductions. *Math. Comupt. Model.* **2013**, *57*, 1625–1632. [CrossRef]
7. Özişik, M.N.; Tzou, D.Y. On the wave theory in heat conduction. *J. Heat Transf.* **1994**, *116*, 526–535. [CrossRef]
8. Orlande, R.B.; Özişik, M.N.; Tzou, D.Y. Inverse analysis for estimating the electron-phonon coupling factor in thin metal films. *J. Appl. Phys.* **1995**, *78*, 1843–1849. [CrossRef]
9. Al Nimr, M.A. Heat transfer mechanisms during short duration laser heating of thin metal films. *Int. J. Thermophys.* **1997**, *18*, 1257–1268. [CrossRef]
10. Zhang, Z.M. *Nano/Microscale Heat Transfer*; McGraw-Hill: New York, NY, USA, 2007.
11. Tzou, D.Y. *Macro- to Microscale Heat Transfer. The Lagging Behavior*; John Wiley & Sons: Hoboken, NJ, USA, 2015.
12. Faghri, A.; Zhang, Y.; Howell, J. *Advanced Heat and Mass Transfer*; Global Digital Press: Columbia, MO, USA, 2010.
13. Smith, A.N.; Norris, P.M. Microscale Heat Transfer, Chapter 18. In *Heat Transfer Handbook*; John Willey & Sons: Hoboken, NJ, USA, 2003.
14. Ciesielski, M. Analytical solution of the dual phase lag equation describing the laser heating of thin metal film. *J. Appl. Math. Comput. Mech.* **2017**, *16*, 33–40. [CrossRef]
15. Kumar, S.; Srivastava, A. Finite integral transform-based analytical solutions of dual phase lag bio-heat transfer equation. *Appl. Math. Model.* **2017**, *52*, 378–403. [CrossRef]
16. Liu, K.-C.; Wang, J.-C. Analysis of thermal damage to laser irradiated tissue based on the dual-phase-lag model. *Int. J. Heat Mass Transf.* **2014**, *70*, 621–628. [CrossRef]
17. Mohammadi-Fakhar, V.; Momeni-Masuleh, S.H. An approximate analytic solution of the heat conduction equation at nanoscale. *Phys. Lett. A* **2010**, *374*, 595–604. [CrossRef]
18. Ramadan, K. Semi-analytical solutions for the dual phase lag heat conduction in multi-layered media. *Int. J. Therm. Sci.* **2009**, *48*, 14–25. [CrossRef]
19. Wang, H.; Dai, W.; Melnik, R. A finite difference method for studying thermal deformation in a double-layered thin film exposed to ultrashort pulsed lasers. *Int. J. Therm. Sci.* **2006**, *45*, 1179–1196. [CrossRef]
20. Dai, W.; Nassar, R. A compact finite difference scheme for solving a three-dimensional heat transport equation in a thin film. *Numer. Methods Partial Differ. Equ.* **2000**, *16*, 441–458. [CrossRef]

21. Mochnacki, B.; Majchrzak, E. Numerical model of thermal interactions between cylindrical cryoprobe and biological tissue using the dual-phase lag equation. *Int. J. Heat Mass Transf.* **2017**, *108*, 1–10. [CrossRef]
22. Majchrzak, E.; Mochnacki, B. Dual-phase-lag equation. Stability conditions of a numerical algorithm based on the explicit scheme of the finite difference method. *J. Appl. Math. Comput. Mech.* **2016**, *15*, 89–96. [CrossRef]
23. Ciesielski, M. Application of the alternating direction implicit method for numerical solution of the dual phase lag equation. *J. Theor. Appl. Mech. Pol.* **2017**, *55*, 839–852. [CrossRef]
24. Majchrzak, E.; Mochnacki, B. Dual-phase lag model of thermal processes in a multi-layered microdomain subjected to a strong laser pulse using the implicit scheme of FDM. *Int. J. Therm. Sci.* **2018**, *133*, 240–251. [CrossRef]
25. Mochnacki, B.; Paruch, M. Estimation of relaxation and thermalization times in microscale heat transfer model. *J. Theor. Appl. Mech. Pol.* **2013**, *51*, 837–845.
26. Kumar, P.; Kumar, D.; Rai, K.N. A numerical study of dual-phase-lag model of bio-heat transfer during hyperthermia treatment. *J. Therm. Biol.* **2015**, *49–50*, 98–105. [CrossRef] [PubMed]
27. Kumar, D.; Kumar, P.; Rai, K.N. A study on DPL model of heat transfer in bi-layer tissues during MFH Treatment. *Comput. Biol. Med.* **2016**, *75*, 160–172. [CrossRef] [PubMed]
28. Kumar, D.; Rai, K.N. A study on thermal damage during hyperthermia treatment based on DPL model for multilayer tissues using finite element Legendre wavelet Galerkin approach. *J. Therm. Biol.* **2016**, *62*, 170–180. [CrossRef] [PubMed]
29. Majchrzak, E. Numerical solution of dual phase lag model of bioheat transfer using the general boundary element method. *CMES Comput. Model. Eng. Sci.* **2010**, *69*, 43–60.
30. Mochnacki, B.; Ciesielski, M. Micro-scale heat transfer. Algorithm basing on the Control Volume Method and the identification of relaxation and thermalization times using the search method. *CMMS* **2015**, *15*, 353–361.
31. Patidar, S.; Kumar, S.; Srivastava, A.; Singh, S. Dual phase lag model-based thermal analysis of tissue phantoms using lattice Boltzmann method. *Int. J. Therm. Sci.* **2016**, *103*, 41–56. [CrossRef]
32. Ho, J.R.; Kuo, C.P.; Jiaung, W.S. Study of heat transfer in multilayered structure within the framework of dual-phase-lag heat conduction model using lattice Boltzmann method. *Int. J. Heat Mass Transf.* **2003**, *46*, 55–69. [CrossRef]
33. Castro, M.A.; Rodríques, F.; Cabrera, J.; Martín, J.A. A compact difference scheme for numerical solution of second order dual-phase-lagging models of microscale heat transfer. *J. Comput. Appl. Math.* **2016**, *291*, 432–440. [CrossRef]
34. Deng, D.; Jiang, Y.; Liang, D.L. High-order finite difference methods fora second order dual-phase-lagging models of microscale heat transfer. *Appl. Math. Comput.* **2017**, *309*, 31–48.
35. Majchrzak, E.; Mochnacki, B. Implicit scheme of the finite difference method for a second-order dual phase lag equation. *J. Theor. Appl. Mech. Pol.* **2018**, *56*, 393–402. [CrossRef]
36. Majchrzak, E.; Mochnacki, B. Numerical solutions of the second-order dual-phase lag equation using the explicit and implicit schemes of the finite difference method. *Int. J. Numer. Methods Heat Fluid Flow* **2019**, *20*, 2099–2120. [CrossRef]
37. Askarizadeh, H.; Baniasadi, E.; Ahmadikia, H. Equilibrium and nonequilibrium thermodynamic analysis of high-order dual-phase-lag heat conduction. *Int. J. Heat Mass Transf.* **2017**, *104*, 301–309. [CrossRef]
38. Ciesielski, M.; Mochnacki, B. Comparison of approaches to the numerical modelling of pure metals solidification using the control volume method. *Int. J. Cast Metals Res.* **2019**, *32*, 213–220. [CrossRef]
39. Grigoropoulos, C.P.; Chimmalgi, A.; Hwang, D.J. *Laser Ablation and Its Applications*; Springer Series in Optical Sciences; Springer: New York, NY, USA, 2007; pp. 473–504.
40. Datta, B.N. *Numerical Linear Algebra and Applications*, 2nd ed.; SIAM: Philadelphia, PA, USA, 2010; p. 162.

© 2020 by the authors. Licensee MDPI, Basel, Switzerland. This article is an open access article distributed under the terms and conditions of the Creative Commons Attribution (CC BY) license (http://creativecommons.org/licenses/by/4.0/).

Article

Two New Strategies for Pricing Freight Options by Means of a Valuation PDE and by Functional Bounds

Lourdes Gómez-Valle [1,†], **Miguel Angel López-Marcos** [2,†] **and Julia Martínez-Rodríguez** [1,*,†]

[1] Departamento de Economía Aplicada e IMUVA, Facultad de Ciencias Económicas y Empresariales, Universidad de Valladolid, 47011 Valladolid, Spain; lourdes@eco.uva.es
[2] Departamento de Matemática Aplicada e IMUVA, Facultad de Ciencias, Universidad de Valladolid, 47011 Valladolid, Spain; malm@mac.uva.es
* Correspondence: julia@eco.uva.es
† These authors contributed equally to this work.

Received: 14 February 2020; Accepted:14 April 2020; Published: 17 April 2020

Abstract: Freight derivative prices have been modeled assuming that the spot freight follows a particular stochastic process in order to manage them, like freight futures, forwards and options. However, an explicit formula for pricing freight options is not known, not even for simple spot freight processes. This is partly due to the fact that there is no valuation equation for pricing freight options. In this paper, we deal with this problem from two independent points of view. On the one hand, we provide a novel theoretical framework for pricing these Asian-style options. In this way, we build a partial differential equation whose solution is the freight option price obtained from stochastic delay differential equations. On the other hand, we prove lower and upper bounds for those freight options which enables us to estimate the option price. In this work, we consider that the spot freight rate follows a general stochastic diffusion process without restrictions in the drift and volatility functions. Finally, using recent data from the Baltic Exchange, we compare the described bounds with the freight option prices.

Keywords: spot freight rates; freight options; stochastic diffusion process; stochastic delay differential equation; risk-neutral measure; arbitrage arguments; partial differential equations

1. Introduction

In this global economy, the transport of every kind of goods around the world has become of great importance. In fact, more than 95% of the world trade is carried by marine vessels, see [1].

The freight (transport by vessels) market is usually considered as a part of the commodity market. However, there are important differences between them. Most commodities are real products while a freight is a service and, as a result, it is not storable. Freight rates also present remarkable properties such as high volatility and risk. The cost of sea transport is affected by fleet supply and commodity demand, but also by external factors such as the price of bunker fuel or seasonal pressures, see [2]. As a consequence, freight derivatives were initially provided to protect ship-owners and charterers against risk. Besides, more recently financial institutions also found great opportunities in it.

There are different types of freight derivatives such as futures, forwards or options, but all of them depend on the freight rate in a settlement period before the maturity, see [3] for more detail. Traded freight options are contracts whose payoffs are the difference between the average of freight rates in a settlement period and the strike price. That is, they are arithmetic Asian-style options. This procedure avoids the possible manipulation, by large participants in the market, of the price just at maturity time. Moreover, the transportation of goods usually takes several days and the freight rates change along this time period.

Taking into account that the freight market is very recent, at the moment, not much scientific research has been done yet. When the spot freight follows a geometric process, a framework for its valuation is developed by Koekebakker et al. [4]. Tvedt [5] models the log spot freight rate in shipping by means of a geometric mean reversion process. Prokopczuk [1] considers that the log spot freight follows an Ornstein–Uhlenbeck process and studies the pricing and hedging of freight futures contract. When the log spot follows a jump-diffusion stochastic process, an accurate valuation of freight options is developed by Nomikos et al. [6] and Kyriakou et al. [7].

In general, in order to obtain a freight option price, it is necessary to use the conditional expectation under the risk-neutral measure because there is no valuation partial differential equation (PDE) for pricing this kind of options, unlike what happens with other derivatives (bonds, futures, European options, etc). Therefore, the Monte Carlo method is used to approximate this conditional expectation, see for example [8]. However, this method is very expensive and inaccurate from a computational point of view.

In this paper, we deal with the freight option valuation problem in two ways. On the one hand, we provide a novel partial differential equation whose solution is the freight option price. This PDE depends on three independent state variables: the spot freight rate, its delay and the continuous version of the average of the spot freight rate over a time period. This framework opens a new way to address this valuation problem. For example, this PDE could be used to obtain a partial explicit solution of the freight option in some models. In other cases, the solution could be approximated by using numerical methods for PDE. We obtained lower and upper bounds of the freight option price. These bounds provide valuable estimations to the option prices.

Our contributions need no restrictive conditions on the model: the spot freight follows a general stochastic diffusion process without restrictions in the drift and volatility functions.

The paper is arranged as follows. In Section 2, a one-factor diffusion model to price freight options is introduced. In Section 3, we provide a novel PDE for pricing these kind of options. In particular, for the geometric model, we obtained a partial solution for this price. In Section 4, we provide lower and upper bounds for the freight option prices. In Section 5, we compare these bounds with the freight option prices in a test problem using data from the Baltic Exchange. Finally, Section 6 concludes.

2. The Option Pricing Model

In this section, we consider a general one-factor diffusion model, which we use to price freight derivatives.

Define $(\Omega, \mathcal{F}, \{\mathcal{F}\}_{t\geq 0}, \mathcal{P})$ as a complete filtered probability space which satisfies the usual conditions and $\{\mathcal{F}\}_{t\geq 0}$ is a filtration, see [9,10].

We assume that the spot freight rate follows the diffusion process, under the risk-neutral measure $\mathcal{Q}$,

$$dS(t) = \mu(S(t))dt + \sigma(S(t))dW(t), \tag{1}$$

where $\mu(S)$ and $\sigma(S)$ are the drift and volatility of the process, respectively, and W is a Wiener process. We suppose that the functions μ and σ satisfy suitable regularity conditions as follows (see [11]):

Assumption 1. *Functions μ and σ are measurable and there exists a constant C such that, for all $x \in \mathbb{R}$,*

$$|\mu(x)| + |\sigma(x)| \leq C(1 + |x|),$$

Assumption 2. *There exists a constant D such that, for all $x, y \in \mathbb{R}$,*

$$|\mu(x) - \mu(y)| + |\sigma(x) - \sigma(y)| \leq D|x - y|.$$

The freight call option price at time t, with settlement period $[T_1, T_N]$, $t \leq T_N$, and strike price K, can be expressed as $C(t, S; K, T_1, \ldots, T_N)$, and at maturity it is

$$C(T_N, S; K, T_1, \ldots, T_N) = \left(\frac{1}{N}\sum_{i=1}^{N} S(T_i) - K\right)^+. \tag{2}$$

On the other hand, we consider a discount factor $D(t) = e^{-\int_0^t r(u)\,du}$. If we assume that the riskless interest rate r is constant, then $D(t) = e^{-rt}$. According to the fundamental theorem of asset pricing (see [10]), the price of a freight call option, at time t, and strike price K, is given by the following conditional expectation

$$C(t, S; K, T_1, \ldots, T_N) = e^{-r(T_N - t)} E^Q\left[\left(\frac{1}{N}\sum_{i=1}^{N} S(T_i) - K\right)^+ \Big| S(t) = S\right]. \tag{3}$$

This price can be represented, taking into account that it is a European call option on a forward freight agreement (FFA), by means of the following expectation (see [4])

$$C(t, S; K, T_1, \ldots, T_N) = e^{-r(T_N - t)} E^Q\left[(F(T_N, S; T_1, \ldots, T_N) - K)^+ \Big| S(t) = S\right], \tag{4}$$

where $F(t, S; T_1, \ldots, T_N) = E^Q\left[\frac{1}{N}\sum_{i=1}^N S(T_i) | S(t) = S\right]$ is the FFA price with settlement period $[T_1, T_N]$. Finally, note that $F(T_N, S; T_1, \ldots, T_N) = E^Q\left[\frac{1}{N}\sum_{i=1}^N S(T_i) | S(T_N) = S\right] = \frac{1}{N}\sum_{i=1}^N S(T_i)$.

3. Valuation Partial Differential Equation

As we have seen in previous sections, freight options are arithmetic Asian-style options, where the average is calculated over a fixed settlement period. Even though in the standard Asian options the settlement period is the total period until maturity, in freight options it is a fixed period close to maturity.

With respect to the standard Asian options, geometric ones usually have an exact pricing formula, however, for arithmetic Asian options such a price does not exist. In the literature, this fact has led to use different methodologies for acceptable and tractable valuation: Monte Carlo simulation approach (see [12,13]) and numerical methods for the PDE provided in [14], as in [15,16], where the spot freight rate follows a geometric process. Moreover, when the arithmetic average is calculated on a fixed period lower than in the standard Asian options, there is not a valuation equation for pricing these freight options.

Therefore, Equation (3) is, nowadays, the main available method to price this kind of derivatives. Unfortunately, in general, it is not an easily manageable form for the empirical application. In order to provide a new framework that allows us to price the freight options in a different way, here we develop a PDE for pricing freight options when the spot freight follows a general diffusion stochastic process. To this end, we will make a similar reasoning for pricing standard Asian options, as in [14], but we need to incorporate a new variable, the delayed spot freight rate. Moreover, when the spot freight rate follows a geometric process, we obtain a partial solution to this PDE.

First, we consider a settlement period $[T_1, T_N]$ such that $d = T_N - T_1$ is a fixed time span, for example, one month. Then, we introduce a continuous version of the average of the spot price, for $t \leq T_N$, as the process $A(t)$:

$$A(t) = \begin{cases} \int_0^t S(z)\,dz, & \text{if } 0 \leq t \leq d, \\ \int_{t-d}^t S(z)\,dz, & \text{if } t > d. \end{cases} \tag{5}$$

We write Equation (5) in differential form and obtain the following stochastic delay differential equation

$$dA(t) = \begin{cases} S(t)\,dt, & \text{if } 0 \leq t \leq d, \\ (S(t) - S(t-d))\,dt, & \text{if } t > d, \end{cases} \quad (6)$$

In order to obtain the equation that verifies the freight option price, we introduce a new variable which is a delay of the spot freight rate along a time period d. We denote this delayed spot freight rate as the new variable

$$X(t) = \begin{cases} S(0), & \text{if } 0 \leq t \leq d, \\ S(t-d), & \text{if } t > d. \end{cases}$$

Therefore, we can rewrite Equation (6) as

$$dA(t) = \begin{cases} S(t)\,dt, & \text{if } 0 \leq t \leq d, \\ (S(t) - X(t))\,dt, & \text{if } t > d. \end{cases}$$

Then, the process $A(t)$ depends on the spot freight rate and on its delay value X as a new variable. In this case, we can approximate the average value of the spot freight rate in the discrete Equation (2), by means of Equation (5), in a continuous way as

$$C(T_N, S, X, A; K, T_1, \ldots, T_N) = \left(\frac{1}{d}A(T_N) - K\right)^+, \quad (7)$$

and the expectation in Equation (3) as

$$C(t, S, X, A; K, T_1, \ldots, T_N) = e^{-r(T_N - t)} E^Q\left[\left(\frac{1}{d}A(T_N) - K\right)^+ \Big| S(t) = S, X(t) = X, A(t) = A\right]. \quad (8)$$

The following theorem provides a PDE satisfied by the freight call option price.

Theorem 1. *The freight call option price function $C(t, S, X, A; K, T_1, \ldots, T_N)$ in Equation (8) satisfies, when $d < t < T_N$, the following PDE*

$$C_t + \mu(S)C_S + \mu(X)C_X + (S-X)C_A + \frac{1}{2}\sigma^2(S)C_{SS} + \frac{1}{2}\sigma^2(X)C_{XX} - rC = 0, \quad (9)$$
$$S > 0, \quad X > 0, \quad A > 0.$$

However, when $0 < t < d$, the function C in Equation (8) verifies the PDE

$$C_t + \mu(S)C_S + SC_A + \frac{1}{2}\sigma^2(S)C_{SS} - rC = 0, \quad (10)$$
$$S > 0, \quad X > 0, \quad A > 0.$$

Proof of Theorem 1. Applying arbitrage arguments in the market, the discounted freight option price is a martingale under the risk-neutral measure Q, see [10]. That is,

$$E^Q\left[D(T_N)C(T_N, S, X, A; K, T_1, \ldots, T_N) | S(t) = S, X(t) = X, A(t) = A\right]$$
$$= D(t)C(t, S, X, A; K, T_1, \ldots, T_N).$$

Then, in the development of $d(D(t)C(t, S, X, A; K, T_1, \ldots, T_N))$, the dt term must be zero.

Note that $dSdS = \sigma^2(S)dt$ and

$$dXdX = \begin{cases} 0, & \text{if } 0 < t < d, \\ \sigma^2(X)\,dt, & \text{if } t > d. \end{cases}$$

Moreover, $dSdX = 0$, because $dW(t)dW(t-d) = 0$, and $dAdA = dSdA = dXdA = 0$. Therefore, by means of Ito Lemma, for $d < t < T_N$, we obtain

$$d(e^{-rt}C) = e^{-rt}\left(-rC + C_t + \mu(S)C_S + \mu(X)C_X + (S-X)C_A + \tfrac{1}{2}\sigma^2(S)C_{SS} + \tfrac{1}{2}\sigma^2(X)C_{XX}\right)dt \\ + e^{-rt}\left(C_S\sigma(S)dW(t) + C_X\sigma(X)dW(t-d)\right), \tag{11}$$

and for $0 < t < d$,

$$d(e^{-rt}C) = e^{-rt}\left(-rC + C_t + \mu(S)C_S + SC_A + \tfrac{1}{2}\sigma^2(S)C_{SS}\right)dt + e^{-rt}C_S\sigma(S)dW(t). \tag{12}$$

Finally, the vanishing of the dt terms in Equations (11) and (12) leads to Equations (9) and (10), respectively. □

Remark 1. *This result allows us to address the valuation problem of freight options in a new way: We obtain a pure final value problem associated to a PDE whose solution gives the freight option price. However, it is very difficult to solve this problem, except in some particular cases. Next, we will consider one of these situations.*

In the freight options literature, some stochastic processes are commonly used to describe the dynamic of the spot freight rate. In particular, it is usual to consider a geometric process where the functions in Equation (1) are $\mu(S) = \mu S$ and $\sigma(S) = \sigma S$, with constants μ and σ. In such a case, in the literature there exist some techniques to approximate the freight option prices although none of them are exact solutions. However, in a similar way to [14], here we value the option on the FFA when the average of the spot freight verifies $A \geq dK$, by solving the PDEs Equations (9) and (10) in Theorem 1.

Proposition 1. *Let $\mu(S) = \mu S$ and $\sigma(S) = \sigma S$ be the drift and volatility of the process (Equation (1)), respectively, with μ and σ constants. Then, the following function is solution to the PDEs, seen in Equations (9) and (10) and verifies the final condition of Equation (7) when $A \geq dK$:*

$\widetilde{C}(t, S, X, A; K, T_1, \ldots, T_N) =$

$$\begin{cases} \left(\tfrac{1}{d}A - K\right)e^{-r(T_N - t)} + \dfrac{e^{-r(T_N - t)}}{d\mu}\left(S(e^{\mu(T_N - t)} - 1) - X(e^{\mu(T_N - d)} - 1)\right), & 0 \leq t \leq d, \\[2mm] \left(\tfrac{1}{d}A - K\right)e^{-r(T_N - t)} + \dfrac{e^{-r(T_N - t)}}{d\mu}(S - X)(e^{\mu(T_N - t)} - 1), & d \leq t \leq T_N. \end{cases} \tag{13}$$

Proof of Proposition 1. First of all, we change the time variable by considering $\tau = T_N - t$. Then, from Equations (7) and (9) we have the initial value problem

$$C_\tau = \mu S C_S + \mu X C_X + (S-X)C_A + \tfrac{1}{2}\sigma^2 S^2 C_{SS} + \tfrac{1}{2}\sigma^2 X^2 C_{XX} - rC, \quad 0 < \tau < T_N - d, \tag{14}$$

$$C(0, S, X, A; K) = \left(\tfrac{1}{d}A - K\right)^+. \tag{15}$$

For $A \geq dK$, as in [14], we look for a linear solution to this problem as:

$$C(\tau, S, X, A; K) = \left(\frac{1}{d}A - K\right) B_1(\tau) + (S - X) B_2(\tau), \quad 0 \leq \tau \leq T_N - d, \tag{16}$$

where B_1 and B_2 are functions depending only of time. Replacing Equation (16) in the PDE Equation (14), we obtain that the functions B_1 and B_2 must verify the following system of ordinary differential equations

$$B_1'(\tau) = -r B_1(\tau),$$
$$B_2'(\tau) = (\mu - r) B_2(\tau) + \frac{1}{d} B_1(\tau).$$

From Equation (15), we get the initial conditions $B_1(0) = 1$ and $B_2(0) = 0$. Solving this system, we obtain the solution to the problem Equations (14) and (15)

$$C(\tau, S, X, A; K) = \left(\frac{1}{d}A - K\right) e^{-r\tau} + (S - X) \frac{e^{-r\tau}}{d\mu} (e^{\mu\tau} - 1), \quad 0 \leq \tau \leq T_N - d. \tag{17}$$

Now, the same change of variable τ in Equation (10), and the value of Equation (17) in $T_N - d$, provide the initial value problem

$$C_\tau = \mu S C_S + S C_A + \frac{1}{2}\sigma^2 S^2 - rC, \quad T_N - d < \tau < T_N, \tag{18}$$

$$C(T_N - d, S, X, A; K) = \left(\frac{1}{d}A - K\right) e^{-r(T_N - d)} + (S - X) \frac{e^{-r(T_N - d)}}{d\mu} (e^{\mu(T_N - d)} - 1). \tag{19}$$

Again, we look for a linear solution as

$$C(\tau, S, X, A; K) = \left(\frac{1}{d}A - K\right) A_1(\tau) + S A_2(\tau) + X A_3(\tau), \quad T_N - d \leq \tau \leq T_N, \tag{20}$$

where A_1, A_2 and A_3 are functions of time.

If we replace Equation (20) into the PDE Equation (18) we obtain that A_1, A_2 and A_3 verify the system of ordinary differential equations:

$$\begin{aligned} A_1'(\tau) &= -r A_1(\tau), \\ A_2'(\tau) &= (\mu - r) A_2(\tau) + \frac{1}{d} A_1(\tau), \\ A_3'(\tau) &= -r A_3(\tau), \end{aligned} \tag{21}$$

and from Equation (19) we derive the initial conditions

$$A_1(T_N - d) = e^{-r(T_N - d)},$$
$$A_2(T_N - d) = \frac{e^{-r(T_N - d)}}{d\mu} (e^{\mu(T_N - d)} - 1),$$
$$A_3(T_N - d) = -\frac{e^{-r(T_N - d)}}{d\mu} (e^{\mu(T_N - d)} - 1).$$

Solving the system Equation (21) with the previous initial conditions, we obtain the solution

$$C(\tau, S, X, A; K) = \left(\frac{1}{d}A - K\right) e^{-r\tau} + S \frac{e^{-r\tau}}{d\mu} (e^{\mu\tau} - 1) - X \frac{e^{-r\tau}}{d\mu} (e^{\mu(T_N - d)} - 1), \quad T_N - d \leq \tau \leq T_N.$$

Finally, if we return to the original time variable t, we obtain the expression in Equation (13) for $\tilde{C}$ which provides the call freight option price when $A \geq dK$. □

Remark 2. *Note that the solution that provides Equation (13) is only valid for $A \geq dK$. Unfortunately, for other values of the continuous average of the spot rate we do not have an explicit expression for the freight call option price. Therefore, even in this simple case, the partial solution to the PDE that we get is not sufficiently to price the freight option. However, it could be useful for the numerical solution of the problem, as we remark in a later section.*

Remark 3. *Although knowing the PDE problem previously described is not sufficient, in general, to get the exact price of the option, we could use numerical methods in order to approximate its solution. However, this is a very hard problem. On the one hand, the PDE involves four independents variables: the time, the spot rate, its delay and the average of the spot rate in the settlement period. Then, it is necessary to design suitable specific numerical methods for this expensive multidimensional problem. On the other hand, the application of numerical methods for a pure final problem requires appropriate boundary conditions. In this sense, for the specific stochastic processes considered in Proposition 1, we can use Equation (13) to obtain such boundary conditions (in a similar way to [14] for Asian options). In any case, the numerical approach of this problem is beyond the scope of this work.*

4. Lower and Upper Bounds for Freight Options

For arithmetic standard Asian options, several bounds have been presented in the literature (see for example [14,17]) and they are obtained in terms of European options. Therefore, assuming a specific dynamics of the spot rate in order to know its probability distribution, these bounds can be valuated. For example, in [18] and [7], an optimal lower bound for freight options is provided when the log spot freight price follows a jump-diffusion process with mean reversion.

In this section, we obtain lower and upper bounds for freight options but, unlike what happens in the Asian case, we do not assume a particular expression of the functions in the spot freight stochastic process.

In the following theorem, as in Equation (4), we consider that the freight option is a European option on an FFA.

Theorem 2. *Let $C(t, S; K, T_1, \ldots, T_N)$ be a freight call option price with settlement period $[T_1, T_N]$ and strike price K. Then,*

$$e^{-r(T_N-t)} \left(F(t,S;T_1,\ldots,T_N) - K\right)^+ \leq C(t,S;K,T_1,\ldots,T_N) \leq \frac{1}{N} \sum_{i=1}^N e^{-r(T_N-T_i)} C_E(t,S;K,T_i), \quad (22)$$

where $F(t, S; T_1, \ldots, T_N)$ is an FFA with settlement period $[T_1, T_N]$, and $C_E(t, S; K, T_i)$ is a European plain vanilla call option with maturity T_i.

Proof of Theorem 2. First of all, note that for a convex function ϕ, $E[\phi(X)] \geq \phi(E[X])$. Therefore, starting with Equation (3) for the freight call option price and taking into account that the maximum function is convex, then

$$C(t,S;K,T_1,\ldots,T_N) = e^{-r(T_N-t)} E^Q \left[\left(\frac{1}{N} \sum_{i=1}^N S(T_i) - K \right)^+ \Big| S(t) = S \right]$$

$$\geq e^{-r(T_N-t)} \left(E^Q \left[\frac{1}{N} \sum_{i=1}^N S(T_i) - K \Big| S(t) = S \right] \right)^+$$

$$= e^{-r(T_N-t)} \left(E^Q \left[\frac{1}{N} \sum_{i=1}^N S(T_i) \Big| S(t) = S \right] - K \right)^+$$

$$= e^{-r(T_N-t)} \left(F(t,S;T_1,\ldots,T_N) - K \right)^+,$$

arriving at the lower bound in Equation (22) which depends on the FFA price $F(t, S; T_1, \ldots, T_N)$.

In order to deduce the upper bound, we use the following relation

$$\left(\sum_{i=1}^{N} a_i\right)^+ \leq \sum_{i=1}^{N} (a_i)^+,$$

which is satisfied for every collection of real numbers $\{a_i\}_{i=1}^N$.

If we apply this relation to the option price formula (Equation (3)), we obtain

$$C(t, S; K, T_1, \ldots, T_N) = e^{-r(T_N-t)} E^Q \left[\left(\frac{1}{N} \sum_{i=1}^{N} S(T_i) - K\right)^+ \bigg| S(t) = S \right]$$

$$= e^{-r(T_N-t)} \frac{1}{N} E^Q \left[\left(\sum_{i=1}^{N} S(T_i) - NK\right)^+ \bigg| S(t) = S \right]$$

$$= e^{-r(T_N-t)} \frac{1}{N} E^Q \left[\left(\sum_{i=1}^{N} (S(T_i) - K)\right)^+ \bigg| S(t) = S \right]$$

$$\leq e^{-r(T_N-t)} \frac{1}{N} \sum_{i=1}^{N} E^Q \left[(S(T_i) - K)^+ \bigg| S(t) = S \right]$$

$$= \frac{1}{N} \sum_{i=1}^{N} e^{-r(T_N-t)} e^{r(T_i-t)} C_E(t, S; K, T_i)$$

$$= \frac{1}{N} \sum_{i=1}^{N} e^{-r(T_N-T_i)} C_E(t, S; K, T_i).$$

In this case, we obtain the upper bound in Equation (22) which depends on the European plain vanilla call options on the spot freight rate $C_E(t, S; K, T_i)$, with maturities at the different dates of the settlement period, T_i, $i = 1, \ldots, N$. □

Remark 4. *As we mentioned in the previous section, pricing the freight call option (Equation (3)) is a complex task. However, its lower and upper bounds, presented in Equation (22), are easier to obtain: FFA and European vanilla option prices with several maturities are required. Therefore, the values of these bounds can be used as an estimation of the window where the freight call option price lies.*

5. Empirical Application

In this section, we analyze the accuracy of the bounds obtained in Section 4. To this end, we consider that the spot freight rate follows a geometric process, which is widely used in the literature (as, for example, in [4]). That is, we assume that the spot freight rate follows a geometric stochastic process

$$dS(t) = \mu S(t) dt + \sigma S(t) dW(t). \tag{23}$$

We estimated the parameters in Equation (23) using Baltic Dry Index data from 2013 to 2019. This index, daily issued by the London-based Baltic Exchange, is mostly used in the freight market. As the spot freight rate follows a geometric Brownian process, we use maximum likelihood obtaining the values $\mu = 0.0041$ and $\sigma = 0.3738$.

Assuming that the market price of risk is $\lambda = 0$, then, the drift under the risk-neutral measure is equal to the drift under physical measure

$$\mu S(t) - \lambda = \mu S(t).$$

The bounds in Theorem 2 are obtained in the following way. The lower bound is computed using the FFA price obtained by [4] for a geometric Brownian motion. As far as the upper bound is concerned, the prices of the European plain vanilla call options on the spot freight rate are obtained in a similar way that in [19], but considering that $\lambda = 0$.

In order to compare the bounds in Equation (22) with the freight option price, we approximate the latter using the Monte Carlo simulation technique, which has been proved to be a flexible and handy method to price options (see, for example, [13]). We approximate the expectation in Equation (3) using a daily time step ($\Delta t = \frac{1}{252}$) and the previously established parameters. We generated 100,000 paths and consider that the settlement period is 1 month and the interest rate is 0.5%. We assumed that the spot freight rate is $S_0 = 1034.6$, which is the average of the Baltic Dry Index from January 2013 to January 2019, and different strike prices from 70% to 130% of this spot freight rate. In order to increase the precision of this technique, we used the antithetic variable method as a variance reduction technique, see [13].

Tables 1 and 2 show several option prices and their corresponding bounds for different maturities (1 and 3 months, respectively) and strike prices (as percentages of the spot price). Both tables confirm the validity of the bounds in Equation (22).

We conclude that the window defined by the bounds, when the maturity is 1 month, is narrower than the one obtained with a maturity of 3 months. In both cases the maximum width of the window is for options at the money (30.56 monetary units for 1 month and 70.05 for 3 months). Moreover, around the spot price the upper bound is closer to the option price than the lower bound. This fact can be observed clearly in Figure 1 that plots the option prices (solid line) and their corresponding lower and upper bounds (dotted and dashed lines, respectively) for several strike prices with maturities of 1, 3, 6 and 12 months. Note that, the higher the maturity the wider the window but, in all cases the behavior of the upper bound fits the option price better than the lower bound.

Finally, Table 3 shows the differences, absolute and relative, between each bound and the freight option price for different values of the strike price, and with a maturity of 3 months. The last row provides the mean of the differences presented in the previous rows. As we can see, when the option is out of the money, the differences in both bounds are higher than when the option is in the money. In fact, these differences reach a maximum when the option is at the money. If we compare the mean of the differences, we observe that the upper bound is much more accurate than the lower bound. Therefore, in this case the upper bound is a good estimation of the option price.

Table 1. European freight call option prices with a maturity of 1 month, for several strikes and their corresponding lower and upper bounds.

Strike	LB	Option Price	UB
70%	310.45	310.44	310.45
75%	258.74	258.73	258.76
80%	207.03	207.03	207.19
85%	155.32	155.42	156.20
90%	103.61	104.86	107.22
95%	51.90	59.63	63.50
100%	0.18	26.68	30.74
105%	0	8.90	12.92
110%	0	2.28	5.05
115%	0	0.42	1.85
120%	0	0.06	0.63
125%	0	0.01	0.21
130%	0	0	0.06

Table 2. European freight call option prices with a maturity of 3 months, for several strikes and their corresponding lower and upper bounds.

Strike	LB	Option Price	UB
70%	310.90	311.64	311.93
75%	259.23	261.69	262.19
80%	207.56	213.70	214.69
85%	155.83	169.37	170.73
90%	104.22	129.67	131.58
95%	52.56	95.78	98.19
100%	0.89	68.69	70.94
105%	0	47.48	49.69
110%	0	31.66	33.79
115%	0	20.94	22.36
120%	0	12.87	14.43
125%	0	8.11	9.10
130%	0	4.86	5.63

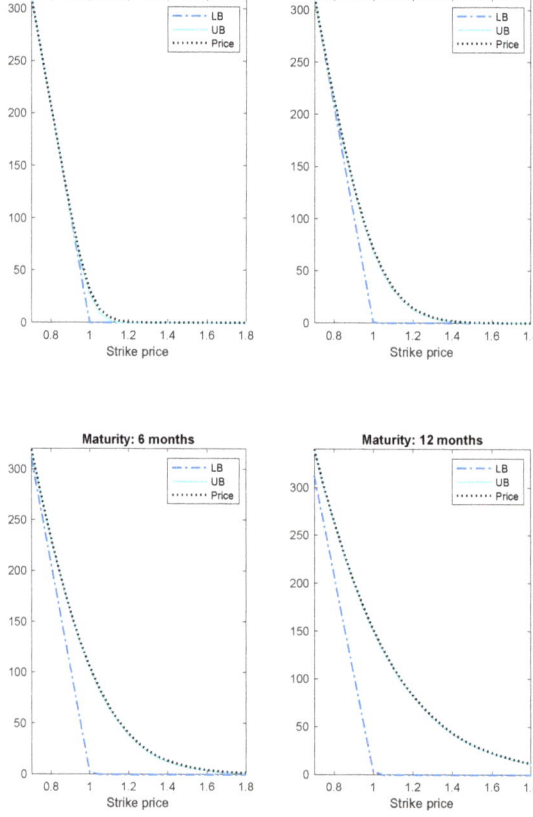

Figure 1. The lower and upper bounds and the option prices according to the strike prices. Maturities: 1, 3, 6 and 12 months.

Table 3. Absolute and relative differences between the freight call option prices with a maturity of 3 months and their lower and upper bounds, for several strike prices.

Strike	Price-LB	UB-Price	(Price-LB)/Price	(UB-Price)/Price
70%	0.7437	0.2871	0.0024	0.0009
75%	2.4605	0.4989	0.0094	0.0019
80%	6.1367	0.9898	0.0287	0.0046
85%	13.4723	1.3604	0.0795	0.0080
90%	25.4416	1.9107	0.1962	0.0147
95%	43.2186	2.4072	0.4512	0.0251
100%	67.7966	2.2578	0.9871	0.0329
105%	47.4780	2.2100	1.0000	0.0465
110%	31.6570	2.1323	1.0000	0.0673
115%	20.9388	1.4180	1.0000	0.0677
120%	12.8660	1.5607	1.0000	0.1213
125%	8.1064	0.9952	1.0000	0.1228
130%	4.8566	0.7709	1.0000	0.1587
Mean	21.9364	1.4461	0.5965	0.0517

6. Discussion and Conclusions

The freight market is a relatively new market but a very important one nowadays. Therefore, more scientific research is necessary in this area. In the freight market, in order to avoid price manipulations by large participants, setting is against the average value of a freight index. As a consequence, freight derivatives have, in general, average-style payoffs which makes them more difficult to price.

In the freight markets literature, we can find few models and methods to price this kind of derivatives. More precisely, to this end, it is usual to consider very specific parametric models. Here we propose new strategies that open a path to price these freight derivatives with general models, which will facilitate its application in the market by practitioners.

The contribution of this paper is twofold. On the one hand, we prove that the freight option price verifies PDEs with three independent state variables: the spot rate, its delay and the average of the spot rate in the settlement period. This result is notable because it offers a new approach to deal with the freight option valuation problem. Moreover, it opens the door to apply numerical methods for pricing freight options. On the other hand, we find and prove some lower and upper bounds for freight options which allow us to approximate its price. Finally, as an empirical application, we calculate these bounds using the Baltic Dry Index, issued by the Baltic Exchange in London, for freight options with different maturities. In such a case, we observe that the upper bound is close to the option price and then, it could be used as approximation to the price, especially for options in the money.

Author Contributions: All authors contributed equally to this article. All authors have read and agreed to the published version of the manuscript.

Funding: This research was funded in part by Consejería de Educación, Junta de Castilla y León, grant numbers VA148G18 and VA138G18, and by Ministerio de Economía y Competitividad, grant number MTM2017-85476-C2-P.

Conflicts of Interest: The authors declare no conflict of interest.

Abbreviations

The following abbreviations are used in this manuscript:

PDE   Partial Differential Equation
FFA   Forward Freight Agreement
LB   Lower Bound
UB   Upper Bound

References

1. Prokopczuk, M. Pricing and hedging in the freight futures market. *J. Fut. Mark.* **2011**, *31*, 440–464. [CrossRef]
2. Goulas, L.; Skiadopoulos, G. Are freight futures markets efficient? Evidence from IMAREX. *Marit. Policy Mag.* **2012**, *28*, 644–659. [CrossRef]
3. Kavussanos, M.G.; Visvikis, I.D. Shipping freight derivatives: A survey of recent evidence. *Marit. Policy Manag.* **2006**, *33*, 233–255. [CrossRef]
4. Koekebakker, S.; Roar, A.; Sodal, S. Pricing freight rate options. *Transp. Res. Part E Logist. Transp. Rev.* **2007**, *43*, 535–548. [CrossRef]
5. Tvedt, J. Valuation of a european futures option in the BIFFEX market. *J. Fut. Mark.* **1998**, *18*, 167–175. [CrossRef]
6. Nomikos, N.K.; Kyriakou, I.; Papapostolou, N.C.; Pouliasis, P.K. Freight options: Price modelling and empirical analysis. *Transp. Res. Part E Logist. Transp. Rev.* **2013**, *51*, 82–94. [CrossRef]
7. Kyriakou, I.; Pouliasis, P.K.; Papapostolou, N.C.; Nomikos, N.K. Income uncertainty and the decision to invest in bulk shipping. *Eur. Financ. Manag.* **2018**, *24*, 387–417. [CrossRef]
8. Ortiz, A.; Martinez, M. Comparison of European and Asian valuation of options with underlying average and stochastic interest rate by Monte Carlo simulation. *Ecorfan J.* **2014**, *5*, 2057–2072.
9. Protter, C. *Stochastic Integration and Differential Equations*; Springer: Berlin/Heidelberg, Germany, 2007.
10. Shreve, S.E. *Stochastic Calculus for Finance II: Continuous Time Models*; Springer: New York, NY, USA, 2004.
11. Oksendal, B. *Stochastic Differential Equations. An Introduction with Applications*; Springer: Berlin/Heidelberg, Germany, 2003.
12. Fusai, G.; Roncoroni, A. *Implementing Models in Quantitative Finance: Methods and Cases*; Springer: Berlin/Heidelberg, Germany, 2008.
13. Huynh, H.T.; Lai, V.S.; Soumaré, I. *Stochastic Simulation and Applications in Finance with MATLAB Programs*; John Wiley & Sons: Chichester, UK, 2008.
14. Kemna, A.G.Z.; Vorst, A.C.F. A pricing method for options based on average asset values. *J. Bank. Financ.* **1990**, *14*, 113–129. [CrossRef]
15. Rogers, L.C.G.; Shi, Z. The value of an Asian option. *J. Appl. Probab.* **1995**, *32*, 1077–1088. [CrossRef]
16. Zvan, R.; Forsyth, P.A.; Vetzal, K.R. Robust numerical methods for PDE models of Asian options. *J. Comp. Financ.* **1998**, *1*, 39–78. [CrossRef]
17. Vanmaele, M.; Deelstra, G.; Liinev, J.; Dhaene, J.; Goovaerts, M.J. Bounds for the price of discrete arithmetic Asian options. *J. Comput. Appl. Math.* **2006**, *185*, 51–90. [CrossRef]
18. Kyriakou, I.; Pouliasis, P.K.; Papapostolou, N.C.; Andriosopoulos, K. Freight derivatives pricing for decoupled mean-reverting diffusion and jumps. *Transp. Res. Part E Logist. Transp. Rev.* **2017**, *108*, 80–96. [CrossRef]
19. Black, F.; Scholes, M. Pricing of options and corporate liabilities. *J. Political Econ.* **1973**, *81*, 637–654. [CrossRef]

© 2020 by the authors. Licensee MDPI, Basel, Switzerland. This article is an open access article distributed under the terms and conditions of the Creative Commons Attribution (CC BY) license (http://creativecommons.org/licenses/by/4.0/).

Article

Stability Analysis of an Age-Structured SEIRS Model with Time Delay

Zhe Yin, Yongguang Yu * and Zhenzhen Lu

Department of mathematics, Beijing Jiaotong University, Beijing 100044, China; 17121567@bjtu.edu.cn (Z.Y.); 18121571@bjtu.edu.cn (Z.L.)
* Correspondence: ygyu@bjtu.edu.cn

Received: 20 February 2020; Accepted: 19 March 2020; Published: 23 March 2020

Abstract: This paper is concerned with the stability of an age-structured susceptible–exposed–infective–recovered–susceptible (SEIRS) model with time delay. Firstly, the traveling wave solution of system can be obtained by using the method of characteristic. The existence and uniqueness of the continuous traveling wave solution is investigated under some hypotheses. Moreover, the age-structured SEIRS system is reduced to the nonlinear autonomous system of delay ODE using some insignificant simplifications. It is studied that the dimensionless indexes for the existence of one disease-free equilibrium point and one endemic equilibrium point of the model. Furthermore, the local stability for the disease-free equilibrium point and the endemic equilibrium point of the infection-induced disease model is established. Finally, some numerical simulations were carried out to illustrate our theoretical results.

Keywords: SEIRS model; age structure; time delay; traveling wave solution; local asymptotic stability; Hopf bifurcation

1. Introduction

In recent years, the study of epidemiology has been a vital problem in ecology. The research of population dynamics has developed rapidly, and many mathematical models have been used to analyze various infectious diseases. Many results have been established in the stability analysis of different epidemic models. The first susceptible–infective–recovered (SIR) epidemic model about disease transmission was established by Kermack and McKendrick in 1927 [1]. Since then, the population dynamics of infectious diseases have attracted the attention of scientists. In 2012, the field of mathematical biology was expanded, particularly in the context of the spread of infectious diseases by Fred Brauer et al. [2]. Nowadays, there are some research work devoted to study the stability of steady states of the SIR, SIRS, SEIR, etc. models [3–7]. It is well known, in the spread of infectious diseases, some infective individuals of population are immune after being recovered (e.g., measles, smallpox, mumps, and others). Meanwhile, some recovered individuals have no immunity (e.g., AIDS, hyperthyroidism, lupus erythematosus, and others), who will return to the susceptible population and continue to be infected. In fact, the probability of becoming infected is different among different individuals, which may depend on the type of infectious diseases and the status of individuals. Therefore, it is necessary to discuss the SEIRS model, which can more clearly describe the spread of infectious diseases in real life.

Time delay is ubiquitous and can be applied in many epidemiology related studies [8,9]. For example, measles has an incubation period of 8–13 days and the incubation period of canine madness is a few months or several years after infection. Sharma et al. [10] developed a five compartmental infection model to describe the spread of avian influenza A (H7N9) virus with two discrete time delays. In addition, Xu et al. [11] analyzed the stability of a SIRS model with time delay. Similarly, Shu [12] and

De la Sena [13] discussed the stability of the SEIR epidemic models with distributed delay respectively. Actually, many authors, such as Cooke [14], Gao [15] and Wang [16], have studied various SEIRS models with time delay.

Besides, age structure is also an important consideration in infectious diseases modeling such as rubella, poliomyelitis, and pertussis, which are transmitted only among children, and venereal diseases, which are transmitted only among adults. Besides, tuberculosis virus carriers in the early incubation period have a higher risk of becoming infective individuals than ones in the late incubation period [17]. Age-structured models have been applied in the epidemic dynamics for decades. In 1986, the dynamics of structured populations was discussed by Metz et al. [18]. Then, the mathematical theory of the age-structured population dynamics was proposed by Iannelli [19]. Afterwards, more and more epidemic models with age structure were studied in [20–28]. Recently, a new age-structured malaria model incorporating the age of latent period and the age of prevention period was formulated by Guo et al. [29]. A new SIRS epidemic model with relapse and infection age on a scale-free network was introduced Huo et al. [30]. However, as far as we can tell, there have been no results on an age-structured SEIRS model with time delay.

The main aim of this paper is to study the stability of an age-structured SEIRS model with time delay. The well-known method of characteristics [25–28] for first-order hyperbolic equations is used to solve this epidemic model. The explicit traveling wave solution is calculated at the preceding moment of time and is described in integral form. Under some hypotheses, the existence and uniqueness of the continuous traveling wave solution of the age-structured SEIRS model is investigated. Moreover, an age-structured SEIRS model with time delay is reduced to the nonlinear ordinary differential equation under some insignificant simplifications. After that, the dimensionless indexes are derived for the existence of the disease-free equilibrium point and the endemic equilibrium point. The local asymptotic stability of the disease-free equilibrium point is studied. By using Hurwitz's criterion and Descartes' rule of signs, the local asymptotic stability of the endemic equilibrium point of system is obtained.

The rest of the paper is organized as follows. In Section 2, an age-structured SEIRS model with time delay is proposed. In Section 3, the traveling wave solution is obtained and some sufficient conditions are established to guarantee the existence and uniqueness of the solution. In Section 4, the stability and Hopf bifurcation analysis of the proposed model are discussed. In Section 5, numerical simulations are provided to illustrate the effectiveness of our main results. Finally, some conclusions are given in Section 6.

2. An Age-Structured SEIRS Model with Time Delay

Motivated by the referred works [28], we discuss two age stages of each subpopulation of an age-structured SEIRS model with time delay in this paper, which include immature stage and mature stage. In the immature stage, individuals can be born, grow up, die, or survive until the maximum age a_1, but, in this case, the individuals are not proliferate until the maximum age a_1. The age a_1 is considered to be the maximum age in the immature stage and it is also considered to be the initial age in the mature stage. In the mature stage, individuals have reached maturity $(a > a_1)$ and can grow up, proliferate, die, or survive to the maximum age A. Here, we consider susceptible, exposed, infective, and recovered individuals of two age stages in an age-structured SEIRS model. Let $S(a,t)$, $E(a,t)$, $I(a,t)$, and $R(a,t)$ be the distribution densities of susceptible, exposed, infective ,and recovered individuals of age x at time t, and they take place in the domain $\Omega = \{(a,t) : 0 \leq a \leq A, 0 \leq t \leq T\}$. The integrals $N_s(t) = \int_0^A S(a,t)da$, $N_e(t) = \int_0^A E(a,t)da$, $N_i(t) = \int_0^A I(a,t)da$, and $N_r(t) = \int_0^A R(a,t)da$ are considered as the number of susceptible, exposed, infective and recovered individuals, respectively. The total population $N(t)$ is: $N(t) = N_s(t) + N_e(t) + N_i(t) + N_r(t)$. What needs to be added is that $S(a,t)$, $E(a,t)$, $I(a,t)$, and $R(a,t)$ should belong to $L_1(\Omega)$ because we assume that the initial total population is limited. Then, the following differential equations with time delay are

$$\begin{cases} \dfrac{\partial S}{\partial t}+\dfrac{\partial S}{\partial a}=-\left(\hat{\alpha}_1(a,t)+\sigma\hat{\theta}_1(a,t)\right)S(a,t)-\left(\gamma_1(a,t-\tau)\int_0^A\beta(a,a',t-\tau)I(a',t-\tau)da'\right)S(a,t)\\ \quad+\rho(a,t)R(a,t),(a,t)\in\Omega,\\ \dfrac{\partial E}{\partial t}+\dfrac{\partial E}{\partial a}=-\left(\hat{\alpha}_2(a,t)+\sigma\hat{\theta}_2(a,t)\right)E(a,t)+\left(\gamma_1(a,t-\tau)\int_0^A\beta(a,a',t-\tau)I(a',t-\tau)da'\right)S(a,t)\\ \quad-\gamma_2(a,t)E(a,t),(a,t)\in\Omega,\\ \dfrac{\partial I}{\partial t}+\dfrac{\partial I}{\partial a}=-\left(\hat{\alpha}_3(a,t)+\sigma\hat{\theta}_3(a,t)\right)I(a,t)+\gamma_2(a,t)E(a,t)-\gamma_3(a,t)I(a,t),(a,t)\in\Omega,\\ \dfrac{\partial R}{\partial t}+\dfrac{\partial R}{\partial a}=-\left(\hat{\alpha}_4(a,t)+\sigma\hat{\theta}_4(a,t)\right)R(a,t)+\gamma_3(a,t)I(a,t)-\rho(a,t)R(a,t),(a,t)\in\Omega, \end{cases} \quad (1)$$

with initial conditions

$$\begin{cases} S(a,0)=S_0(a),a\in[0,A],\\ E(a,0)=E_0(a),a\in[0,A],\\ I(a,t)=I_0(a,t),a\in[0,A],t\in[-\tau,0],\\ R(a,0)=0,x\in[0,A], \end{cases} \quad (2)$$

and boundary conditions

$$\begin{cases} S(0,t)=\int_{a_1}^A\Big(\mu_1\theta_1(a,t)S(a,t)+\mu_2\theta_2(a,t)\left(1-p(a,t)\right)E(a,t)+\mu_3\theta_3(a,t)\left(1-q(a,t)\right)I(a,t)\\ \quad+\mu_4\theta_4(a,t)R(a,t)\Big)da,t\in(0,T),\\ E(0,t)=\mu_2\int_{a_1}^A p(a,t)\theta_2(a,t)E(a,t)da,t\in(0,T),\\ I(0,t)=\mu_3\int_{a_1}^A q(a,t)\theta_3(a,t)I(a,t)da,t\in(0,T),\\ R(0,t)=0,t\in(0,T), \end{cases} \quad (3)$$

where $\hat{\theta}_i(a,t)(i=1,2,3,4)$ denote fertility rates of females of each subpopulation of age a; $\hat{\theta}_i(a,t)=\theta_i(a,t)$, if $a\in[a_1,A]$; $\hat{\theta}_i(a,t)=0$, if $a\notin[a_1,A]$; $\hat{\alpha}_i(a,t)$ are natural death rates of each subpopulation of age a; σ is the provided parameter: $\sigma=1$ if the individuals die when they produce and $\sigma=0$ if the individuals continue to survive when they produce; $\gamma_1(a,t)$ is a transmission coefficient which describes the varying probability of infectiousness and it is related to a great many social, environmental, and epidemiological factors; $\beta(a,a',t)$ is the contact rate between infected population (age a') and susceptible population (age a) per unit time; τ ($\tau>0$) is a fixed incubation period of infection; $\gamma_2(a,t)$ is the conversion rate from exposed population to infected population of age a; $\gamma_3(a,t)$ is the recovery rate of age a; $\rho(a,t)$ is the conversion rate from recovered population losing immunity to the susceptible population of age a; $\mu_i(i=1,2,3,4)$ are reproductive rates of each subpopulation in the proliferating stage; $p(a,t)$ is part of exposed individuals which procreate new exposed individuals of age a; and, similarly, $q(a,t)$ is part of infective individuals which procreate new infective individuals of age a.

3. Traveling Wave Solution

In this section, we mainly use the method of characteristics [2,18,25–28] for first-order hyperbolic partial differential equations (Equation (1)). Then, the system in Equation (1) is reduced to nonlinear delayed integro-differential equations along the characteristics curve $a-t=constant$ [25,26]. Let $u=a-t$, the following system with time delay is obtained:

$$\begin{cases} S_t = -\left(\hat{\alpha}_1(u+t,t) + \sigma\hat{\theta}_1(u+t,t)\right)S(u+t,t) - \left(\gamma_1(u+t,t-\tau)\int_0^A \beta(a,a',t-\tau)I(a',t-\tau)da'\right) \\ \quad \times S(u+t,t) + \rho(u+t,t)R(u+t,t), \\ E_t = -\left(\hat{\alpha}_2(u+t,t) + \sigma\hat{\theta}_2(u+t,t)\right)E(u+t,t) + \left(\gamma_1(u+t,t-\tau)\int_0^A \beta(a,a',t-\tau)I(a',t-\tau)da'\right) \\ \quad \times S(u+t,t) - \gamma_2(u+t,t)E(u+t,t), \\ I_t = -\left(\hat{\alpha}_3(u+t,t) + \sigma\hat{\theta}_3(u+t,t)\right)I(u+t,t) + \gamma_2(u+t,t)E(u+t,t) - \gamma_3(u+t,t)I(u+t,t), \\ R_t = -\left(\hat{\alpha}_4(u+t,t) + \sigma\hat{\theta}_4(u+t,t)\right)R(u+t,t) + \gamma_3(u+t,t)I(u+t,t) - \rho(u+t,t)R(u+t,t), \end{cases} \quad (4)$$

with initial functions

$$\begin{cases} S(u,0) = S_0(u), \\ E(u,0) = E_0(u), \\ I(u+t,t) = I_0(u+t,t), t \in [-\tau,0], \\ R(u,0) = R_0(u) = 0. \end{cases} \quad (5)$$

In general, we divide time interval $[0, T]$ into K intervals $[t_{k-1}, t_k]$, where $k = 1, ..., K$, $t_0 = 0$, $t_K = T$, $t_k = kA$. Then, Ω can be grouped into two sets:

$$\Omega_k^{(1)} = \{(a,t) | t \in [(k-1)A, (a+(k-1)A)], a \in [0,A]\}, \quad (6)$$

$$\Omega_k^{(2)} = \{(a,t) | t \in [(a+(k-1)A), kA], a \in [0,A]\}, \quad (7)$$

$$\Omega = \cup_{k=1}^{K}(\Omega_k^{(1)} \cup \Omega_k^{(2)}). \quad (8)$$

We also need to define an auxiliary set for $k = 1, ..., K$:

$$\bar{\Omega}^{(k)} = \left\{[-u_n^{(k)}, -u_{n+1}^{(k)}] \mid u_n^{(k)} = na_1 + (k-1)A, n = 1, ..., N-1, u_N^{(k)} = kA\right\}, \quad (9)$$

where $N = [\frac{A}{a_1}] + 1, if \frac{A}{a_1} - [\frac{A}{a_1}] > 0; N = \frac{A}{a_1}, if \frac{A}{a_1} - [\frac{A}{a_1}] > 0$. $[x]$ denotes a whole part of real number x. The following hypotheses are given:

(H_1): $S_0(a)$, $E_0(a)$ and $I_0(a,0)$ are non-negative and continuous when $a \in [0, A]$; when $a \to A - 0$, $S_0(A) = 0$, $E_0(A) = 0$ and $I_0(A, 0) = 0$.

(H_2): $\hat{\alpha}_i(a,t), \theta_i(a,t), \gamma_i(a,t), \beta(a,t), \rho(a,t), p(a,t), q(a,t) \in C(\Omega)$.

(H_3): $\hat{\alpha}_i(a,t) > 0, \theta_i(a,t) \geq 0, \gamma_i(a,t) > 0, \beta(a,t) \geq 0$. In addition, $0 < \int_{a_1}^A \theta_i(a,t)da \leq 1$, $0 < \int_0^A \beta(a,t)dx \leq 1, 0 \leq \rho(a,t) \leq 1, 0 \leq p(a,t) \leq 1, 0 \leq q(a,t) \leq 1$.

(H_4): zero-order compatibility conditions are

$$\begin{cases} S_0(0) = \int_{a_1}^A \left(\mu_1\theta_1(a,0)S_0(a) + \mu_2\theta_2(a,0)(1-p(a,0))E_0(a) + \mu_3\theta_3(a,0)(1-q(a,0))I_0(a,0)\right)da, \\ E_0(0) = \mu_2 \int_{a_1}^A \theta_2(a,0)p(a,0)E_0(a)da, \\ I_0(0,0) = \mu_3 \int_{a_1}^A \theta_3(a,0)q(a,0)I_0(a,0)da, \\ R_0(0) = 0. \end{cases} \quad (10)$$

For convenience, we assume Hypotheses (H_1)–(H_4) are satisfied.

Then, we solve the system in Equation (4) by using the well-known steps method [8,9]. For the first step $h = 1$, i.e., for the first time interval $t \in [0, \tau]$, the solution to the initial value problem in Equation (4) can be obtained according to the known values of function $I(x, t - \tau)$. Repeating the pervious step for $h = 2, 3, 4, ..., H$, for the time interval $t \in [(h-1)\tau, h\tau]$, we can obtain the solution to the initial value problem in Equation (4) for the whole time interval $t \in [0, T]$:

$$\begin{cases} S_1(u,t) = S_0(u)W_1(u,0,t,\tau) + \int_0^t W_1(u,\phi,t,\tau)\rho(u+\phi,\phi)R(u+\phi,\phi)d\phi, t \in [0,\tau], \\ S_h(u,t) = S_{h-1}(u)W_1(u,(h-1)\tau,t,\tau) + \int_{(h-1)\tau}^t W_1(u,\phi,t,\tau)\rho(u+\phi,\phi)R(u+\phi,\phi)d\phi \\ \quad = S_0(u)W_1(v,0,t,\tau) + \int_0^t W_1(u,\phi,t,\tau)\rho(v+\phi,\phi)R(u+\phi,\phi)d\phi, t \in [(h-1)\tau, h\tau], \\ W_1(u,t_0,t,\tau) = exp\left(-\int_{t_0}^t (\alpha_1(u+\phi,\phi) + \gamma_1(u+\phi,\phi-\tau)D(u+\phi,\phi-\tau))d\phi\right), \end{cases} \quad (11)$$

$$\begin{cases} E_1(u,t) = E_0(u)W_2(u,0,t) + \int_0^t W_2(u,\phi,t)\gamma_1(u+\phi,\phi-\tau)S(u+\phi,\phi)D(u+\phi,\phi-\tau)d\phi, \\ t \in [0,\tau], \\ E_h(u,t) = E_{h-1}(u)W_2(u,(h-1)\tau,t) + \int_{(h-1)\tau}^t W_2(u,\phi,t)\gamma_1(u+\phi,\phi-\tau)S(u+\phi,\phi) \\ \times D(u+\phi,\phi-\tau)d\phi = E_0(u)W_2(u,0,t) + \int_0^t W_2(u,\phi,t)\gamma_1(u+\phi,\phi-\tau) \\ \times S(u+\phi,\phi)D(u+\phi,\phi-\tau)d\phi, t \in [(h-1)\tau, h\tau], \\ W_2(u,t_0,t) = exp\left(-\int_{t_0}^t (\alpha_2(u+\phi,\phi) + \gamma_2(u+\phi,\phi))d\phi\right), \end{cases} \quad (12)$$

$$\begin{cases} I_1(u,t) = I_0(u,0)W_3(u,0,t) + \int_0^t W_3(u,\phi,t)\gamma_2(u+\phi,\phi-\tau)E(u+\phi,\phi)d\phi, t \in [0,\tau], \\ I_h(u,t) = I_{h-1}(u,(h-1)\tau)W_3(u,(h-1)\tau,t) + \int_{(h-1)\tau}^t W_3(u,\phi,t)\gamma_2(u+\phi,\phi-\tau) \\ \times E(u+\phi,\phi)d\phi = I_0(u,0)W_3(u,0,t) + \int_0^t W_3(u,\phi,t)\gamma_2(u+\phi,\phi-\tau) \\ \times E(u+\phi,\phi)d\phi, t \in [(h-1)\tau, h\tau], \\ W_3(u,t_0,t) = exp\left(-\int_{t_0}^t (\alpha_3(u+\phi,\phi) + \gamma_3(u+\phi,\phi))d\phi\right), \end{cases} \quad (13)$$

$$\begin{cases} R_1(u,t) = \int_0^t W_4(u,\phi,t)\gamma_3(u+\phi,\phi-\tau)I(u+\phi,\phi)d\phi, t \in [0,\tau] \\ R_h(u,t) = R_{h-1}(u,(h-1)\tau)W_4(u,(h-1)\tau),t) + \int_{(h-1)\tau}^t W_4(u,\phi,t)\gamma_3(u+\phi,\phi-\tau) \\ \times I(u+\phi,\phi)d\phi = \int_0^t W_4(u,\phi,t)\gamma_3(u+\phi,\phi-\tau)I(u+\phi,\phi)d\phi, t \in [(h-1)\tau, h\tau], \\ W_4(u,t_0,t) = exp\left(-\int_{t_0}^t (\alpha_4(u+\phi,\phi) + \rho(u+\phi,\phi))d\phi\right), \end{cases} \quad (14)$$

where $\alpha_i(a,t) = \hat{\alpha}_i(a,t) + \sigma\hat{\theta}_i(a,t)$, $i = 1, 2, 3, 4$, $D(u+\phi, \phi-\tau) = \int_0^A \beta(u+\phi, \eta, \phi-\tau)I(\eta, \phi-\tau)d\eta$.

The exact solution to Equations (11)–(14) of the system in Equation (4) can be expressed in terms of the original variables. Let $k = 1$ and $u = a - t$; we have the solution of the system in Equation (1) in the sets $\Omega_k^{(1)}$ and $\Omega_k^{(2)}$. Repeating this process for $k = 2, 3, ...K$, the form of explicit traveling wave solution, i.e., numerical solution of the system in Equation (1), can be obtained in all domains $\Omega_k^{(1)}$ and $\Omega_k^{(2)}$ for the values of $k = 1, ..., K$:

$$(S(a,t), E(a,t), I(a,t), R(a,t)) = \begin{cases} \left.\begin{array}{l} S_h^{(k-1)}(u,t) = S_h^{(k-1)}(a-t,t), \\ E_h^{(k-1)}(u,t) = E_h^{(k-1)}(a-t,t), \\ I_h^{(k-1)}(u,t) = I_h^{(k-1)}(a-t,t), \\ R_h^{(k-1)}(u,t) = R_h^{(k-1)}(a-t,t), \end{array}\right\} if (a,t) \in \Omega_k^{(1)}, \\ \left.\begin{array}{l} S_h^{(k)}(u,t) = S_h^{(k)}(a-t,t), \\ E_h^{(k)}(u,t) = E_h^{(k)}(a-t,t), \\ I_h^{(k)}(u,t) = I_h^{(k)}(a-t,t), \\ R_h^{(k)}(u,t) = R_h^{(k)}(a-t,t), \end{array}\right\} if (a,t) \in \Omega_k^{(2)}, \end{cases} \quad (15)$$

$$\begin{cases} S^{(0)}(u,t) = S_0(u)W_1(u,0,t,\tau) + \int_0^t W_1(u,\phi,t,\tau)\rho(u+\phi,\phi)R^{(0)}(u+\phi,\phi)d\phi, \\ S^{(k)}(u,t) = F^{(k)}(u)W_1(u,-u,t,\tau) + \int_{-u}^t W_1(u,\phi,t,\tau)\rho(u+\phi,\phi)R^{(k)}(u+\phi,\phi)d\phi, \end{cases} \quad (16)$$

$$\begin{cases} E^{(0)}(u,t) = E_0(u)W_2(u,0,t) + \int_0^t W_2(u,\phi,t)\gamma_1(u+\phi,\phi-\tau)S^{(0)}(u+\phi,\phi)D(u+\phi,\phi-\tau)d\phi, \\ E^{(k)}(u,t) = G^{(k)}(u)W_2(u,-u,t) + \int_{-u}^t W_2(u,\phi,t)\gamma_1(u+\phi,\phi-\tau)S^{(k)}(u+\phi,\phi)D(u+\phi,\phi-\tau)d\phi, \end{cases} \quad (17)$$

$$\begin{cases} I^{(0)}(u,t) = I_0(u,0)W_3(u,0,t) + \int_0^t W_3(u,\phi,t)\gamma_2(u+\phi,\phi-\tau)E^{(0)}(u+\phi,\phi)d\phi, \\ I^{(k)}(u,t) = Q^{(k)}(u)W_3(u,-u,t) + \int_{-u}^t W_3(u,\phi,t)\gamma_2(u+\phi,\phi-\tau)E^{(k)}(u+\phi,\phi)d\phi, \end{cases} \quad (18)$$

$$\begin{cases} R^{(0)}(u,t) = \int_0^t W_4(u,\phi,t)\gamma_3(u+\phi,\phi-\tau)I^{(0)}(u+\phi,\phi)d\phi, \\ R^{(k)}(u,t) = \int_{-u}^t W_4(u,\phi,t)\gamma_3(u+\phi,\phi-\tau)I^{(k)}(u+\phi,\phi)d\phi, \end{cases} \quad (19)$$

where $W_1(u,t_0,t,\tau)$, $W_2(u,t_0,t)$, $W_3(u,t_0,t)$, and $W_4(u,t_0,t)$ are shown in Equations (11)–(14), respectively. $F^{(k)}(u)$, $G^{(k)}(u)$, and $Q^{(k)}(u)$ are given by defining functions $F_n^{(k)}(u)$, $G_n^{(k)}(u)$, $Q_n^{(k)}(u)$, $S_n^{(k)}(u)$, $E_n^{(k)}(u)$, $I_n^{(k)}(u)$, and $R_n^{(k)}(u)$, $k = 1, ..., K$:

$$(F^{(k)}(u)), G^{(k)}(u), Q^{(k)}(u)) = (F_n^{(k)}(u)), G_n^{(k)}(u), Q_n^{(k)}(u)), u \in [-u_n^{(k)}, -u_{n-1}^{(k)}], n = 1, ..., N_\tau,$$

thus we have

$$\begin{cases} S_n^{(k)}(u,t) = F_n^{(k)}(u)W_1(u,-u,t,\tau) + \int_{-u}^t W_1(u,\phi,t,\tau)\rho(u+\phi,\phi)R_n^{(k)}(u+\phi,\phi)d\phi, \\ E_n^{(k)}(u,t) = G_n^{(k)}(u)W_2(u,-u,t) + \int_{-u}^t W_2(u,\phi,t)\gamma_1(u+\phi,\phi-\tau)S_n^{(k)}(u+\phi,\phi)D(u+\phi,\phi-\tau)d\phi, \\ I_n^{(k)}(u,t) = Q_n^{(k)}(u)W_3(u,-u,t) + \int_{-u}^t W_3(u,\phi,t)\gamma_2(u+\phi,\phi-\tau)E_n^{(k)}(u+\phi,\phi)d\phi, \\ R_n^{(k)}(u,t) = \int_{-u}^t W_4(u,\phi,t)\gamma_3(u+\phi,\phi-\tau)I_n^{(k)}(u+\phi,\phi)d\phi, \end{cases}$$

where $n = 1, \ldots, N_\tau - 1$ and functions $F_n^{(k)}(u)$, $G_n^{(k)}(u)$, and $Q_n^{(k)}(u)$ can be defined according to the recurrent algorithm as follows:

$$F_1^{(k)}(v) = \int_{a_1+v}^{A+v} [\mu_1 \theta_1(u-v,-v) S^{(k-1)}(u,-v) + \mu_2 \theta_2(u-v,-v)(1-p(u-v,-v)) E^{(k-1)}(u,-v)$$
$$+ \mu_3 \theta_3(u-v,-v)(1-q(u-v,-v)) I^{(k-1)}(u,-v) + \mu_4 \theta_4(u-v,-v) R^{(k-1)}(u,-v)] du,$$
$$v \in [-u_1^{(k)}, -u_0^{(k)}],$$

$$G_1^{(k)}(v) = \mu_2 \int_{a_1+v}^{A+v} p(u-v,-v) \theta_2(u-v,-u) E^{(k-1)}(u,-v) du, v \in [-u_1^{(k)}, -u_0^{(k)}],$$

$$Q_1^{(k)}(v) = \mu_3 \int_{a_1+v}^{A+v} q(u-v,-v) \theta_3(u-v,-u) I^{(k-1)}(u,-v) du, v \in [-u_1^{(k)}, -u_0^{(k)}],$$

$$F_2^{(k)}(v) = \int_{a_1+v}^{-u_0^{(k)}} [\mu_1 \theta_1(u-v,-v) S^{(k)}(u,-v) + \mu_2 \theta_2(u-v,-v)(1-p(u-v,-v)) E^{(k)}(u,-v)$$
$$+ \mu_3 \theta_3(u-v,-v)(1-q(u-v,-v)) I^{(k)}(u,-v) + \mu_4 \theta_4(u-v,-v) R^{(k)}(u,-v)] du$$
$$+ \int_{-u_0^{(k)}}^{A+v} [\mu_1 \theta_1(u-v,-v) S^{(k-1)}(u,-v) + \mu_2 \theta_2(u-v,-v)(1-p(u-v,-v))$$
$$\times E^{(k-1)}(u,-v) + \mu_3 \theta_3(u-v,-v)(1-q(u-v,-v)) I^{(k-1)}(u,-v) + \mu_4 \theta_4(u-v,-v)$$
$$\times R^{(k-1)}(u,-v)] du, v \in [-u_2^{(k)}, -u_1^{(k)}],$$

$$G_2^{(k)}(v) = \mu_2 \int_{a_1+v}^{-u_0^{(k)}} p(u-v,-v) \theta_2(u-v,-u) E^{(k)}(u,-v) du + \mu_2 \int_{-u_0^{(k)}}^{A+v} p(u-v,-v)$$
$$\times \theta_2(u-v,-u) E^{(k-1)}(u,-v) du, v \in [-u_2^{(k)}, -u_1^{(k)}],$$

$$Q_2^{(k)}(v) = \mu_3 \int_{a_1+v}^{-u_0^{(k)}} q(u-v,-v) \theta_3(u-v,-u) I^{(k)}(u,-v) du + \mu_3 \int_{-u_0^{(k)}}^{A+v} q(u-v,-v)$$
$$\times \theta_3(u-v,-u) I^{(k-1)}(u,-v) du, v \in [-u_2^{(k)}, -u_1^{(k)}],$$

$$F_n^{(k)}(v) = \int_{a_1+v}^{-u_{n-2}^{(k)}} [\mu_1 \theta_1(u-v,-v) S^{(k)}(u,-v) + \mu_2 \theta_2(u-v,-v)(1-p(u-v,-v)) E^{(k)}(u,-v)$$
$$+ \mu_3 \theta_3(u-v,-v)(1-q(u-v,-v)) I^{(k)}(u,-v) + \mu_4 \theta_4(u-v,-v) R^{(k)}(u,-v)] du$$
$$+ \sum_{i=0}^{n-3} \int_{-u_{i+1}^{(k)}}^{-u_i^{(k)}} [\mu_1 \theta_1(u-v,-v) S_{i+1}^{(k)}(u,-v) + \mu_2 \theta_2(u-v,-v)(1-p(u-v,-v))$$
$$\times E_{i+1}^{(k)}(u,-v) + \mu_3 \theta_3(u-v,-v)(1-q(u-v,-v)) I_{i+1}^{(k)}(u,-v) + \mu_4 \theta_4(u-v,-v)$$
$$+ R_{i+1}^{(k)}(u,-v)] du + \int_{-u_0^{(k)}}^{A+v} [\mu_1 \theta_1(u-v,-v) S^{(k-1)}(u,-v) + \mu_2 \theta_2(u-v,-v)$$
$$\times (1-p(u-v,-v)) E^{(k-1)}(u,-v) + \mu_3 \theta_3(u-v,-v)(1-q(u-v,-v)) I^{(k-1)}(u,-v)$$
$$+ \mu_4 \theta_4(u-v,-v) R^{(k-1)}(u,-v)] du, v \in [-u_n^{(k)}, -u_{n-1}^{(k)}], n = 3, \ldots, N_\tau,$$

$$G_n^{(k)}(v) = \mu_2 \int_{a_1+v}^{-u_{n-2}^{(k)}} p(u-v,-v) \theta_2(u-v,-u) E_{n-1}^{(k)}(u,-v) du + \mu_2 \sum_{i=0}^{n-3} \int_{-u_{i+1}^{(k)}}^{-u_i^{(k)}} p(u-v,-v)$$
$$\times \theta_2(u-v,-u) E_{i+1}^{(k)}(u,-v) du + \mu_2 \int_{-u_0^{(k)}}^{A+v} p(u-v,-v) \theta_2(u-v,-u) E^{(k-1)}(u,-v) du,$$
$$v \in [-u_n^{(k)}, -u_{n-1}^{(k)}], n = 3, \ldots, N_\tau,$$

$$Q_n^{(k)}(v) = \mu_3 \int_{a_1+v}^{-u_{n-2}^{(k)}} q(u-v,-v)\theta_3(u-v,-u)I_{n-1}^{(k)}(u,-v)du + \mu_3 \sum_{i=0}^{n-3} \int_{-u_{i+1}^{(k)}}^{-u_i^{(k)}} q(u-v,-v)$$
$$\times \theta_3(u-v,-u)I_{i+1}^{(k)}(u,-v)du + \mu_3 \int_{-u_0^{(k)}}^{A+v} q(u-v,-v)\theta_3(u-v,-u)I^{(k-1)}(u,-v)du,$$
$$v \in [-u_n^{(k)}, -u_{n-1}^{(k)}], n = 3, \ldots, N_\tau,$$
$$F_1^{(k)}(u) = \int_{a_1}^{A} [\mu_1\theta_1(v,-u)S^{(k-1)}(v,-u) + \mu_2\theta_2(v,-u)(1-p(v,-u))E^{(k-1)}(v,-u)$$
$$+ \mu_3\theta_3(v,-u)(1-q(v,-u))I^{(k-1)}(v,-u) + \mu_4\theta_4(v,-u)R^{(k-1)}(v,-u)]dv,$$
$$G^{(k)}(u) = \mu_2 \int_{a_1}^{A} p(v,-u)\theta_2(v,-u)E^{(k-1)}(v,-u)dv,$$
$$Q^{(k)}(u) = \mu_3 \int_{a_1}^{A} q(v,-u)\theta_3(v,-u)I^{(k-1)}(v,-u)dv.$$

Theorem 1. *If Assumptions (H_1)–(H_4) are satisfied, there exists a unique continuous traveling wave solution to Equations (15)–(19) of the system in Equation (1).*

Proof. By analogy with [28], if $S_0(a)$, $E_0(a)$, $I_0(a,0)$, and $R_0(a)$ satisfy compatibility conditions (i.e., Hypothesis H_4), then two parts of the traveling wave in Equation (15) can be combined continuously. If the parameters of the system in Equation (1) fulfill Hypotheses (H_1)–(H_3), there exists a unique continuous traveling wave solution to Equations (15)–(19) of the system in Equation (1). The method is very similar, thus we omit it. □

4. Stability Analysis of System

Consider the nonlinear autonomous system in Equation (1) where the following parameters are constants: $\theta_i(a,t) = \theta_{i0}(i = 1,2,3,4)$, $\alpha_i(a,t) = \alpha_{i0} = \alpha_i + \sigma\theta_{i0}(i = 1,2,3,4)$, $\gamma_1(a,t) = \gamma_{10}$, $\gamma_2(a,t) = \gamma_{20}$, $\gamma_3(a,t) = \gamma_{30}$, $\beta(a,a',t) = \beta_0$, $p(a,t) = p_0$, $q(a,t) = q_0$, and $\rho(a,t) = \rho_0$, where α_{i0}, θ_{i0}, γ_{i0}, β_0, p_0, q_0, and ρ_0 are positive constants. In this paper, a partial differential equation (Equation (1)) with the initial-boundary values in Equations (2) and (3) reduced to a nonlinear ordinary differential equation. Take the maturity age $a_1 \to 0$, which is not an essential simplification. Integrating Equation (1) in regard to age a from 0 to A, and using the real conditions $S(A,t) = 0$, $E(A,t) = 0$, $I(A,t) = 0$ and $R(A,t) = 0$, the initial value problem of the nonlinear ordinary differential equation autonomous system that describes the population dynamics of the number of population $N_s(t)$, $N_e(t)$, $N_i(t)$, and $N_r(t)$ is:

$$\begin{cases} N_s'(t) = -(\alpha_{10} - \mu_1\theta_{10})N_s(t) - \gamma_{10}\beta_0 N_s(t)N_i(t-\tau) + \mu_2\theta_{20}(1-p_0)N_e(t) \\ \quad + \mu_3\theta_{30}(1-q_0)N_i(t) + (\mu_4\theta_{40}+\rho_0)N_r(t), \\ N_e'(t) = -(\alpha_{20} + \gamma_{20} - \mu_2\theta_{20}p_0)N_e(t) + \gamma_{10}\beta_0 N_s(t)N_i(t-\tau), \\ N_i'(t) = -(\alpha_{30} + \gamma_{30} - \mu_3\theta_{30}q_0)N_i(t) + \gamma_{20}N_e(t), \\ N_r'(t) = -(\alpha_{40} + \rho_0)N_r(t) + \gamma_{30}N_i(t), \end{cases} \quad (20)$$

with initial functions

$$\begin{cases} N_s(0) = \int_0^A S_0(a)da, \\ N_e(t) = \int_0^A E_0(a,t)da, t \in [-\tau, 0], \\ N_i(0) = \int_0^A I_0(a)da, \\ N_r(0) = 0. \end{cases} \quad (21)$$

The basic reproduction number [31] is defined as

$$R_0 = \frac{N\gamma_{20}\gamma_{10}\beta_0}{(\alpha_{20}+\gamma_{20}-\mu_2\theta_{20}p_0)(\alpha_{30}+\gamma_{30}-\mu_3\theta_{30}q_0)}. \quad (22)$$

According to the analysis of [28], if $R_0 > 1$, it can lead to the outbreak of infectious diseases. Hence, the following theorem is obtained.

Theorem 2. *If $\gamma_{10} = 0$, the system in Equation (20) has only a disease-free equilibrium point $H_0 = (0, 0, 0, 0)$; if $R_0 > 1$ and the parameters of the system in Equation (20) satisfy $R_1 > 1$, $R_2 < 1$, $R_3 < 1$, and $R_4 < 1$, where $R_1 = \frac{\mu_1\theta_{10}}{\alpha_{10}}$, $R_2 = \frac{\mu_2\theta_{20}p_0}{\alpha_{20}+\gamma_{20}}$, $R_3 = \frac{\mu_3\theta_{30}q_0}{\alpha_{30}+\gamma_{30}}$, and $R_4 = \frac{\mu_3\theta_{30}(1-q_0)\gamma_{20}+\gamma_{20}\gamma_{30}(\mu_4\theta_{40}+\rho_0)(\alpha_{40}+\rho_0)^{-1}}{(\alpha_{20}+\gamma_{20}-\mu_2\theta_{20})(\alpha_{30}+\gamma_{30}-\mu_3\theta_{30}q_0)}$. The system in Equation (20) has only a positive endemic equilibrium point $H^* = (N_s^*, N_e^*, N_i^*, N_r^*)$, where $N_s^*, N_e^*, N_i^*,$ and N_r^* are given in the proof.*

Proof. When $\gamma_{10} = 0$, that is no infectious diseases, there exists one disease-free equilibrium point $H_0 = (0, 0, 0, 0)$. When the fertility rate and the death rate of susceptible population are satisfied by

$$R_1 = \frac{\mu_1\theta_{10}}{\alpha_{10}} > 1, \quad (23)$$

where R_1 is a dimensionless index for the existence of the disease-free equilibrium point. It can be seen that $R_1 > 1$ presents that death rate of susceptible population is less than their reproductive rate. We can take it further into consideration with the dynamic behavior of susceptible population. The endemic equilibrium point $H^* = (N_s^*, N_e^*, N_i^*, N_r^*)$ is the solution of nonlinear system

$$-(\alpha_{10} - \mu_1\theta_{10})N_s^* - \gamma_{10}\beta_0 N_s^* N_i^* + \mu_2\theta_{20}(1-p_0)N_e^* + \mu_3\theta_{30}(1-q_0)N_i(t) + (\mu_4\theta_{40}+\rho_0)N_r^* = 0, \quad (24)$$

$$-(\alpha_{20}+\gamma_{20}-\mu_2\theta_{20}p_0)N_e^* + \gamma_{10}\beta_0 N_s^* N_i^* = 0, \quad (25)$$

$$-(\alpha_{30}+\gamma_{30}-\mu_3\theta_{30}q_0)N_i^* + \gamma_{20}N_e^* = 0, \quad (26)$$

$$-(\alpha_{40}+\rho_0)N_r^* + \gamma_{30}N_i^* = 0. \quad (27)$$

On the basis of the parameters of the system in Equation (1) satisfying Hypothesis (H_1)–(H_3), the endemic equilibrium point $N_s^* > 0$, $N_e^* > 0$, $N_i^* > 0$ and $N_r^* > 0$ exist if and only if the parameters of Equations (24)–(27) satisfy

$$R_2 = \frac{\mu_2\theta_{20}p_0}{\alpha_{20}+\gamma_{20}} < 1, \quad (28)$$

$$R_3 = \frac{\mu_3\theta_{30}q_0}{\alpha_{30}+\gamma_{30}} < 1, \quad (29)$$

$$R_4 = \frac{\mu_3\theta_{30}(1-q_0)\gamma_{20} + \gamma_{20}\gamma_{30}(\mu_4\theta_{40}+\rho_0)(\alpha_{40}+\rho_0)^{-1}}{(\alpha_{20}+\gamma_{20}-\mu_2\theta_{20})(\alpha_{30}+\gamma_{30}-\mu_3\theta_{30}q_0)} < 1, \quad (30)$$

where R_2, R_3, and R_4 are dimensionless indexes for the existence of the endemic equilibrium point H^*. It can be known that $R_2 < 1$ presents that the death rate and conversion rate of exposed population outweigh their reproductive rate. The density of exposes cannot increase indefinitely because of the balance of death rate, conversion rate, and reproductive rate. From the biological point of view, higher death rate and conversion rate are the consequence of infectious disease. $R_3 < 1$ denotes that reproductive rate of infected population is less than their death rate and conversion rate. From the biological point of view, higher death rate and conversion rate in both cases are the fundamental results of infectious diseases. Let $M_0 = (\alpha_{20}+\gamma_{20}-\mu_2\theta_{20})(\alpha_{30}+\gamma_{30}-\mu_3\theta_{30}q_0) - \mu_3\theta_{30}(1-q_0)\gamma_{20} - \gamma_{20}\gamma_{30}(\mu_4\theta_{40}+\rho_0)(\alpha_{40}+\rho_0)^{-1}$. The inequality $R_4 < 1$ is equivalent to $M_0 > 0$. Then, the endemic equilibrium point $H^* = (N_s^*, N_e^*, N_i^*, N_r^*)$ exists and can be explicitly expressed as

$$N_s^* = \frac{(\alpha_{20} + \gamma_{20} - \mu_2\theta_{20}p_0)(\alpha_{30} + \gamma_{30} - \mu_3\theta_{30}q_0)}{\gamma_{20}\gamma_{10}\beta_0}, N_e^* = \frac{(\alpha_{30} + \gamma_{30} - \mu_3\theta_{30}q_0)}{\gamma_{20}}N_i^*,$$

$$N_i^* = \frac{(\mu_1\theta_{10} - \alpha_{10})\gamma_{20}}{M_0}N_s^*, N_r^* = \frac{\gamma_{30}}{(\alpha_{40} + \rho_0)}N_i^*.$$

□

First, the local stability of the disease-free equilibrium point H_0 is analyzed.

Theorem 3. *For all $\tau \geq 0$, the disease-free equilibrium point $H_0 = (0, 0, 0, 0)$ is locally asymptotically stable if $R_1 < 1$, $R_2 < 1$ and $R_3 < 1$.*

Proof. The characteristic equation of the system in Equation (20) for the disease-free equilibrium point H_0 is

$$(\lambda + \alpha_{10} - \mu_1\theta_{10})(\lambda + \alpha_{20} + \gamma_{20} - \mu_2\theta_{20}p_0)(\lambda + \alpha_{30} + \gamma_{30} - \mu_3\theta_{30}q_0)(\lambda + \alpha_{40} + \rho_0) = 0. \quad (31)$$

Four roots $\lambda_{1,2,3,4}$ of Equation (31) are given:

$$\lambda_1 = \mu_1\theta_{10} - \alpha_{10}, \lambda_2 = -(\alpha_{20} + \gamma_{20} - \mu_2\theta_{20}p_0), \lambda_3 = -(\alpha_{30} + \gamma_{30} - \mu_3\theta_{30}q_0), \lambda_4 = -(\alpha_{40} + \rho_0).$$

If $\alpha_{10} > \mu_1\theta_{10}(R_1 < 1)$, $\alpha_{20} + \gamma_{20} > \mu_2\theta_{20}p_0$ ($R_2 < 1$), and $\alpha_{30} + \gamma_{30} > \mu_3\theta_{30}q_0$ ($R_3 < 1$), characteristic equation in Equation (31) has four real negative roots. Then, the disease-free equilibrium point $H_0 = (0, 0, 0, 0)$ is locally asymptotically stable. □

Next, we study the stability of the endemic equilibrium point H^* by linearizing the nonlinear autonomous system and calculating characteristic equations. Accordingly, we can obtain polynomial equations which can analyze the stability of system. After linearization of the system in Equation (20), we get

$$\begin{cases} L_s'(t) = -(\alpha_{10} - \mu_1\theta_{10})L_s(t) - \gamma_{10}\beta_0[L_s(t)N_i^* + N_s^*L_i(t-\tau)] + \mu_2\theta_{20}(1-p_0)L_e(t) \\ \quad + \mu_3\theta_{30}(1-q_0)L_i(t) + (\mu_4\theta_{40} + \rho_0)L_r(t), \\ L_e'(t) = -(\alpha_{20} + \gamma_{20} - \mu_2\theta_{20}p_0)L_e(t) + \gamma_{10}\beta_0[L_s(t)N_i^* + N_s^*L_i(t-\tau)], \\ L_i'(t) = -(\alpha_{30} + \gamma_{30} - \mu_3\theta_{30}q_0)L_i(t) + \gamma_{20}L_e(t), \\ L_r'(t) = -(\alpha_{40} + \rho_0)L_r(t) + \gamma_{30}L_i(t). \end{cases} \quad (32)$$

Consider the form of the exponential solution of the linearized system in Equation (32) as

$$L_s(t) = \tilde{L}_s e^{\lambda\tau}, L_e(t) = \tilde{L}_e e^{\lambda\tau}, L_i(t) = \tilde{L}_i e^{\lambda\tau}, L_r(t) = \tilde{L}_r e^{\lambda\tau}, \quad (33)$$

where λ is a parameter. Substituting these solutions into Equation (32), we have

$$\begin{cases} (\lambda + \alpha_{10} - \mu_1\theta_{10} + \gamma_{10}\beta_0 N_i^*)\tilde{L}_s + [\gamma_{10}\beta_0 N_s^* e^{-\lambda\tau} - \mu_3\theta_{30}(1-q_0)]\tilde{L}_i - \mu_2\theta_{20}(1-p_0)\tilde{L}_e \\ \quad - (\mu_4\theta_{40} + \rho_0)\tilde{L}_r = 0, \\ (\lambda + \alpha_{20} + \gamma_{20} - \mu_2\theta_{20}p_0)\tilde{L}_e - \gamma_{10}\beta_0\tilde{L}_s N_i^* - \gamma_{10}\beta_0 N_s^* \tilde{L}_i e^{-\lambda\tau} = 0, \\ (\lambda + \alpha_{30} + \gamma_{30} - \mu_3\theta_{30}q_0)\tilde{L}_i - \gamma_{20}\tilde{L}_e = 0, \\ (\lambda + \alpha_{40} + \rho_0)\tilde{L}_r - \gamma_{30}\tilde{L}_i = 0. \end{cases}$$

Denote $M_1 = \mu_1\theta_{10} - \alpha_{10}, M_2 = \alpha_{20} + \gamma_{20} - \mu_2\theta_{20}, M_3 = \alpha_{40} + \rho_0, M_4 = \alpha_{20} + \gamma_{20} - \mu_2\theta_{20}p_0$ and $M_5 = \alpha_{30} + \gamma_{30} - \mu_3\theta_{30}q_0$. Hence $M_1 > 0, M_2 > 0, M_3 > 0, M_4 > 0$ and $M_5 > 0$. The characteristic equation of the system in Equation (32) for the endemic equilibrium point $H^*=(N_s^*, N_e^*, N_i^*, N_r^*)$ is given:

$$\lambda^4 - (\lambda^2 + D_1\lambda - D_2)D_3 e^{-\lambda\tau} + D_5\lambda^3 + D_6\lambda^2 + D_7\lambda = 0, \tag{34}$$

where $D_1 = M_3 - M_1$, $D_2 = M_1 M_3$, $D_3 = M_4 M_5$, $D_4 = M_4 + M_5$, $D_5 = D_1 + D_4 + M_1 D_3 M_0^{-1}$, $D_6 = D_1 D_4 - D_2 + D_3 + [D_4 - \mu_2\theta_{20}(1-p_0) + M_3] M_1 D_3 M_0^{-1}$, $D_7 = D_1 D_3 - D_2 D_4 + [(M_2 + M_5)M_3 + M_2 M_5 - \mu_3\theta_{30}(1-q_0)\gamma_{20}] M_1 D_3 M_0^{-1}$.

Theorem 4. *For the system in Equation (20), if $D_5(D_6 - D_3)(D_7 - D_1 D_3) - D_5^2 D_2 D_3 - (D_7 - D_1 D_3)^2 > 0$, the endemic equilibrium point $H^*=(N_s^*, N_e^*, N_i^*, N_r^*)$ is locally asymptotically stable when $\tau = 0$.*

Proof. When $\tau = 0$, the characteristic equation (Equation (34)) becomes

$$\lambda^4 + D_5\lambda^3 + (D_6 - D_3)\lambda^2 + (D_7 - D_1 D_3)\lambda + D_2 D_3 = 0. \tag{35}$$

It is evident that $D_2 > 0, D_3 > 0$ and $D_4 > 0$ when $N_s^*, N_e^*, N_i^*, N_r^*$ satisfy the nonnegativity conditions in Equation (23)–(30). The following equations are also satisfied:

$D_3 M_0^{-1} - 1 = [\mu_3\theta_{30}(1-q_0)\gamma_{20} - \gamma_{20}\gamma_{30}(\mu_4\theta_{40} + \rho_0)M_3^{-1}]M_0^{-1} > 0$,
$D_5 = D_4 + M_1(D_3 M_0^{-1} - 1) + M_3 > 0$,
$D_6 - D_3 = D_2(D_3 M_0^{-1} - 1) + M_1 M_5 D_3 M_0^{-1} + M_4[\mu_3\theta_{30}(1-q_0)\gamma_{20} + \gamma_{20}\gamma_{30}(\mu_4\theta_{40} + \rho_0)M_3^{-1}] > 0$,
$D_7 - D_1 D_3 = M_4[\mu_3\theta_{30}(1-q_0)\gamma_{20} + \gamma_{20}\gamma_{30}(\mu_4\theta_{40} + \rho_0)M_3^{-1}]D_2 M_0^{-1} + M_5 D_2(D_3 M_0^{-1} - 1) + M_1 D_3 + \gamma_{20}\gamma_{30}(\mu_4\theta_{40} + \rho_0)M_3^{-1} M_1 D_3 M_0^{-1} > 0$,
$D_2 D_3 = M_1 M_3 M_4 M_5 > 0$.

Suppose that

$$D_5(D_6 - D_3)(D_7 - D_1 D_3) - D_5^2 D_2 D_3 - (D_7 - D_1 D_3)^2 > 0, \tag{36}$$

Then, according to the Routh–Hurwitz criterion, the roots of the characteristic equations of system are all negative real. Thus, the endemic equilibrium point of the autonomous system in Equation (20) is asymptotically stable with $\tau = 0$. □

Next, the characteristic equation (Equation (34)) with $\tau > 0$ is presented for discussion.

Theorem 5. *For the system in Equation (20), if $D_6^2 - 2D_5 D_7 - D_3^2 > 0$ and $D_7^2 - 2D_2 D_3^2 - D_1^2 D_3^2 > 0$, the results can be obtained:*

(i) When $0 < \tau < \tau_0$, the endemic equilibrium point $H^*=(N_s^*, N_e^*, N_i^*, N_r^*)$ is locally asymptotically stable.
(ii) When $\tau > \tau_0$, the endemic equilibrium point $H^*=(N_s^*, N_e^*, N_i^*, N_r^*)$ is unstable.
(iii) A Hopf bifurcation occurs at $H^* = (N_s^*, N_e^*, N_i^*, N_r^*)$ when $\tau = \tau_k$ $(k = 0, 1, 2, \cdots)$.

Proof. When $\tau > 0$, suppose that Equation (34) has a purely imaginary root $\lambda = i\omega(\omega > 0)$. The characteristic equation (Equation (34)) converses to the following form:

$$\omega^4 - (-\omega^2 + iD_1\omega - D_2)D_3(\cos(\omega\tau) - i\sin(\omega\tau)) - iD_5\omega^3 - D_6\omega^2 + iD_7\omega = 0.$$

Separating the real and imaginary parts yields two corresponding equations:

$$\omega D_1 D_3 \sin(\omega\tau) - (\omega^2 D_3 + D_2 D_3)\cos(\omega\tau) = \omega^4 - D_6\omega^2, \quad (37)$$

$$(\omega^2 D_3 + D_2 D_3)\sin(\omega\tau) + \omega D_1 D_3 \cos(\omega\tau) = D_7\omega - D_5\omega^3. \quad (38)$$

Thus,

$$sin\omega\tau = \frac{(\omega^2 D_3 + D_2 D_3)(D_7\omega - D_5\omega^3) + \omega D_1 D_3(\omega^4 - D_6\omega^2)}{(\omega D_1 D_3)^2 + (\omega^2 D_3 + D_2 D_3)^2}, \quad (39)$$

$$cos\omega\tau = \frac{\omega D_1 D_3(D_7\omega - D_5\omega^3) - (\omega^4 - D_6\omega^2)(\omega^2 D_3 + D_2 D_3)}{(\omega D_1 D_3)^2 + (\omega^2 D_3 + D_2 D_3)^2}. \quad (40)$$

Let $x = \omega^2$. Squaring Equations (37) and (38) and adding them together, we can get

$$x^4 + (D_5^2 - 2D_6)x^3 + (D_6^2 - 2D_5 D_7 - D_3^2)x^2 + (D_7^2 - 2D_2 D_3^2 - D_1^2 D_3^2)x - D_2^2 D_3^2 = 0. \quad (41)$$

Let $l(x) = x^4 + (D_5^2 - 2D_6)x^3 + (D_6^2 - 2D_5 D_7 - D_3^2)x^2 + (D_7^2 - 2D_2 D_3^2 - D_1^2 D_3^2)x - D_2^2 D_3^2$. According to the previous restrictions in Equation (23)–(30), we can also get

$$D_5^2 - 2D_6 = M_3^2 + M_4^2 + M_5^2 + M_1^2(D_3 M_0^{-1} - 1)^2 + 2\mu_2\theta_{20}(1 - p_0)M_1 D_3 M_0^{-1} > 0. \quad (42)$$

Suppose that

$$D_6^2 - 2D_5 D_7 - D_3^2 > 0, \quad (43)$$

$$D_7^2 - 2D_2 D_3^2 - D_1^2 D_3^2 > 0. \quad (44)$$

In accordance with Descartes' rule of signs, Equation (41) has only one changed sign. Thus, it only has one positive root. Let x^* be the small unique positive root and it always exists. The unknown parameter ω of Equations (37) and (38) is defined as $\pm i\omega_0 = \pm i\sqrt{x_0}$. Combining with Equations (37) and (38), the form of time delay τ_k is gained:

$$\tau_k = \frac{1}{\omega_0}\arcsin\left(\frac{(\omega_0^2 D_3 + D_2 D_3)(D_7\omega_0 - D_5\omega_0^3) + \omega_0 D_1 D_3(\omega_0^4 - D_6\omega_0^2)}{(\omega_0 D_1 D_3)^2 + (\omega_0^2 D_3 + D_2 D_3)^2}\right) + 2k\pi), k = 0, 1, 2, \cdots. \quad (45)$$

Therefore, when $\tau \in (0, \tau_0)$, all roots of Equation (34) have strictly negative real parts. The endemic equilibrium point $H^* = (N_s^*, N_e^*, N_i^*, N_r^*)$ of the system in Equation (20) is locally asymptotically stable. When $\tau = \tau_0$, the roots of Equation (34) have strictly negative real parts except for $\pm i\omega_0$. When $\tau > \tau_0$, the endemic equilibrium point $H^* = (N_s^*, N_e^*, N_i^*, N_r^*)$ of the system in Equation (20) is unstable. Then, differentiating both sides of Equation (34) with respect to τ, we obtain

$$\left(\frac{d\lambda}{d\tau}\right)^{-1} = \frac{(2\lambda + D_1)D_3 - (4\lambda^3 + 3D_5\lambda^2 + 2D_6\lambda + D_7)e^{\lambda\tau}}{(\lambda^3 + D_1\lambda^2 - D_2\lambda)D_3} - \frac{\tau}{\lambda}. \quad (46)$$

Further, one leads to

$$Re\left\{\left(\frac{d\lambda}{d\tau}\right)^{-1}\right\}_{\lambda=i\omega_0} = \frac{Q_1 + Q_2 - Q_3}{\omega_0^2[(\omega_0 D_1 D_3)^2 + (\omega_0^2 D_3 + D_2 D_3)^2]}, \quad (47)$$

where

$Q_1 = \omega_0^2 D_1 D_3(4\omega_0^3 - 2D_6\omega_0)sin\omega_0\tau_0 + \omega_0^2 D_1 D_3(D_7 - 3D_5\omega_0^2)cos\omega_0\tau_0$,
$Q_2 = \omega_0(\omega_0^2 D_3 + D_2 D_3)(D_7 - 3D_5\omega_0^2)sin\omega_0\tau_0 - \omega_0(\omega_0^2 D_3 + D_2 D_3)(4\omega_0^3 - 2D_6\omega_0)cos\omega_0\tau_0$,
$Q_3 = 2\omega_0^2 D_3(\omega_0^2 D_3 + D_2 D_3) + (\omega_0 D_1 D_3)^2$.

Since conditions in Equation (42)–(44) hold, then

$$sign\left\{Re(\frac{d\lambda}{d\tau})\right\}_{\tau=\tau_0} = sign\left\{Re(\frac{d\lambda}{d\tau})^{-1}\right\}_{\lambda=i\omega_0} = sign\left\{\frac{l'(x_0)}{(\omega_0 D_1 D_3)^2 + (\omega_0^2 D_3 + D_2 D_3)^2}\right\}_{x_0=\omega_0^2} > 0, \quad (48)$$

where $l'(x_0) = 4x_0^3 + 3(D_5^2 - 2D_6)x_0^2 + 2(D_6^2 - 2D_5 D_7 - D_3^2)x_0 + (D_7^2 - 2D_2 D_3^2 - D_1^2 D_3^2) > 0$. Hence, based on the properties of the Hopf bifurcation discussed in [32], the transversal condition holds and a Hopf bifurcation occurs at $\tau = \tau_0$. The proof is complete. □

5. Simulation

To affirm the stability analysis above, we numerically simulated the disease-free equilibrium point H_0 and the endemic equilibrium point H^* of the system in Equation (20). We are more concerned with the indicators and conditions for outbreaks of "local" population or local asymptotic stability of equilibrium points. All programs were developed using the software Matlab R2016a (see Program S1).

First, we considered the stability of the disease-free equilibrium point $H_0 = (0,0,0,0)$. Let $\alpha_{i0} = 0.13, i = 1,2,3,4; \mu_1 = 0.3; \mu_2 = 0.5; \mu_3 = 0.5; \mu_4 = 0.3; \theta_{10} = 0.4; \theta_{20} = 0.1; \theta_{30} = 0.1; \theta_{40} = 0.2; \gamma_{10} = 0.1; \gamma_{20} = 0.15; \gamma_{30} = 0.1; \beta_0 = 0.0023; p_0 = 0.1; q_0 = 0.15; \rho_0 = 0.02$; and the initial value was $(N_s(0), N_e(0), N_i(0), N_r(0)) = (1500, 1000, 500, 0)$. The parameters of the system in Equation (20) satisfy conditions $R_1 < 1$, $R_2 < 1$, and $R_3 < 1$ and the conditions in Theorem 3 are satisfied in Figure 1. The disease-free equilibrium point H_0 of the system in Equation (20) is locally asymptotically stable.

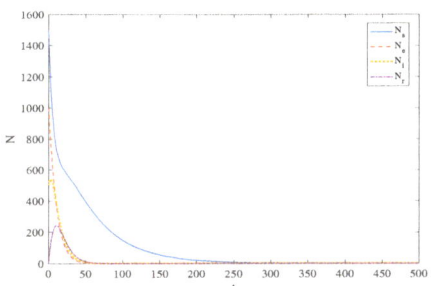

Figure 1. Local asymptotic stability of the disease-free equilibrium point $H_0 = (0,0,0,0)$.

Next, we considered the endemic equilibrium point $H^* = (N_s^*, N_e^*, N_i^*, N_r^*)$ with $\tau = 0$. Let $\alpha_{i0} = 0.13, i = 1,2,3,4; \mu_1 = 0.6; \mu_2 = 0.5; \mu_3 = 0.5; \mu_4 = 0.3; \theta_{10} = 0.4; \theta_{20} = 0.1; \theta_{30} = 0.1; \theta_{40} = 0.2; \gamma_{10} = 0.1; \gamma_{20} = 0.15; \gamma_{30} = 0.1; \beta_0 = 0.0023; p_0 = 0.1; q_0 = 0.15; \rho_0 = 0.02$; and the initial value was $(N_s(0), N_e(0), N_i(0), N_r(0)) = (1500, 1000, 500, 0)$. The parameters of the system in Equation (20) satisfy the condition in Equation (36). The result is presented in Figure 2, which comes from Theorem 4; the system in Equation (20) has a unique endemic equilibrium point $H^* = (N_s^*, N_e^*, N_i^*, N_r^*) \approx (1723, 1108, 797, 531)$ and the system in Equation (20) is locally asymptotically stable.

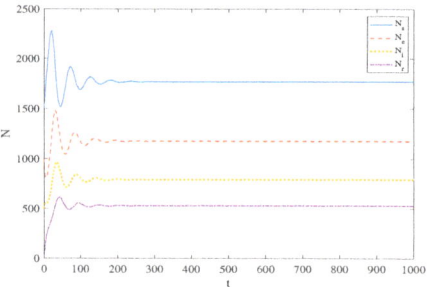

Figure 2. Local asymptotic stability of the endemic equilibrium point $H^* = (N_s^*, N_e^*, N_i^*, N_r^*)$ when $\tau = 0$.

Similarly, let $\alpha_{i0} = 0.13, i = 1,2,3,4; \mu_1 = 0.6; \mu_2 = 0.5; \mu_3 = 0.5; \mu_4 = 0.3; \theta_{10} = 0.4; \theta_{20} = 0.1; \theta_{30} = 0.1; \theta_{40} = 0.2; \gamma_{10} = 0.1; \gamma_{20} = 0.16; \gamma_{30} = 0.1; \beta_0 = 0.0023; p_0 = 0.1; q_0 = 0.15; \rho_0 = 0.02$; and the initial value be $(N_s(0), N_e(0), N_i(0), N_r(0)) = (1500, 1000, 500, 0)$. In other words, $N_s(0) < N_s^*$, $N_e(0) < N_e^*$, $N_i(0) < N_i^*$, and $N_r(0) < N_r^*$. The parameters of the system in Equation (20) satisfy the conditions in Equations (43) and (44). In this case, we have the roots of Equation (41) are $x_1 = -0.1398$, $x_2 = 0.0073$, and $x_{3,4} = -0.213 \pm 0.0248i$. Thus, we get $x_0 = 0.0073$, $\omega_0 = 0.0855$, $\tau_0 = 5.3526$, and $\tau_1 = 73.4875$. The population dynamic behaviors of the endemic equilibrium point H^* are shown in Figure 3 ($\tau = 4.3526$) and Figure 4 ($\tau = 6.3526$), respectively. It can been shown that the endemic equilibrium point H^* spends longer time to become locally asymptotically stable only when $0 < \tau < \tau_0$ (see Figure 3). When $\tau > \tau_0$, it is quite clear that larger values of the time delay causes periodic oscillations. The equilibrium point loses its stability. Thus, the solution of the system in Equation (20) is unstable with the increase of time t (see Figure 4).

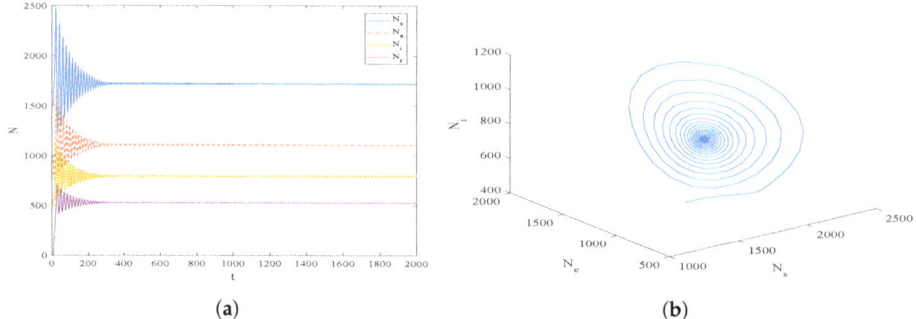

(a) (b)

Figure 3. Local asymptotic stability of the endemic equilibrium point $H^* = (N_s^*, N_e^*, N_i^*, N_r^*)$ when $\tau = 4.3526 < \tau_0$. (**a**) Time response diagram of N_s, N_e, N_i and N_r; (**b**) Phase diagram of N_s, N_e and N_i.

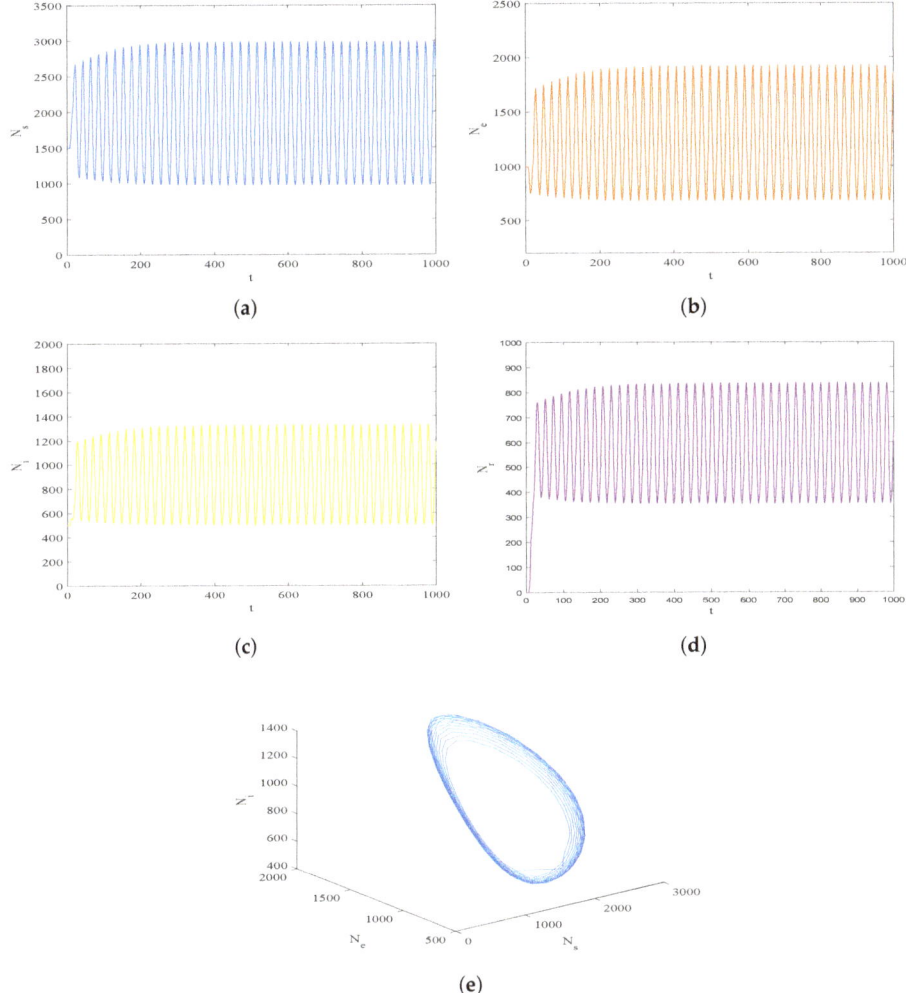

Figure 4. Instability of the endemic equilibrium point $H^* = (N_s^*, N_e^*, N_i^*, N_r^*)$ when $\tau = 6.3526 > \tau_0$. (a) Time response diagram of N_s; (b) Time response diagram of N_e; (c) Time response diagram of N_i; (d) Time response diagram of N_r; (e) Phase diagram of N_s, N_e and N_i.

6. Concluding Remarks

In this article, the theory of an age-structured SEIRS model with time delay is analyzed. The model is based on the delayed nonlinear partial differential equation of initial-boundary value problems. The traveling wave solution of the system in Equation (1) is obtained using the method of characteristic and the recurrent algorithm. Then, we can obtain the existence and uniqueness of the continuous traveling wave solution of system according to hypotheses. The age-structured SEIRS model with time delay is reduced to a nonlinear ordinary differential equation under some insufficient simplifications. This allows us to obtain some sufficient conditions of existence of two equilibrium points of an age-structured SEIRS system: R_1 is a dimensionless index for the existence of the disease-free equilibrium point H_0. R_2, R_3, and R_4 are dimensionless indexes for the existence of the endemic equilibrium point H^*. From the biological point of view, the endemic equilibrium point H^* only exists in the case of high values of death and conversion rate of exposed and infected population. The disease-free equilibrium point H_0 and the endemic equilibrium point H^* are given. The disease-free equilibrium point $H_0 = (0,0,0,0)$ is locally asymptotically stable if $R_1 < 1$, $R_2 < 1$, and $R_3 < 1$. The stability of the endemic equilibrium point $H^* = (N_s^*, N_e^*, N_i^*, N_r^*)$ with $\tau = 0$ and $\tau > 0$ are analyzed: for $R_0 > 1$, if the condition in Equation (36) holds, the endemic equilibrium point H^* is locally asymptotically stable when $\tau = 0$; if the conditions in Equations (43) and (44) hold, the endemic equilibrium point H^* is locally asymptotically stable when time delay $0 < \tau < \tau_0$; if the conditions in Equations (43)–(44) hold, the endemic equilibrium point H^* is unstable when τ satisfies $\tau > \tau_0$; Hopf bifurcation occurs at $\tau = \tau_k (k = 0, 1, 2, ...)$. When time delay exceeds the critical value τ_0, the system in Equation (20) loses its stability and Hopf bifurcation occurs. In this case, the susceptible, exposed, infected, and recovered population in the model will coexist in an oscillating mode and infectious diseases will get out of control. It can be seen that time delay has an important effect on the spread of infectious diseases. Therefore, we should shorten the time delay as much as possible in order to predict and eliminate infectious diseases. Our further research is using some bifurcation control strategies to control the occurrence of the Hopf bifurcation so as to control the occurrence of infectious diseases.

In general, this study provides the practical understanding of the different dynamic behaviors of an age-structured susceptible–exposed–infected–recovered–susceptible model with time delay, which is helpful for us to understand the application of epidemiology better in real life.

Supplementary Materials: The following are available online at http://www.mdpi.com/2227-7390/8/3/455/s1, Program S1: Software source code for Figures 1–4.

Author Contributions: Conceptualization, Z.Y.; Funding acquisition, Y.Y.; Methodology, Z.Y.; Software, Z.Y. and Z.L.; and Supervision, Y.Y. All authors have read and agreed to the published version of the manuscript.

Funding: This work was supported by the National Nature Science Foundation of China (No. 61772063) and Beijing Natural Science Foundation (No. Z180005).

Acknowledgments: The authors would like to thank for the support funds.

Conflicts of Interest: The authors declare that there are no conflicts of interest regarding the publication of this paper.

References

1. Kermack, W.O.; McKendrick, A.G. A contribution to the mathematical theory of epidemics. *Proc. R. Soc. Lond. A* **1927**, *115*, 700–721. [CrossRef]
2. Brauer, F.; Castillo-Chavez, C. *Mathematical Models in Population Biology and Epidemiology*; Springer: New York, NY, USA, 2012.
3. Zhou, X.; Cui, J.A. Analysis stability and bifurcation for an SEIR epidemic model with saturated recovery rate. *Commun. Nonliear Sci. Numer. Simul.* **2011**, *16*, 4438–4450. [CrossRef]
4. Chauhan, S.; Misra, O.P.; Dhar, J. Stability Analysis of SIR Model with Vaccination. *J. Comput. Appl. Math.* **2014**, *4*, 17–23.
5. Vargas-De-Len, C. On the global stability of SIS, SIR and SIRS epidemic model with standard incidence. *Chaos Solitons Fractals* **2011**, *44*, 1106–1110. [CrossRef]
6. Shaikhet, L. *Stability of SIR Epidemic Model Equilibrium Points*; Springer: Berlin/Heidelberg, Germany, 2013.
7. Chen, L.J.; Sun, J.T. Global stability and optimal control of an SIRS epidemic model on heterogeneous networks. *Physica A* **2014**, *15*, 196–204. [CrossRef]
8. Lakshmanan, M.; Senthilkumar, D.V. *Dynamics of Nonlinear Time-Delay Systems*; Springer: Berlin/Heidelberg, Germany, 2010; p. 313.
9. Smith, H. *An Introduction to Delay Differential Equations with Application to the Life Sciences*; Springer: Berlin/Heidelberg, Germany, 2011; p. 172.
10. Sharma, S.; Mondal, A.; Pal, A.K.; Samanta, G.P. Stability analysis and optimal control of avian influenza virus A with time delays. *Int. J. Dyn. Control* **2018**, *6*, 1351–1366. [CrossRef]

11. Xu, R.; Ma, Z.E.; Wang, Z.P. Global stability of a delayed SIRS epidemic model with saturation incidence and temporary immunity. *J. Comput. Appl. Math.* **2010**, *59*, 3211–3221. [CrossRef]
12. Shu, H.Y.; Fan, D.J.; Wei J.J. Global stability of multi-group SEIR epidemic models with distributed delays and nonlinear transmission. *Nonlinear Anal. RWA* **2012**, *13*, 1581–1592. [CrossRef]
13. De la Sena, M.; Alonso-Quesadaa, S.; Ibeasb, A. On the stability of an SEIR epidemic model with distributed time-delay and a general class of feedback vaccination rules. *Appl. Math. Comput.* **2015**, *270*, 953–976. [CrossRef]
14. Cooke, K.L.; Driessche, P. Analysis of an SEIRS epidemic model with two delays. *J. Math. Biol.* **1996**, *35*, 240–260. [CrossRef]
15. Gao, S.J.; Chen, L.S.; Teng, Z.d. Impulsive Vaccination of an SEIRS Model with Time Delay and Varying Total Population Size. *Bull. Math. Biol.* **2007**, *69*, 731–745. [CrossRef] [PubMed]
16. Wang, W.D. Global behavior of an SEIRS epidemic model with time delays. *Appl. Math. Lett.* **2002**, *15*, 423–428. [CrossRef]
17. Schenzle, D. An age-structured model of pre-and post-vaccination measles transmission. *IMA J. Appl. Math.* **1984**, *1*, 169–191. [CrossRef] [PubMed]
18. Metz, J.A.J.; Diekmann, O. The Dynamics of Physiologically Structured Populations. In *Lecture Notes in Biomathematics*; Springer: Berlin, Germany, 1986; p. 68.
19. Iannelli, M. *Mathematical Theory of Age-Structured Population Dynamics*; Giardini Editorie Stampatori: Pisa, Italy, 1995.
20. Hethcote, H.W. The mathematics of infectious diseases. *SIAM Rev.* **2000**, *42*, 599–653. [CrossRef]
21. Mccluskey, C. Global stability for an SEI epidemiological model with continuous age-structure in the exposed and infectious classes. *Math. Biosci. Eng.* **2012**, *9*, 819–841.
22. Zou, L.; Ruan, S.; Zhang, W. An age-structured model for the transmission dynamics of hepatitis B. *SIAM J. Math. Appl.* **2010**, *70*, 3121–3139. [CrossRef]
23. Gandolfi, A.; Iannelli, M.; Marinoschi, G. An age-structured model of epidermis growth. *J. Math. Biol.* **2011**, *62*, 111–141. [CrossRef]
24. Alexanderian, A.; Gobbert, M.K.; Fister, K.R.; Gaff, H.; Lenhart, S.; Schaefer, E. An age-structured model for the spread of epidemic cholera. *Nonlinear Anal. RWA* **2011**, *12*, 3483–3498. [CrossRef]
25. Akimenko, V.V. Asymptotically stable states of non-linear age-structured monocyclic cell population model I. Travelling wave solution. *Math. Comput. Simul.* **2017**, *133*, 2–23. [CrossRef]
26. Akimenko, V.V. Asymptotically stable states of non-linear age-structured monocyclic cell population model II. Numerical simulation. *Math. Comput. Simul.* **2017**, *133*, 24–38. [CrossRef]
27. Akimenko, V.V. Nonlinear age-structured models of polycyclic population dynamics with death rates as a power functions with exponent n. *Math. Comput. Simul.* **2017**, *133*, 175–205. [CrossRef]
28. Akimenko, V.V. An age-structured SIR epidemic model with fixed incubation period of infection. *Math. Comput. Appl.* **2017**, *73*, 1485–1504. [CrossRef]
29. Guo, Z.K.; Huo, H.F.; Xiang, H. Global dynamics of an age-structured malaria model with prevention. *Math. Biosci. Eng.* **2019**, *16*, 1625–1653. [CrossRef] [PubMed]
30. Huo, H.F.; Yang, P.; Xiang, H. Dynamics for an SIRS epidemic model with infection age and relapse on a scale-free network. *J. Frankl. Inst.* **2019**, *356*, 7411–7443.
31. Diekmann, O.; Heesterbeek, J.A.P.; Metz, J.A.J. On the definition and the computation of the basic reproduction ratio R_0 in models for infectious diseases in heterogeneous populations. *J. Math. Biol.* **1990**, *28*, 365. [CrossRef] [PubMed]
32. Hassard, B.; Kazarinoff, N.; Wan, Y.H. *Theory and Applications of Hopf Bifurcation*; Cambridge University Press: London, UK, 1981.

© 2020 by the authors. Licensee MDPI, Basel, Switzerland. This article is an open access article distributed under the terms and conditions of the Creative Commons Attribution (CC BY) license (http://creativecommons.org/licenses/by/4.0/).

Article

Bounded Solutions of Semilinear Time Delay Hyperbolic Differential and Difference Equations

Allaberen Ashyralyev [1,2,3],* and Deniz Agirseven [4]

1. Department of Mathematics, Near East University, Mersin 10 99138, Turkey
2. Department of Mathematics, Peoples' Friendship University Russia, Moscow 117198, Russia
3. Institute of Mathematics and Mathematical Modeling, Almaty 050010, Kazakhstan
4. Department of Mathematics, Trakya University, Edirne 22030, Turkey; denizagirseven@trakya.edu.tr or denizagirseven@gmail.com
* Correspondence: allaberen.ashyralyev@neu.edu.tr or aallaberen@gmail.com

Received: 28 October 2019; Accepted: 27 November 2019; Published: 2 December 2019

Abstract: In this paper, we study the initial value problem for a semilinear delay hyperbolic equation in Hilbert spaces with a self-adjoint positive definite operator. The mean theorem on the existence and uniqueness of a bounded solution of this differential problem for a semilinear hyperbolic equation with unbounded time delay term is established. In applications, the existence and uniqueness of bounded solutions of four problems for semilinear hyperbolic equations with time delay in unbounded term are obtained. For the approximate solution of this abstract differential problem, the two-step difference scheme of a first order of accuracy is presented. The mean theorem on the existence and uniqueness of a uniformly bounded solution of this difference scheme with respect to time stepsize is established. In applications, the existence and uniqueness of a uniformly bounded solutions with respect to time and space stepsizes of difference schemes for four semilinear partial differential equations with time delay in unbounded term are obtained. In general, it is not possible to get the exact solution of semilinear hyperbolic equations with unbounded time delay term. Therefore, numerical results for the solution of difference schemes for one and two dimensional semilinear hyperbolic equation with time delay are presented. Finally, some numerical examples are given to confirm the theoretical analysis.

Keywords: semilinear problems with delay; hyperbolic equations; difference scheme; stability; Hilbert space

MSC: 39A30; 35L20; 65M06

---

1. Introduction

Delay differential equations are used to model biological, physical, system engineering, and sociological processes as well as naturally occurring oscillatory systems (see, for examples, [1–9]).

It is known that in differential and difference equations, the involvement of the delay term causes deep difficulties in the analysis of these equations. Lu [10] studies modified iterative schemes by combing the method of upper-lower solutions and the Jacobi method or the Gauss–Seidel method for finite-difference solutions of reaction-diffusion systems with time delays. Ashyralyev and Sobolevskii [11] study the initial-value problem for linear delay parabolic equations in a Banach space and present a sufficient condition for the stability of the solution of this initial-value problem. The stability estimates in Hölder norms for the solutions of the initial-boundary value problem for delay parabolic equations were established.

Ashyralyev and Agirseven [12–18] studied some initial-boundary value problems for linear delay parabolic differential equations. Theorems on stability and convergence of difference schemes for the numerical solution of initial and boundary value problems for linear parabolic equations with time delay were proved. Moreover, Ashyralyev, Agirseven, and Ceylan [19] investigated finding sufficient conditions for the existence and uniqueness of a bounded solution of the initial value problem for the semilinear delay parabolic equation in a Banach space. The main theorem on the existence and uniqueness of a bounded solution of this problem was established. In applications, existence and uniqueness of a bounded solution of problems for four semilinear delay parabolic equations were obtained. Numerical results were given.

Henriquez, Cuevas, and Caicedo [20] investigated the existence of almost periodic solutions for linear retarded functional differential equations with finite delay. The existence of almost periodic solutions with the stabilization of distributed control systems was obtained.

Hao, Fan, Cao, and Sun [21] proposed a linearized quasi-compact finite difference scheme for semilinear space-fractional diffusion equations with a fixed time delay. Under the local Lipschitz conditions, they proved the solvability and convergence of the scheme in the discrete maximum norm by the energy method.

Liang [22] studied the convergence and asymptotic stability of semidiscrete and full discrete schemes for linear parabolic equations with time delay. She proved that the semidiscrete scheme, backward Euler and Crank–Nicolson full discrete schemes can unconditionally preserve the delay-independent asymptotic stability with some additional restrictions on time and spatial stepsizes of the forward Euler full discrete scheme.

Bhrawy, Abdelkawy, and Mallawi [23] investigated the Chebyshev Gauss–Lobatto pseudospectral scheme in spatial directions for the approximate solution of one-dimensional, coupled, and two-dimensional parabolic equations with time delays. They also develop an efficient numerical algorithm based on the Chebyshev pseudospectral algorithm to obtain the two spatial variables in approximate solving the two-dimensional parabolic equations with time delay.

Applying Vishik's results and methods of operator tools, Ismailov, Guler, and Ipek [24] described all solvable extensions of a minimal operator generated by linear delay differential-operator expression of first order in the Hilbert space of vector-functions in finite interval. Sharp formulas for the spectrums of these solvable extensions were obtained. Theoretical results have been supported by applications.

Piriadarshani and Sengadir [25] obtained an existence theorem for a semilinear partial differential equation with infinite delay employing a phase space in which discretizations can naturally be performed. For linear partial differential equations with infinite delay they show that the solutions of the ordinary differential equation with infinite delay obtained by the semi-discretization converge to the original solution.

Castro, Rodriguez, Cabrera, and Martin [26] developed an explicit finite difference scheme for a model with coefficients variable in time and studied their properties of convergence and stability.

Hyperbolic equations without time delay arise in many branches of science and engineering, for example, electrodynamics, thermodynamics, hydrodynamics, fluid dynamics, wave propagation, hyperbolic geometry, and discrete mathematics (see, e.g., [8,9,27–34], and the references given therein). The geometry of complex networks is closely related with their structure and function. Shang Yilun [34] investigated the Gromov-hyperbolicity of the Newman–Watts model of small-world networks. It is known that asymptotic Erdős–Rényi random graphs are not hyperbolic. We show that the Newman–Watts ones built on top of them by adding lattice-induced clustering are not hyperbolic as the network size goes to infinity. Numerical simulations are provided to illustrate the effects of various parameters on hyperbolicity in this model. The geometry of complex networks has a close relationship with their structure and function. Shang Yilun [33] investigated Gromov-hyperbolicity of inhomogeneous random networks modeled by the Chung-Lu model $G(w)$. His numerical simulations illustrate this non-hyperbolicity of $G(w)$ for power law degree distributions among others.

In numerical methods for solving hyperbolic equations, the problem of stability has received a great deal of importance and attention. The method of operators as a tool for the study of the stability of the solution of local and nonlocal problems to hyperbolic differential and difference equations in Hilbert and Banach spaces has been systematically developed by many authors (see, e.g., [27–32,35–37]).

A large cycle of works on difference schemes for hyperbolic equations (see, e.g., [38–42] and the references given therein), in which stability was established under the assumption that the magnitude of the grid steps τ and h with respect to the time and space variables, were connected. In abstract terms that means that the condition $\tau||A_h|| \to 0$ when $\tau \to 0$ holds.

Of course there is great interest in the study of absolute stable difference schemes of a high order of accuracy for hyperbolic equations, in which stability was established without any assumptions in respect to the grid steps τ and h. Such type of stability inequalities for the solutions of the first order of accuracy difference scheme for the abstract hyperbolic equation in Hilbert spaces were established for the first time in [43]. The first and second order of accuracy difference schemes generated by integer power of space operator of approximate solutions of the abstract initial value problem for the abstract hyperbolic equation in Hilbert spaces were presented in [11]. The stability estimates for the solution of these difference schemes were established.

The survey paper [44] contains the recent results on the local and nonlocal well-posed problems for second order differential and difference equations. Stability of differential problems for hyperbolic equations and of difference schemes for approximate solution of these problems were presented.

However, the stability theory of problems for a hyperbolic equation with unbounded time delay term is not well-investigated. A few researchers are interested in these kinds of problems. Bounded solutions of semilinear one dimensional hyperbolic differential equations with time delay term have been investigated in earlier papers [45–48]. In the paper [49] the existence and uniqueness of a bounded solution of nonlinear hyperbolic differential equations with bounded time delay term were established. The generality of the operator approach allows for treating a wider class of delay nonlinear hyperbolic differential equations with bounded time delay term. In general, hyperbolic differential equations with unbounded time delay term are blown up [7]. Therefore, the boundedness solution of problems for hyperbolic equations with unbounded time delay term is not well-investigated.

Our goal in the present paper is to investigate the boundedness solution of problems for semilinear hyperbolic equations with unbounded time delay term. We study the initial value problem for the semilinear hyperbolic differential equation with time delay

$$\begin{cases} \frac{d^2u(t)}{dt^2} + Au(t) = f(t, u(t), u_t(t-w), u(t-w)), \ t > 0, \\ u(t) = \varphi(t), \ -w \leq t \leq 0 \end{cases} \tag{1}$$

in a Hilbert space H with a self-adjoint positive definite operator A. Here $\varphi(t)$ is a continuously differentiable abstract function defined on the interval $[-w, 0]$ with values in H and $f(t)$ is continuous abstract function defined on the interval $[0, \infty)$ with values in H. Assume that A is unbounded operator and $(Ax, x) > \delta(x, x)$, for $x \neq 0, x \in H$ and $\delta > 0$.

A function $u(t)$ is called a solution of problem (1), if the following conditions are fulfilled:

i. $u(t)$ is twice continuously differentiable function on the interval $[0, \infty)$, $f(t, u(t), u_t(t-w), u(t-w))$ is continuous and bounded function on $[0, \infty)$

ii. The element $u(t)$ belongs to $D(A)$ for all $t \in [0, \infty)$, and the function $Au(t)$ is continuous on the interval $[0, \infty)$.

iii. $u(t)$ satisfies the equation and initial conditions in Equation (1).

In the present paper, the main theorem on the existence and uniqueness of a bounded solution of the differential problem (1) is established. In applications, the existence and uniqueness of a bounded solution of four problems for semilinear hyperbolic equations with time delay are obtained. A first order of accuracy difference scheme for the numerical solution of this problem is presented. The theorem on the existence and uniqueness of a uniformly bounded solution of this difference scheme with respect to τ is established. In applications, existence and uniqueness of a bounded solution of a problem for four semilinear delay parabolic equations were established. Numerical results for the solution of difference schemes for one and two dimensional nonlinear hyperbolic equation with time delay are presented.

2. Main Existence and Uniqueness Theorem of the Differential Problem

Throughout this paper, $c(t)$ and $s(t)$ are operator-functions defined by formulas

$$c(t)u = \frac{e^{iA^{\frac{1}{2}}t} + e^{-iA^{\frac{1}{2}}t}}{2}u, \quad s(t)u = \int_0^t c(y)u\,dy. \tag{2}$$

We will give some auxiliary statements which will be useful in the sequel.

Recall that the norm $\|A\|_{H \to H}$ of a bounded operator $A : H \to H$ is by definition the smallest number C for which estimate

$$\|Au\|_H \leq C\|u\|_H$$

holds. Equivalently,

$$\|A\|_{H \to H} = \sup_{\|u\|_H = 1} \|Au\|_H.$$

Lemma 1. *For $t \geq 0$, the following estimates hold:*

$$\|A^{-\frac{1}{2}}\|_{H \to H} \leq \delta^{-\frac{1}{2}}, \quad \|c(t)\|_{H \to H} \leq 1, \quad \|As(t)\|_{H \to H} \leq 1. \tag{3}$$

Proof. Applying formulas (2) and the spectral representation of the self-adjoint positive definite operator A in a Hilbert space H, we can write (see [50])

$$\left\|A^{-\frac{1}{2}}\right\|_{H \to H} \leq \sup_{\delta \leq \lambda < \infty} \left|\lambda^{-\frac{1}{2}}\right| \leq \delta^{-\frac{1}{2}},$$

$$\|c(t)\|_{H \to H} \leq \sup_{\delta \leq \lambda < \infty} \left|\frac{e^{i\lambda^{\frac{1}{2}}t} + e^{-i\lambda^{\frac{1}{2}}t}}{2}\right| = \sup_{\delta \leq \lambda < \infty} \left|\cos\left(\lambda^{\frac{1}{2}}t\right)\right| \leq 1,$$

$$\left\|A^{\frac{1}{2}}s(t)\right\|_{H \to H} \leq \sup_{\delta \leq \lambda < \infty} \left|\frac{e^{i\lambda^{\frac{1}{2}}t} - e^{-i\lambda^{\frac{1}{2}}t}}{2i}\right| = \sup_{\delta \leq \lambda < \infty} \left|\sin\left(\lambda^{\frac{1}{2}}t\right)\right| \leq 1,$$

for any $t \geq 0$. Lemma 1 is proved. □

The approach of proof of main theorem is based on reducing problem (1) to an integral equation of Volterra type

$$u(t) = c(t - (m-1)w)\,u((m-1)w) + s(t - (m-1)w)\frac{du((m-1)w)}{dt}$$

$$+ \int_{(m-1)w}^t s(t-y)\,f(y, u(y), u_y(y-w), u(y-w))\,dy,$$

$$(m-1)w \leq t \leq mw, \quad m = 1, 2, \ldots, \quad u(t) = \varphi(t), \quad -w \leq t \leq 0$$

in $[0, \infty) \times H$ and the application of successive approximations. Note that on $(m-1)w \leq t \leq mw$, $m = 1, 2, \ldots$, $u_t(t-w)$ and $u(t-w)$ are given. Therefore, the recursive formula for the solution of problem (1) is

$$u_i(t) = c(t-(m-1)w) u((m-1)w) + s(t-(m-1)w) \frac{du((m-1)w)}{dt} \qquad (4)$$
$$+ \int_{(m-1)w}^{t} s(t-y) f(y, u_{i-1}(y), u_y(y-w), u(y-w)) dy,$$

$$u_0(t) = c(t-(m-1)w) u((m-1)w) + s(t-(m-1)w) \frac{du((m-1)w)}{dt},$$

$$(m-1)w \leq t \leq mw, \quad m = 1, 2, ...,$$

$$u_i(t) = \varphi(t), \quad i = 1, 2, ..., \quad -w \leq t \leq 0.$$

Theorem 1. *Suppose that $\varphi(t)$ is a continuously differentiable function on $[-\omega, 0]$ and $\varphi(t) \in D(A)$, $\varphi'(t) \in D(A^{\frac{1}{2}})$ and*

$$\|A\varphi(t)\|_H \leq M, \quad \|A^{1/2}\varphi'(t)\|_H \leq \tilde{M}. \qquad (5)$$

Besides let $f : [0, \infty) \times H \times H \times H \longrightarrow H$ be continuous and bounded function, that is

$$\|f(t, u, v, z)\|_H \leq \bar{M} \qquad (6)$$

in $[0, \infty) \times H \times H \times H$ and Lipschitz condition holds uniformly with respect to t, v and z

$$\|f(t, u, v, z) - f(t, w, v, z)\|_H \leq L\|u - w\|_H. \qquad (7)$$

Here, M, $\tilde{M}$, $\bar{M}$, Ł are positive constants. Then there exists a unique solution to problem (1) which is bounded in $C(H)$. Here, $C(H) = C([0, \infty), H)$ stands for the Banach space of the abstract continuous and bounded functions $v(t)$ defined on $[0, \infty)$ with values in H, equipped with the norm

$$\|v\|_{C(H)} = \sup_{0 \leq t < \infty} \|v(t)\|_H.$$

Proof. Let $0 \leq t \leq \omega$. Then, according to Equation (4), we get

$$u_i(t) = c(t) \varphi(0) + s(t) \varphi'(0) + \int_0^t s(t-y) f(y, u_{i-1}(y), \varphi_y(y-w), \varphi(y-w)) dy, \qquad (8)$$

$$u_i'(t) = -As(t) \varphi(0) + c(t) \varphi'(0) + \int_0^t c(t-y) f(y, u_{i-1}(y), \varphi_y(y-w), \varphi(y-w)) dy \qquad (9)$$

for all $i = 1, 2, \ldots$. Therefore,

$$u(t) = u_0(t) + \sum_{i=0}^{\infty} (u_{i+1}(t) - u_i(t)), \qquad (10)$$

$$u'(t) = u_0'(t) + \sum_{i=0}^{\infty} (u_{i+1}'(t) - u_i'(t)), \qquad (11)$$

where

$$u_0(t) = c(t) \varphi(0) + s(t) \varphi'(0), \quad u_0'(t) = -As(t) \varphi(0) + c(t) \varphi'(0).$$

Applying the triangle inequality and estimates (3) and (5), we get

$$\|u_0(t)\|_H \leq \|A^{-1}\|_{H \to H}$$

$$\times \left[\|c(t)\|_{H \to H} \|A\varphi(0)\|_H + \|A^{1/2} s(t)\|_{H \to H} \|A^{1/2}\varphi'(0)\|_H \right]$$

$$\leq \delta^{-1} \left[\|A\varphi(0)\|_H + \|A^{1/2}\varphi'(0)\|_H \right] \leq \delta^{-1} \left[M + \tilde{M} \right],$$

$$\|u_0'(t)\|_H \leq \|A^{-\frac{1}{2}}\|_{H \to H}$$

$$\times \left[\|A^{1/2}s(t)\|_{H\to H}\|A\varphi(0)\|_H + \|c(t)\|_{H\to H}\|A^{1/2}\varphi'(0)\|_H \right]$$

$$\leq \delta^{-\frac{1}{2}} \left[\|A\varphi(0)\|_H + \|A^{1/2}\varphi'(0)\|_H \right] \leq \delta^{-\frac{1}{2}} \left[M + \widetilde{M} \right].$$

Applying formulas (8) and (9) and the triangle inequality and estimates (3) and (6), we get

$$\|u_1(t) - u_0(t)\|_H \leq \|A^{-\frac{1}{2}}\|_{H\to H}$$

$$\times \int_0^t \|A^{1/2}s(t-y)\|_{H\to H}\|f(y, u_0(y), \varphi_y(y-w), \varphi(y-w))\|_H dy \leq \delta^{-\frac{1}{2}}\bar{M}t,$$

$$\|u_1'(t) - u_0'(t)\|_H \leq \int_0^t \|c(t-y)\|_{H\to H}\|f(y, u_0(y), \varphi_y(y-w), \varphi(y-w))\|_H dy \leq \bar{M}t.$$

Using the triangle inequality, we get

$$\|u_1(t)\|_H \leq \delta^{-1}\left[M + \widetilde{M}\right] + \delta^{-\frac{1}{2}}\bar{M}t,$$

$$\|u_1'(t)\|_H \leq \delta^{-\frac{1}{2}}\left[M + \widetilde{M}\right] + \bar{M}t.$$

Applying formulas (8) and (9) and estimates (3), (6), and (7), we get

$$\|u_2(t) - u_1(t)\|_H \leq \|A^{-\frac{1}{2}}\|_{H\to H}$$

$$\times \int_0^t \|A^{1/2}s(t-y)\|_{H\to H}\|f(y, u_1(y), \varphi_y(y-w), \varphi(y-w)) - f(y, u_0(y), \varphi_y(y-w), \varphi(y-w))\|_H dy$$

$$\leq \delta^{-\frac{1}{2}}L\int_0^t \|u_1(y) - u_0(y)\|_H dy \leq \delta^{-1}L\bar{M}\int_0^t y\, dy = \frac{\bar{M}}{L}\frac{(\delta^{-\frac{1}{2}}Lt)^2}{2!},$$

$$\|u_2'(t) - u_1'(t)\|_H$$

$$\leq \int_0^t \|c(t-y)\|_{H\to H}\|f(y, u_1(y), \varphi_y(y-w), \varphi(y-w)) - f(y, u_0(y), \varphi_y(y-w), \varphi(y-w))\|_H dy$$

$$\leq L\int_0^t \|u_1(y) - u_0(y)\|_H dy \leq \delta^{-\frac{1}{2}}L\bar{M}\int_0^t y\, dy = \delta^{-\frac{1}{2}}\frac{\bar{M}}{L}\frac{(Lt)^2}{2!}.$$

Then

$$\|u_2(t)\|_H \leq \delta^{-1}\left[M + \widetilde{M}\right] + \delta^{-\frac{1}{2}}\bar{M}t + \frac{\bar{M}}{L}\frac{(\delta^{-\frac{1}{2}}Lt)^2}{2!},$$

$$\|u_2'(t)\|_H \leq \delta^{-\frac{1}{2}}\left[M + \widetilde{M}\right] + \bar{M}t + \frac{\bar{M}}{L\delta^{-\frac{1}{2}}}\frac{(Lt)^2}{2!}.$$

Let

$$\|u_n(t) - u_{n-1}(t)\|_H \leq \frac{\bar{M}}{L}\frac{(\delta^{-\frac{1}{2}}Lt)^n}{n!},\ \|u_n'(t) - u_{n-1}'(t)\|_H \leq \frac{\bar{M}}{L\delta^{-\frac{1}{2}}}\frac{(Lt)^n}{n!}$$

and

$$\|u_n(t)\|_H \leq \delta^{-1}\left[M + \widetilde{M}\right] + \delta^{-\frac{1}{2}}\bar{M}t + \frac{\bar{M}}{L}\frac{(\delta^{-\frac{1}{2}}Lt)^2}{2!} + \cdots + \frac{\bar{M}}{L}\frac{(\delta^{-\frac{1}{2}}Lt)^n}{n!},$$

$$\|u_n'(t)\|_H \leq \delta^{-\frac{1}{2}}\left[M + \widetilde{M}\right] + \bar{M}t + \frac{\bar{M}}{L\delta^{-\frac{1}{2}}}\frac{(L\delta^{-\frac{1}{2}}t)^2}{2!} + \cdots + \frac{\bar{M}}{L\delta^{-\frac{1}{2}}}\frac{(L\delta^{-\frac{1}{2}}t)^n}{n!}.$$

Then, we obtain

$$\|u_{n+1}(t) - u_n(t)\|_H \leq \|A^{-\frac{1}{2}}\|_{H\to H}$$

$$\times \int_0^t \|A^{1/2} s(t-y)\|_{H\to H} \|f(y,u_n(y),\varphi_y(y-w),\varphi(y-w)) - f(y,u_{n-1}(y),\varphi_y(y-w),\varphi(y-w))\|_H dy$$

$$\le \delta^{-\frac{1}{2}} L \int_0^t \|u_n(y) - u_{n-1}(y)\|_H dy \le \delta^{-\frac{1}{2}} L \int_0^t \frac{\bar{M}}{L} \frac{(\delta^{-\frac{1}{2}}Ly)^n}{n!} y\, dy = \frac{\bar{M}}{L}\frac{(\delta^{-\frac{1}{2}}Lt)^{n+1}}{(n+1)!},$$

$$\|u'_{n+1}(t) - u'_n(t)\|_H$$

$$\le \int_0^t \|c(t-y)\|_{H\to H}\|f(y,u_n(y),\varphi_y(y-w),\varphi(y-w)) - f(y,u_{n-1}(y),\varphi_y(y-w),\varphi(y-w))\|_H dy$$

$$\le \int_0^t L\|u_n(y) - u_{n-1}(y)\|_H ds \le \int_0^t L \frac{\bar{M}}{L}\frac{(\delta^{-\frac{1}{2}}Ly)^n}{n!} dy = \frac{\bar{M}}{L\delta^{-\frac{1}{2}}}\frac{(L\delta^{-\frac{1}{2}}t)^{n+1}}{(n+1)!}.$$

Therefore,

$$\|u_{n+1}(t) - u_n(t)\|_H \le \frac{\bar{M}}{L}\frac{(\delta^{-\frac{1}{2}}Lt)^{n+1}}{(n+1)!}, \quad \|u'_{n+1}(t) - u'_n(t)\|_H \le \frac{\bar{M}}{L\delta^{-\frac{1}{2}}}\frac{(L\delta^{-\frac{1}{2}}t)^{n+1}}{(n+1)!}$$

and

$$\|u_{n+1}(t)\|_H \le \delta^{-1}\left[M + \tilde{M}\right] + \delta^{-\frac{1}{2}}\bar{M}t + \frac{\bar{M}}{L}\frac{(\delta^{-\frac{1}{2}}Lt)^2}{2!} \cdots + \frac{\bar{M}}{L}\frac{(\delta^{-\frac{1}{2}}Lt)^{n+1}}{(n+1)!},$$

$$\|u'_{n+1}(t)\|_H \le \delta^{-\frac{1}{2}}\left[M + \tilde{M}\right] + \bar{M}t + \frac{\bar{M}}{L\delta^{-\frac{1}{2}}}\frac{(L\delta^{-\frac{1}{2}}t)^2}{2!} + \cdots + \frac{\bar{M}}{L\delta^{-\frac{1}{2}}}\frac{(L\delta^{-\frac{1}{2}}t)^{n+1}}{(n+1)!}$$

are true for any n, $n \ge 1$ by mathematical induction. In a similar manner, for any n, we can obtain

$$\|A^{\frac{1}{2}}u_{n+1}(t) - A^{\frac{1}{2}}u_n(t)\|_H \le \frac{\bar{M}}{L\delta^{-\frac{1}{2}}}\frac{(L\delta^{-\frac{1}{2}}t)^{n+1}}{(n+1)!}$$

and

$$\|A^{\frac{1}{2}}u_{n+1}(t)\|_H \le \delta^{-\frac{1}{2}}\left[M + \tilde{M}\right] + \bar{M}t + \frac{\bar{M}}{L\delta^{-\frac{1}{2}}}\frac{(L\delta^{-\frac{1}{2}}t)^2}{2!} + \cdots + \frac{\bar{M}}{L\delta^{-\frac{1}{2}}}\frac{(L\delta^{-\frac{1}{2}}t)^{n+1}}{(n+1)!}.$$

From that and formulas (10) and (11) it follows that

$$\|u(t)\|_H \le \|u_0(t)\|_H + \sum_{i=0}^{\infty}\|u_{i+1}(t) - u_i(t)\|_H$$

$$\le \delta^{-1}\left[M + \tilde{M}\right] + \sum_{i=0}^{\infty}\frac{\bar{M}}{L}\frac{(\delta^{-\frac{1}{2}}Lt)^{i+1}}{(i+1)!}$$

$$\le \delta^{-1}\left[M + \tilde{M}\right] + \frac{\bar{M}}{L}e^{\delta^{-\frac{1}{2}}Lt}, \quad 0 \le t \le w,$$

$$\|u'(t)\|_H \le \|u'_0(t)\|_H + \sum_{i=0}^{\infty}\|u'_{i+1}(t) - u'_i(t)\|_H$$

$$\le \delta^{-\frac{1}{2}}\left[M + \tilde{M}\right] + \sum_{i=0}^{\infty}\frac{\bar{M}}{L\delta^{-\frac{1}{2}}}\frac{(L\delta^{-\frac{1}{2}}t)^{i+1}}{(i+1)!}$$

$$\le \delta^{-\frac{1}{2}}\left[M + \tilde{M}\right] + \frac{\bar{M}}{L\delta^{-\frac{1}{2}}}e^{L\delta^{-\frac{1}{2}}t}, \quad 0 \le t \le w,$$

$$\|A^{\frac{1}{2}}u(t)\|_H \le \|A^{\frac{1}{2}}u_0(t)\|_H + \sum_{i=0}^{\infty}\|A^{\frac{1}{2}}u_{i+1}(t) - A^{\frac{1}{2}}u_i(t)\|_H$$

$$\leq \delta^{-\frac{1}{2}} \left[M + \tilde{M}\right] + \sum_{i=0}^{\infty} \frac{\bar{M}}{L\delta^{-\frac{1}{2}}} \frac{(L\delta^{-\frac{1}{2}}t)^{i+1}}{(i+1)!}$$

$$\leq \delta^{-\frac{1}{2}} \left[M + \tilde{M}\right] + \frac{\bar{M}}{L\delta^{-\frac{1}{2}}} e^{L\delta^{-\frac{1}{2}}t}, \ 0 \leq t \leq w,$$

which proves the existence of a bounded solution of problem (1) in $[0, w] \times H$.

Let $mw \leq t \leq (m+1)w$, $m = 1, 2, \ldots$. Then, according to Equation (4), we can write

$$u_i(t) = c(t - mw) u(mw) + s(t - mw) \frac{du(mw)}{dt} \\ + \int_{mw}^{t} s(t-y) f(y, u_{i-1}(y), u_y(y-w), u(y-w)) dy, \ i = 1, 2, \ldots, \quad (12)$$

$$u_i'(t) = -As(t - mw) u(mw) + c(t - mw) \frac{du(mw)}{dt} \\ + \int_{mw}^{t} c(t-y) f(y, u_{i-1}(y), u_y(y-w), u(y-w)) dy, \ i = 1, 2, \ldots. \quad (13)$$

Therefore,

$$u(t) = u_0(t) + \sum_{i=0}^{\infty} (u_{i+1}(t) - u_i(t)), \quad (14)$$

$$u_i'(t) = u_0'(t) + \sum_{i=0}^{\infty} (u_{i+1}'(t) - u_i'(t)), \quad (15)$$

where

$$u_0(t) = c(t - mw) u(mw) + s(t - mw) \frac{du(mw)}{dt},$$

$$u_0'(t) = -As(t - mw) u(mw) + c(t - mw) \frac{du(mw)}{dt}.$$

Assume that problem (1) has a bounded solution in $[(m-1)w, mw] \times H$ and

$$\|A^{1/2} u(t)\|_H \leq M_{m-1}, \ \|u'(t)\|_H \leq \tilde{M}_{m-1}. \quad (16)$$

Applying estimates (3) and (16), we get

$$\|u_0(t)\|_H \leq \|A^{-\frac{1}{2}}\|_{H \to H}$$

$$\times \left[\|c(t)\|_{H \to H} \|A^{1/2} u(mw)\|_H + \|A^{1/2} s(t)\|_{H \to H} \|u'(mw)\|_H\right] \leq \delta^{-\frac{1}{2}} \left[M_{m-1} + \tilde{M}_{m-1}\right],$$

$$\|u_0'(t)\|_H \leq \left[\|A^{1/2} s(t)\|_{H \to H} \|A^{1/2} u(mw)\|_H + \|c(t)\|_{H \to H} \|\varphi'(0)\|_H\right] \leq M_{m-1} + \tilde{M}_{m-1}.$$

Applying formulas (12) and (13) and estimates (3) and (6), we get

$$\|u_1(t) - u_0(t)\|_H \leq \|A^{-\frac{1}{2}}\|_{H \to H}$$

$$\times \int_{mw}^{t} \|A^{1/2} s(t-y)\|_{H \to H} \|f(y, u_0(y), u_y(y-w), u(y-w))\|_H dy \leq \delta^{-\frac{1}{2}} \bar{M} (t - mw),$$

$$\|u_1'(t) - u_0'(t)\|_H \leq \int_{mw}^{t} \|c(t-y)\|_{H \to H} \|f(y, u_0(y), u_y(y-w), u(y-w))\|_H dy \leq \bar{M} (t - mw).$$

Using the triangle inequality, we get

$$\|u_1(t)\|_H \leq \delta^{-\frac{1}{2}} \left[M_{m-1} + \tilde{M}_{m-1}\right] + \delta^{-\frac{1}{2}} \bar{M} (t - mw),$$

$$\|u_1'(t)\|_H \leq M_{m-1} + \tilde{M}_{m-1} + \bar{M} (t - mw).$$

Applying formulas (9) and (12) and estimates (3), (6), and (7), we get

$$\|u_2(t) - u_1(t)\|_H \leq \|A^{-\frac{1}{2}}\|_{H \to H}$$

$$\times \int_0^t \|A^{1/2} s(t-y)\|_{H \to H} \|f(y, u_1(y), u_y(y-w), u(y-w)) - f(y, u_0(y), u_y(y-w), u(y-w))\|_H dy$$

$$\leq \delta^{-\frac{1}{2}} L \int_{m\omega}^t \|u_1(y) - u_0(y)\|_H dy \leq \delta^{-1} L \bar{M} \int_{m\omega}^t (y - m\omega) \, dy = \frac{\bar{M}}{L} \frac{(\delta^{-\frac{1}{2}} L (t - m\omega))^2}{2!},$$

$$\|u_2'(t) - u_1'(t)\|_H$$

$$\leq \int_{m\omega}^t \|c(t-y)\|_{H \to H} \|f(y, u_1(y), u_y(y-w), u(y-w)) - f(y, u_0(y), u_y(y-w), u(y-w))\|_H dy$$

$$\leq L \int_{m\omega}^t \|u_1(y) - u_0(y)\|_H dy \leq L \bar{M} \delta^{-\frac{1}{2}} \int_{m\omega}^t (y - m\omega) \, dy = \frac{\bar{M}}{L\delta^{-\frac{1}{2}}} \frac{(L\delta^{-\frac{1}{2}} (t - m\omega))^2}{2!}.$$

Then

$$\|u_2(t)\|_H \leq \delta^{-\frac{1}{2}} \left[M_{m-1} + \widetilde{M}_{m-1} \right] + \delta^{-\frac{1}{2}} \bar{M} (t - m\omega) + \frac{\bar{M}}{L} \frac{(\delta^{-\frac{1}{2}} L (t - m\omega))^2}{2!},$$

$$\|u_2'(t)\|_H \leq M_{m-1} + \widetilde{M}_{m-1} + \bar{M}(t - m\omega) + \frac{\bar{M}}{L\delta^{-\frac{1}{2}}} \frac{(L\delta^{-\frac{1}{2}} (t - m\omega))^2}{2!}.$$

Let

$$\|u_n(t) - u_{n-1}(t)\|_H \leq \frac{\bar{M}}{L} \frac{(\delta^{-\frac{1}{2}} L (t - m\omega))^n}{n!}, \quad \|u_n'(t) - u_{n-1}'(t)\|_H \leq \frac{\bar{M}}{L\delta^{-\frac{1}{2}}} \frac{(L\delta^{-\frac{1}{2}} (t - m\omega))^n}{n!}$$

and

$$\|u_n(t)\|_H \leq \delta^{-\frac{1}{2}} \left[M_{m-1} + \widetilde{M}_{m-1} \right] + \delta^{-\frac{1}{2}} \bar{M}(t - m\omega) + \frac{\bar{M}}{L} \frac{(\delta^{-\frac{1}{2}} L (t - m\omega))^2}{2!} + \cdots + \frac{\bar{M}}{L} \frac{(\delta^{-\frac{1}{2}} L (t - m\omega))^n}{n!},$$

$$\|u_n'(t)\|_H \leq M_{m-1} + \widetilde{M}_{m-1} + \bar{M}(t - m\omega) + \frac{\bar{M}}{L\delta^{-\frac{1}{2}}} \frac{(L\delta^{-\frac{1}{2}} (t - m\omega))^2}{2!} + \cdots + \frac{\bar{M}}{L\delta^{-\frac{1}{2}}} \frac{(L\delta^{-\frac{1}{2}} (t - m\omega))^n}{n!}.$$

Then, we obtain

$$\|u_{n+1}(t) - u_n(t)\|_H \leq \|A^{-\frac{1}{2}}\|_{H \to H}$$

$$\times \int_{m\omega}^t \|A^{1/2} s(t-y)\|_{H \to H} \|f(y, u_n(y), u_y(y-w), u(y-w)) - f(y, u_{n-1}(y), u_y(y-w), u(y-w))\|_H dy$$

$$\leq \delta^{-\frac{1}{2}} L \int_{m\omega}^t \|u_n(y) - u_{n-1}(y)\|_H dy \leq \delta^{-\frac{1}{2}} L \int_{m\omega}^t \frac{\bar{M}}{L} \frac{(\delta^{-\frac{1}{2}} L (y - m\omega))^n}{n!} dy = \frac{\bar{M}}{L} \frac{(\delta^{-\frac{1}{2}} L (t - m\omega))^{n+1}}{(n+1)!},$$

$$\|u_{n+1}'(t) - u_n'(t)\|_H$$

$$\leq \int_{m\omega}^t \|c(t-y)\|_{H \to H} \|f(y, u_n(y), u_y(y-w), u(y-w)) - f(y, u_{n-1}(y), u_y(y-w), u(y-w))\|_H dy$$

$$\leq \int_{m\omega}^t L \|u_n(y) - u_{n-1}(y)\|_H ds \leq \int_{m\omega}^t L \frac{\bar{M}}{L} \frac{(L\delta^{-\frac{1}{2}} (y - m\omega))^n}{n!} dy = \frac{\bar{M}}{L\delta^{-\frac{1}{2}}} \frac{(L\delta^{-\frac{1}{2}} (t - m\omega))^{n+1}}{(n+1)!}.$$

Therefore,

$$\|u_{n+1}(t) - u_n(t)\|_H \leq \frac{\bar{M}}{L} \frac{(\delta^{-\frac{1}{2}} L (t - m\omega))^{n+1}}{(n+1)!}, \quad \|u_{n+1}'(t) - u_n'(t)\|_H \leq \frac{\bar{M}}{L\delta^{-\frac{1}{2}}} \frac{(L\delta^{-\frac{1}{2}} (t - m\omega))^{n+1}}{(n+1)!}.$$

and

$$\|u_{n+1}(t)\|_H \leq \delta^{-\frac{1}{2}}\left[M_{m-1} + \widetilde{M}_{m-1}\right] + \delta^{-\frac{1}{2}}\bar{M}(t-m\omega) + \frac{\bar{M}}{L}\frac{(\delta^{-\frac{1}{2}}L(t-m\omega))^2}{2!} \cdots + \frac{\bar{M}}{L}\frac{(\delta^{-\frac{1}{2}}L(t-m\omega))^{n+1}}{(n+1)!},$$

$$\|u'_{n+1}(t)\|_H \leq M_{m-1} + \widetilde{M}_{m-1} + \bar{M}(t-m\omega) + \frac{\bar{M}}{L\delta^{-\frac{1}{2}}}\frac{(L\delta^{-\frac{1}{2}}(t-m\omega))^2}{2!} + \cdots + \frac{\bar{M}}{L\delta^{-\frac{1}{2}}}\frac{(L\delta^{-\frac{1}{2}}(t-m\omega))^{n+1}}{(n+1)!}$$

are true for any $n, n \geq 1$ by mathematical induction. From that and formulas (14) and (15) it follows that

$$\|u(t)\|_H \leq \|u_0(t)\|_H$$

$$+ \sum_{i=0}^{\infty}\|u_{i+1}(t) - u_i(t)\|_H \leq \delta^{-\frac{1}{2}}\left[M + \widetilde{M}\right] + \sum_{i=0}^{\infty}\frac{\bar{M}}{L}\frac{(\delta^{-\frac{1}{2}}L(t-m\omega))^{i+1}}{(i+1)!}$$

$$\leq \delta^{-\frac{1}{2}}\left[M_{m-1} + \widetilde{M}_{m-1}\right] + \frac{\bar{M}}{L}e^{\delta^{-\frac{1}{2}}L(t-m\omega)}, \quad m\omega \leq t \leq (m+1)\omega,$$

$$\|u'(t)\|_H \leq \|u'_0(t)\|_H + \sum_{i=0}^{\infty}\|u'_{i+1}(t) - u'_i(t)\|_H$$

$$\leq \delta^{-\frac{1}{2}}\left[M + \widetilde{M}\right] + \sum_{i=0}^{\infty}\frac{\bar{M}}{L\delta^{-\frac{1}{2}}}\frac{(L\delta^{-\frac{1}{2}}(t-m\omega))^{i+1}}{(i+1)!}$$

$$\leq M_{m-1} + \widetilde{M}_{m-1} + \frac{\bar{M}}{L\delta^{-\frac{1}{2}}}e^{L\delta^{-\frac{1}{2}}(t-m\omega)}, \quad m\omega \leq t \leq (m+1)\omega$$

which proves the existence of a bounded solution of problem (1) in $[m\omega, (m+1)\omega] \times H$.

Now we will prove uniqueness of the bounded solution of the problem. Suppose that there is a bounded solution $v(t)$ of problem (1) and $v(t) \neq u(t)$. Denoting $z(t) = v(t) - u(t)$ and using Equation (1), we get

$$\begin{cases} \frac{d^2z(t)}{dt^2} + Az(t) = f(t, v(t), v_t(t-w), v(t-w)) - f(t, u(t), u_t(t-w), u(t-w)), & t > 0, \\ z(t) = 0, & -w \leq t \leq 0 \end{cases}$$

for $z(t)$.

Let $0 \leq t \leq w$. Since $v(t-w) = u(t-w) = \varphi(t-w)$, we can write

$$\begin{cases} \frac{d^2z(t)}{dt^2} + Az(t) = f(t, v(t), \varphi_t(t-w), \varphi(t-w)) - f(t, u(t), \varphi_t(t-w), \varphi(t-w)), & t > 0, \\ z(t) = 0, & -w \leq t \leq 0. \end{cases}$$

Therefore,

$$z(t) = \int_0^t s(t-y)\left[f(y, v(y), \varphi_y(y-w), \varphi(y-w)) - f(y, u(y), \varphi_y(y-w), \varphi(y-w))\right] dy.$$

Applying estimates (3) and (6), we get

$$\|z(t)\|_H \leq \|A^{-\frac{1}{2}}\|_{H \to H}$$

$$\times \int_0^t \|A^{1/2}s(t-y)\| \|f(y, v(y), \varphi_y(y-w), \varphi(y-w)) - f(y, u(y), \varphi_y(y-w), \varphi(y-w))\|_H dy$$

$$\leq L\delta^{-\frac{1}{2}}\int_0^t \|v(y) - u(y)\|_H ds \leq L\delta^{-\frac{1}{2}}\int_0^t \|z(y)\|_H dy.$$

Applying the integral inequality, we get

$$\|z(t)\|_H \leq 0.$$

From that it follows that $z(t) = 0$ which proves the uniqueness of a bounded solution of problem (1) in $[0, w] \times H$. Using the same method and mathematical induction, we can establish the uniqueness of a bounded solution of problem (1) in $[0, \infty) \times H$. Theorem 1 is proved. □

3. Applications

First, we consider the initial value problem for a semilinear hyperbolic equation with time delay and with nonlocal conditions

$$\begin{cases} u_{tt}(t,x) - (a(x)u_x(t,x))_x + \delta u(t,x) = f(t,x,u(t,x),u_t(t-w,x),u(t-w,x)), \\ 0 < t < \infty, \ x \in (0,l), \\ u(t,x) = \varphi(t,x), \ -w \leq t \leq 0, \ x \in [0,l], \\ u(t,0) = u(t,l), u_x(t,0) = u_x(t,l), \ 0 \leq t < \infty, \end{cases} \quad (17)$$

where $a(x)$ and $\varphi(t,x)$ are given sufficiently smooth functions, $\delta > 0$ is the sufficiently large number. Suppose that $a(x) \geq a > 0$ and $a(l) = a(0)$.

Theorem 2. *Suppose the following hypotheses:*

1. For any $t, -w \leq t \leq 0$
$$\|\varphi(t,.)\|_{W_2^2[0,l]} \leq M, \ \|\varphi'(t,.)\|_{W_2^1[0,l]} \leq \widetilde{M}. \quad (18)$$

2. The function $f: [0, \infty) \times (0,l) \times L_2[0,l] \times L_2[0,l] \times L_2[0,l] \to L_2[0,l]$ be continuous and bounded, that is
$$\|f(t,\cdot,u,v,w)\|_{L_2[0,l]} \leq \overline{M} \quad (19)$$

and Lipschitz condition is satisfied uniformly with respect to t, z, w

$$\|f(t,\cdot,u,z,w) - f(t,\cdot,v,z,w)\|_{L_2[0,l]} \leq L\|u-v\|_{L_2[0,l]}.$$

3. Here and in future, $M, \widetilde{M}, \overline{M}, L$ are positive constants. Assume that all compatibility conditions are satisfied. Then there exists a unique solution to problem (17) which is bounded in $[0, \infty) \times L_2[0,l]$.

The proof of Theorem 2 is based on the abstract Theorem 1, on the self-adjointness and positivity in $L_2[0,l]$ of a differential operator A defined by the formula

$$Au = -\frac{d}{dx}\left(a(x)\frac{du}{dx}\right) + \delta u \quad (20)$$

with domain $D(A) = \{u \in W_2^2[0,l] : u(0) = u(l), \ u'(0) = u'(l)\}$ [51] and on estimates

$$\|c\{t\}\|_{L_2[0,l] \to L_2[0,l]} \leq 1, \ \|(A)^{\frac{1}{2}} s\{t\}\|_{L_2[0,l] \to L_2[0,l]} \leq 1, \ t \geq 0. \quad (21)$$

Second, we consider the initial value problem for a semilinear hyperbolic equation with time delay and with involution

$$\begin{cases} u_{tt}(t,x) - (a(x)u_x(t,x))_x + \delta u(t,x) - \beta (a(-x)u_x(t,-x))_x \\ = f(t,x,u(t,x),u_t(t-w,x),u(t-w,x)), \ 0 < t < \infty, \ x \in (-l,l), \\ u(t,x) = \varphi(t,x), \ -\omega \leq t \leq 0, \ x \in [-l,l], \\ u(t,-l) = u(t,l) = 0, \ 0 \leq t < \infty, \end{cases} \quad (22)$$

where $a(x)$ and $\varphi(t,x)$ are given sufficiently smooth functions, $\delta > 0$ is the sufficiently large number. Suppose that $a \geq a(x) = a(-x) \geq \delta > 0$, $\delta - a|\beta| \geq 0$.

Theorem 3. *Suppose the following hypotheses:*

1. *For any t, $-w \leq t \leq 0$*

$$\|\varphi(t,.)\|_{W_2^2[-l,l]} \leq M, \ \|\varphi'(t,.)\|_{W_2^1[-l,l]} \leq \widetilde{M}.$$

2. *The function $f : [0,\infty) \times (-l,l) \times L_2[-l,l] \times L_2[-l,l] \times L_2[-l,l] \to L_2[-l,l]$ be continuous and bounded, that is*

$$\|f(t,\cdot,u,v,w)\|_{L_2[-l,l]} \leq \overline{M}$$

and Lipschitz condition is satisfied uniformly with respect to t, z, w

$$\|f(t,\cdot,u,z,w) - f(t,v,z,w)\|_{L_2[-l,l]} \leq L \|u-v\|_{L_2[-l,l]}.$$

Assume that all compatibility conditions are satisfied. Then there exists a unique solution to problem (22) which is bounded in $[0,\infty) \times L_2[-l,l]$.

The proof of Theorem 3 is based on the abstract Theorem 1, on the self-adjointness and positivity in $L_2[-l,l]$ of a differential operator A defined by the formula

$$Av(x) = -(a(x)v_x(x))_x - \beta(a(-x)v_x(-x))_x + \delta v(x)$$

with the domain $D(A) = \{u \in W_2^2[-l,l] : u(-l) = u(l) = 0\}$ [52] and on the estimate

$$\|c\{t\}\|_{L_2[-l,l] \to L_2[-l,l]} \leq 1, \ \|(A)^{\frac{1}{2}}s\{t\}\|_{L_2[-l,l] \to L_2[-l,l]} \leq 1, \ t \geq 0.$$

Third, let $\Omega \subset R^m$ be a bounded open domain with smooth boundary S, $\overline{\Omega} = \Omega \cup S$. In $[0,\infty) \times \Omega$ we consider the initial boundary value problem for a multidimensional semilinear delay differential equation of hyperbolic type

$$\begin{cases} \frac{\partial^2 u(t,x)}{\partial t^2} - \sum_{r=1}^{m}(a_r(x)u_{x_r})_{x_r} + \delta u(t,x) \\ = f(t,x,u(t,x),u_t(t-w,x),u(t-w,x)), \ 0 < t < \infty, \ x = (x_1,...,x_m) \in \Omega, \\ u(t,x) = \varphi(t,x), \ -\omega \leq t \leq 0, \ x \in \overline{\Omega}, \\ u(t,x) = 0, \ x \in S, \ 0 \leq t < \infty, \end{cases} \quad (23)$$

where $a_r(x)$ and $\varphi(t,x)$ are given sufficiently smooth functions and $\delta > 0$ is the sufficiently large number and $a_r(x) > 0$.

Theorem 4. *Suppose the following hypotheses:*

1. For any $t, -w \leq t \leq 0$,

$$\|\varphi(t,.)\|_{W_2^2(\overline{\Omega})} \leq M, \quad \|\varphi'(t,.)\|_{W_2^1(\overline{\Omega})} \leq \tilde{M}.$$

2. The function $f : [0,\infty) \times Q \times L_2(\overline{\Omega}) \times L_2(\overline{\Omega}) \times L_2(\overline{\Omega}) \to L_2(\overline{\Omega})$ be continuous and bounded, that is

$$\|f(t,u,v,w)\|_{L_2(\overline{\Omega})} \leq \overline{M}$$

and Lipschitz condition is satisfied uniformly with respect to t, z, w

$$\|f(t,u,z,w) - f(t,v,z,w)\|_{L_2(\overline{\Omega})} \leq L \|u - v\|_{L_2(\overline{\Omega})}.$$

Assume that all compatibility conditions are satisfied. Then there exists a unique solution to problem (23) which is bounded in $[0,\infty) \times L_2(\overline{\Omega})$.

The proof of Theorem 4 is based on the abstract Theorem 1, on the self-adjointness and positivity in $L_2(\overline{\Omega})$ of a differential operator A defined by the formula

$$Au(x) = -\sum_{r=1}^{m}(a_r(x)u_{x_r})_{x_r} + \delta u(x) \qquad (24)$$

with domain [53]

$$D(A) = \{u(x) : u(x), \ u_{x_r}(x), \ (a_r(x)u_{x_r})_{x_r} \in L_2(\overline{\Omega}), \ 1 \leq r \leq m, \ u(x) = 0, \ x \in S\}$$

and on the estimate

$$\|c\{t\}\|_{L_2(\overline{\Omega}) \to L_2(\overline{\Omega})} \leq 1, \ \|(A)^{\frac{1}{2}} s\{t\}\|_{L_2(\overline{\Omega}) \to L_2(\overline{\Omega})} \leq 1, \ t \geq 0 \qquad (25)$$

and on the following theorem on the coercivity inequality for the solution of the elliptic differential problem in $L_2(\overline{\Omega})$.

Theorem 5. *For the solutions of the elliptic differential problem*

$$\begin{cases} A^x u(x) = \omega(x), \ x \in \Omega, \\ u(x) = 0, \ x \in S, \end{cases}$$

the coercivity inequality [53]

$$\sum_{r=1}^{m} \|u_{x_r x_r}\|_{L_2(\overline{\Omega})} \leq M_1 \|\omega\|_{L_2(\overline{\Omega})}.$$

is satisfied. Here M_1 is independent of $\omega(x)$.

Fourth, in $[0, \infty) \times \Omega$ we consider the initial boundary value problem for a multidimensional semilinear delay hyperbolic equation

$$\begin{cases} \frac{\partial^2 u(t,x)}{\partial t^2} - \sum_{r=1}^{m}(a_r(x)u_{x_r})_{x_r} + \delta u(t,x) \\ = f(t, x, u(t,x), u_t(t-w, x), u(t-w, x)), \quad x = (x_1, ..., x_m) \in \Omega, \\ u(t, x) = \varphi(t, x), \quad -w \leq t \leq 0, \quad x \in \overline{\Omega}, \\ \frac{\partial u}{\partial \overrightarrow{p}}(t, x) = 0, \quad x \in S, \quad 0 \leq t < \infty, \end{cases} \quad (26)$$

where $a_r(x)$ and $\varphi(t, x)$ are given sufficiently smooth functions and $\delta > 0$ is the sufficiently large number and $a_r(x) > 0$. Here, $\overrightarrow{p}$ is the normal vector to Ω.

Theorem 6. *Suppose that assumptions of Theorem 4 hold. Assume that all compatibility conditions are satisfied. Then same stability estimates for the solution of (26) hold.*

The proof of Theorem 6 is based on the abstract Theorem 3, on the self-adjointness and positivity of a differential operator A defined by the formula

$$Au(x) = -\sum_{r=1}^{m}(a_r(x)u_{x_r})_{x_r} + \delta u(x) \quad (27)$$

with domain

$$D(A) = \left\{ u(x) : u(x), \, u_{x_r}(x), \, (a_r(x)u_{x_r})_{x_r} \in L_2(\overline{\Omega}), \, 1 \leq r \leq m, \, \frac{\partial}{\partial \overrightarrow{p}} u(x) = 0, \, x \in S \right\}$$

in $L_2(\overline{\Omega})$ and on the following theorem on the coercivity inequality for the solution of the elliptic differential problem in $L_2(\overline{\Omega})$.

Theorem 7. *For the solutions of the elliptic differential problem*

$$\begin{cases} A^x u(x) = \omega(x), \quad x \in \Omega, \\ \frac{\partial u(x)}{\partial \overrightarrow{p}} = 0, \quad x \in S, \end{cases}$$

the coercivity inequality [53]

$$\sum_{r=1}^{m} \|u_{x_r x_r}\|_{L_2(\overline{\Omega})} \leq M_1(\delta) \|\omega\|_{L_2(\overline{\Omega})}$$

is satisfied. Here $M_1(\delta)$ is independent of $\omega(x)$.

4. The Main Theorem on Existence and Uniqueness of a Uniformly Bounded Solution of the Difference Scheme

In the present section for the approximate solution of Equation (1) we will study the first order of accuracy difference scheme

$$\begin{cases} \frac{u^{k+1}-2u^k+u^{k-1}}{\tau^2} + Au^{k+1} = f(t_k, u^k, \frac{u^{k-N}-u^{k-N-1}}{\tau}, u^{k-N}), \\ t_k = k\tau, \ 1 \leq k < \infty, \ N\tau = \omega, \\ (I+\tau^2 A)\frac{u^{k+1}-u^k}{\tau} = \frac{u^k-u^{k-1}}{\tau}, \ k = mN, \ m = 0, 1, ..., \\ u^k = \varphi_k, \ \varphi_k = \varphi(t_k), \ t_k = k\tau, \ -N \leq k \leq 0. \end{cases} \quad (28)$$

The approach of proof of the theorem on the existence and uniqueness of a bounded solution of difference scheme (28) uniformly with respect to τ is based on reducing this difference scheme to an equivalent nonlinear equations. Equivalent nonlinear equations for the difference scheme (28) is

$$u^k = \begin{cases} \varphi_0 + \tau R\widetilde{R}\frac{\varphi_0-\varphi_{-1}}{\tau}, \ k=1, \\[4pt] \frac{R^{k-1}+\widetilde{R}^{k-1}}{2}\varphi_0 + \tau\left(R-\widetilde{R}\right)^{-1}\left(R^k-\widetilde{R}^k\right)R\widetilde{R}\frac{\varphi_0-\varphi_{-1}}{\tau} \\[4pt] +\tau\left(R-\widetilde{R}\right)^{-1}\sum_{p=1}^{k-1} R\widetilde{R}\left(R^{k-p}-\widetilde{R}^{k-p}\right)f(t_p, ju^p, \frac{\varphi_{p-N}-\varphi_{p-N-1}}{\tau}, \varphi_{p-N})\tau, \\[4pt] 2\leq k \leq N, \\[4pt] u^{mN} + \tau R\widetilde{R}\frac{u^{mN}-u^{mN-1}}{\tau}, \ k=mN+1, \\[4pt] \frac{R^{k-mN-1}+\widetilde{R}^{k-mN-1}}{2}u^{mN} + \left(R-\widetilde{R}\right)^{-1}\left(R^{k-mN}-\widetilde{R}^{k-mN}\right)R\widetilde{R}\frac{u^{mN}-u^{mN-1}}{\tau} \\[4pt] +\tau\left(R-\widetilde{R}\right)^{-1}\sum_{p=mN+1}^{k-1} R\widetilde{R}\left(R^{k-p}-\widetilde{R}^{k-p}\right)f(t_p, u^p, \frac{u^{p-N}-u^{p-N-1}}{\tau}, u^{p-N})\tau, \\[4pt] 2+mN \leq k \leq (m+1)N, \ m=1,2,\cdots \end{cases} \quad (29)$$

in $C_\tau(H)$ and the use of successive approximations. Here and in future $R = (I + \tau i A^{\frac{1}{2}})^{-1}, \widetilde{R} = (I - \tau i A^{\frac{1}{2}})^{-1}$ and $C_\tau(H) = C([0,\infty)_\tau, H)$ stands for the Banach space of the mesh functions $v^\tau = \{v^l\}_{l=0}^\infty$ defined on a grid space

$$[0,\infty)_\tau = \{t_k = k\tau, \ k=0,1,..., \ N\tau = w\}$$

with values in H, equipped with the norm

$$\|v^\tau\|_{C_\tau(H)} = \sup_{0\leq l < \infty} \|v^l\|_H.$$

The recursive formula for the solution of difference scheme (28) is

$$\begin{cases} \frac{j u^{k+1} - 2 j u^k + j u^{k-1}}{\tau^2} + A j u^{k+1} = f(t_k, (j-1) u^k, \frac{u^{k-N} - u^{k-N-1}}{\tau}, u^{k-N}), \\ t_k = k\tau, \ mN + 1 \leq k \leq (m+1)N, \ m = 0, 1, 2, \cdots, \ N\tau = \omega, \\ (I + \tau^2 A) \frac{j u^{k+1} - j u^k}{\tau} = \frac{j u^k - j u^{k-1}}{\tau}, \ k = mN, \ m = 1, \ldots, \\ u^k = \varphi_k, \ t_k = k\tau, \ -N \leq k \leq 0, \\ j = 1, 2, \cdots, \ 0 u^k \text{ is given for any } k. \end{cases} \quad (30)$$

From Equations (29) and (30) it follows

$$j u^k = \begin{cases} \varphi_0 + \tau R \widetilde{R} \frac{\varphi_0 - \varphi_{-1}}{\tau}, k = 1, \\[4pt] \frac{R^{k-1} + \widetilde{R}^{k-1}}{2} \varphi_0 + \tau \left(R - \widetilde{R}\right)^{-1} \left(R^k - \widetilde{R}^k\right) R \widetilde{R} \frac{\varphi_0 - \varphi_{-1}}{\tau} \\[4pt] + \tau \left(R - \widetilde{R}\right)^{-1} \sum_{p=1}^{k-1} R \widetilde{R} \left(R^{k-p} - \widetilde{R}^{k-p}\right) f(t_p, j u^p, \frac{\varphi_{p-N} - \varphi_{p-N-1}}{\tau}, \varphi_{p-N}) \tau, \\[4pt] 2 \leq k \leq N, \\[4pt] u^{mN} + \tau R \widetilde{R} \frac{u^{mN} - u^{mN-1}}{\tau}, \ k = mN + 1, \\[4pt] \frac{R^{k-mN-1} + \widetilde{R}^{k-mN-1}}{2} u^{mN} + \tau \left(R - \widetilde{R}\right)^{-1} \left(R^{k-mN} - \widetilde{R}^{k-mN}\right) R \widetilde{R} \frac{u^{mN} - u^{mN-1}}{\tau} \\[4pt] + \left(R - \widetilde{R}\right)^{-1} R \widetilde{R} \\[4pt] \times \tau \sum_{p=mN+1}^{k-1} \left(R^{k-p} - \widetilde{R}^{k-p}\right) f(t_p, (j-1) u^p, \frac{u^{p-N} - u^{p-N-1}}{\tau}, u^{p-N}) \tau, \\[4pt] mN \leq k \leq (m+1)N, \ m = 1, 2, \cdots, \ j = 1, 2, \ldots, \end{cases} \quad (31)$$

where

$$0 u^k = \begin{cases} \varphi_0 + R \widetilde{R} \frac{\varphi_0 - \varphi_{-1}}{\tau}, \ k = 1, \\[4pt] \frac{R^{k-1} + \widetilde{R}^{k-1}}{2} \varphi_0 + \tau \left(R - \widetilde{R}\right)^{-1} \left(R^k - \widetilde{R}^k\right) R \widetilde{R} \frac{\varphi_0 - \varphi_{-1}}{\tau}, \\[4pt] 2 \leq k \leq N, \\[4pt] u^{mN} + \tau R \widetilde{R} \frac{u^{mN} - u^{mN-1}}{\tau}, \ k = mN + 1, \\[4pt] \frac{R^{k-mN-1} + \widetilde{R}^{k-mN-1}}{2} u^{mN} + \tau \left(R - \widetilde{R}\right)^{-1} \left(R^{k-mN} - \widetilde{R}^{k-mN}\right) R \widetilde{R} \frac{u^{mN} - u^{mN-1}}{\tau} \\[4pt] 2 + mN \leq k \leq (m+1)N, \ m = 1, 2, \cdots. \end{cases}$$

Let us give the lemma that will be needed below.

Lemma 2. *The following estimates hold:*

$$\|R\|_{H \to H} \leq 1, \quad \|\tau AR\|_{H \to H} \leq 1, \quad \|\widetilde{R}\|_{H \to H} \leq 1, \quad \|\tau A \widetilde{R}\|_{H \to H} \leq 1, \tag{32}$$

$$\left\|\tau A(I + \tau^2 A^2)^{-1}\right\|_{H \to H} \leq 1. \tag{33}$$

The proof of Lemma 2 is based on the spectral representation of the self-adjoint positive definite operator in a Hilbert space (see, [50]).

Theorem 8. *Let the assumptions of Theorem 1 be satisfied. Then, there exists a unique solution $u^\tau = \{u^k\}_{k=0}^\infty$ of difference scheme (28) which is bounded in $C_\tau(H)$ of uniformly with respect to τ.*

Proof. Step 1. Uniformly boundedness of solution of difference scheme (28) on $[0, w]_\tau$. Assume that $1 \leq k \leq N$. According to the method of recursive approximation (31), we get

$$u^k = 0u^k + \sum_{i=0}^\infty \left[(i+1)u^k - iu^k\right], \tag{34}$$

where

$$0u^1 = \varphi_0 + \tau R \widetilde{R} \frac{\varphi_0 - \varphi_{-1}}{\tau},$$

$$0u^k = \frac{R^{k-1} + \widetilde{R}^{k-1}}{2} \varphi_0 + \tau \left(R - \widetilde{R}\right)^{-1} \left(R^k - \widetilde{R}^k\right) R \widetilde{R} \frac{\varphi_0 - \varphi_{-1}}{\tau}, \quad 2 \leq k \leq N, \tag{35}$$

$$ju^1 = \varphi_0 + \tau R \widetilde{R} \frac{\varphi_0 - \varphi_{-1}}{\tau}, \quad j = 1, 2, \cdots,$$

$$ju^k = \frac{R^{k-1} + \widetilde{R}^{k-1}}{2} \varphi_0 + \tau \left(R - \widetilde{R}\right)^{-1} \left(R^k - \widetilde{R}^k\right) R \widetilde{R} \frac{\varphi_0 - \varphi_{-1}}{\tau} + \tau \left(R - \widetilde{R}\right)^{-1}$$

$$\times \sum_{p=1}^{k-1} R \widetilde{R} \left(R^{k-p} - \widetilde{R}^{k-p}\right) f(t_p, ju^p, \frac{\varphi_{p-N} - \varphi_{p-N-1}}{\tau}, \varphi_{p-N}) \tau, \quad 2 \leq k \leq N, \; j = 1, 2, \cdots. \tag{36}$$

Applying formula (35), the triangle inequality and estimates (5), (32), and (33), we get

$$\left\|0u^1\right\|_H \leq \|A^{-1}\|_{H \to H} \left[\|A\varphi_0\|_H + \left\|\tau A^{\frac{1}{2}} R \widetilde{R}\right\|_{H \to H} \left\|A^{\frac{1}{2}} \frac{\varphi_0 - \varphi_{-1}}{\tau}\right\|_H\right] \leq \delta^{-1} \left[M + \widetilde{M}\right],$$

$$\left\|\frac{0u^1 - \varphi_0}{\tau}\right\|_H \leq \|A^{-\frac{1}{2}}\|_{H \to H} \|R\widetilde{R}\|_{H \to H} \left\|A^{\frac{1}{2}} \frac{\varphi_0 - \varphi_{-1}}{\tau}\right\|_H \leq \delta^{-\frac{1}{2}} \widetilde{M},$$

$$\left\|0u^k\right\|_H \leq \|A^{-1}\|_{H \to H} \left[\frac{1}{2}\left[\|R\|_{H \to H}^{k-1} + \|\widetilde{R}\|_{H \to H}^{k-1}\right] \|A\varphi_0\|_H + \frac{1}{2} \left[\|R\|_{H \to H}^k + \|\widetilde{R}\|_{H \to H}^k\right]\right.$$

$$\left. \times \left\|A^{\frac{1}{2}} \frac{\varphi_0 - \varphi_{-1}}{\tau}\right\|_H\right] \leq \delta^{-1} \left[\|A\varphi_0\|_H + \left\|A^{\frac{1}{2}} \frac{\varphi_0 - \varphi_{-1}}{\tau}\right\|_H\right] \leq \delta^{-1} \left[M + \widetilde{M}\right],$$

$$\left\|\frac{0u^k - 0u^{k-1}}{\tau}\right\|_H \leq \|A^{-\frac{1}{2}}\|_{H \to H} \left[\frac{1}{2}\left[\|R\|_{H \to H}^{k-1} + \|\widetilde{R}\|_{H \to H}^{k-1}\right] \|A\varphi_0\|_H + \frac{1}{2} \left[\|R\|_{H \to H}^k + \|\widetilde{R}\|_{H \to H}^k\right]\right.$$

$$\left. \times \left\|A^{\frac{1}{2}} \frac{\varphi_0 - \varphi_{-1}}{\tau}\right\|_H\right] \leq \delta^{-\frac{1}{2}} \left[\|A\varphi_0\|_H + \left\|A^{\frac{1}{2}} \frac{\varphi_0 - \varphi_{-1}}{\tau}\right\|_H\right] \leq \delta^{-\frac{1}{2}} \left[M + \widetilde{M}\right]$$

for any $2 \leq k \leq N$. Applying formula (36), and estimates (6), (32), and (33), we get

$$\|1u^1 - 0u^1\|_H = 0,$$

$$\|1u^k - 0u^k\|_H \leq \|A^{-\frac{1}{2}}\|_{H \to H}$$

$$\times \sum_{p=1}^{k-1} \frac{1}{2} \left[\|R\|_{H \to H}^{k-p} + \|\widetilde{R}\|_{H \to H}^{k-p} \right] \|f(t_p, 1u^p, \frac{\varphi_{p-N} - \varphi_{p-N-1}}{\tau}, \varphi_{p-N})\|_H \tau \leq \delta^{-\frac{1}{2}} \bar{M} t_k,$$

$$\left\| \frac{1u^1 - \varphi_0}{\tau} - \frac{0u^1 - \varphi_0}{\tau} \right\|_H = 0,$$

$$\left\| \frac{1u^k - 1u^{k-1}}{\tau} - \frac{0u^k - 0u^{k-1}}{\tau} \right\|_H$$

$$\leq \sum_{p=1}^{k-1} \frac{1}{2} \left[\|R\|_{H \to H}^{k-p} + \|\widetilde{R}\|_{H \to H}^{k-p} \right] \|f(t_p, 1u^p, \frac{\varphi_{p-N} - \varphi_{p-N-1}}{\tau}, \varphi_{p-N})\|_H \tau \leq \bar{M} t_k$$

for any $k = 2, \cdots, N$. Using the triangle inequality, we get

$$\|1u^k\|_H \leq \delta^{-1} \left[M + \widetilde{M} \right] + \delta^{-\frac{1}{2}} \bar{M} t_k,$$

$$\left\| \frac{1u^k - 1u^{k-1}}{\tau} \right\|_H \leq \delta^{-\frac{1}{2}} \left[M + \widetilde{M} \right] + \bar{M} t_k$$

for any $k = 1, \cdots, N$. Applying formula (36), and estimates (7), (32), and (33), we get

$$\|2u^1 - 1u^1\|_H = 0,$$

$$\|2u^k - 1u^k\|_H \leq \|A^{-\frac{1}{2}}\|_{H \to H} \sum_{p=1}^{k-1} \frac{1}{2} \left[\|R\|_{H \to H}^{k-p} + \|\widetilde{R}\|_{H \to H}^{k-p} \right]$$

$$\times \left\| f(t_p, 1u^p, \frac{\varphi_{p-N} - \varphi_{p-N-1}}{\tau}, \varphi_{p-N}) - f(t_p, 0u^p, \frac{\varphi_{p-N} - \varphi_{p-N-1}}{\tau}, \varphi_{p-N}) \right\|_H \tau$$

$$\leq \delta^{-\frac{1}{2}} \sum_{p=1}^{k-1} L \|1u^p - 0u^p\|_H \tau \leq \delta^{-\frac{1}{2}} \sum_{p=1}^{k-1} L \delta^{-\frac{1}{2}} \bar{M} t_p \tau \leq \delta^{-1} L \bar{M} \frac{t_k^2}{2!},$$

$$\left\| \frac{2u^1 - 2u^0}{\tau} - \frac{1u^1 - 1u^0}{\tau} \right\|_H = 0,$$

$$\left\| \frac{2u^k - 2u^{k-1}}{\tau} - \frac{1u^k - 1u^{k-1}}{\tau} \right\|_H \leq \sum_{p=1}^{k-1} \frac{1}{2} \left[\|R\|_{H \to H}^{k-p} + \|\widetilde{R}\|_{H \to H}^{k-p} \right]$$

$$\times \left\| f(t_p, 1u^p, \frac{\varphi_{p-N} - \varphi_{p-N-1}}{\tau}, \varphi_{p-N}) - f(t_p, 0u^p, \frac{\varphi_{p-N} - \varphi_{p-N-1}}{\tau}, \varphi_{p-N}) \right\|_H \tau$$

$$\leq \sum_{p=1}^{k-1} L \|1u^p - 0u^p\|_H \tau \leq \sum_{p=1}^{k-1} L \delta^{-\frac{1}{2}} \bar{M} t_p \tau \leq \frac{\bar{M}}{L \delta^{-\frac{1}{2}}} \frac{\left(\delta^{-\frac{1}{2}} L t_k \right)^2}{2!},$$

for any $k = 2, \cdots, N$. Using the triangle inequality, we get

$$\|2u^k\|_H \leq \delta^{-1} \left[M + \widetilde{M} \right] + \delta^{-\frac{1}{2}} \bar{M} t_k + \frac{\bar{M}}{L} \frac{\left(\delta^{-\frac{1}{2}} L t_k \right)^2}{2!},$$

$$\left\|\frac{2u^k - 2u^{k-1}}{\tau}\right\|_H \leq \delta^{-\frac{1}{2}}\left[M + \widetilde{M}\right] + \bar{M}t_k + \frac{\bar{M}}{L\delta^{-\frac{1}{2}}}\frac{\left(\delta^{-\frac{1}{2}}Lt_k\right)^2}{2!}$$

for any $k = 1, \cdots, N$. Let

$$\|nu^k - (n-1)u^k\|_H \leq \frac{\bar{M}}{L}\frac{(L\delta^{-\frac{1}{2}}t_k)^n}{n!},$$

$$\left\|\frac{nu^k - nu^{k-1}}{\tau} - \frac{(n-1)u^k - (n-1)u^{k-1}}{\tau}\right\|_H \leq \frac{\bar{M}}{L\delta^{-\frac{1}{2}}}\frac{(L\delta^{-\frac{1}{2}}t_k)^n}{n!},$$

$$\|nu^k\|_H \leq \delta^{-1}\left[M + \widetilde{M}\right] + \delta^{-\frac{1}{2}}\bar{M}t_k + \bar{M}\frac{\left(\delta^{-\frac{1}{2}}Lt_k\right)^2}{2!} + \cdots + \frac{\bar{M}}{L}\frac{(L\delta^{-\frac{1}{2}}t_k)^n}{n!},$$

$$\left\|\frac{nu^k - nu^{k-1}}{\tau}\right\|_H \leq \delta^{-\frac{1}{2}}\left[M + \widetilde{M}\right] + \bar{M}t_k$$

$$+ \frac{\bar{M}}{L\delta^{-\frac{1}{2}}}\frac{\left(\delta^{-\frac{1}{2}}Lt_k\right)^2}{2!} + \cdots + \frac{\bar{M}}{L\delta^{-\frac{1}{2}}}\frac{(L\delta^{-\frac{1}{2}}t_k)^n}{n!}$$

for any $k = 1, \cdots, N$. Applying formula (36), and estimates (7), (32), and (33), we get

$$\|(n+1)u^1 - nu^1\|_H = 0,$$

$$\|(n+1)u^k - nu^k\|_H \leq \|A^{-\frac{1}{2}}\|_{H \to H}\sum_{p=1}^{k-1}\frac{1}{2}\left[\|R\|_{H \to H}^{k-p} + \|\widetilde{R}\|_{H \to H}^{k-p}\right]$$

$$\times \left\|f(t_p, nu^p, \frac{\varphi_{p-N} - \varphi_{p-N-1}}{\tau}, \varphi_{p-N}) - f(t_p, (n-1)u^p, \frac{\varphi_{p-N} - \varphi_{p-N-1}}{\tau}, \varphi_{p-N})\right\|_H \tau$$

$$\leq \delta^{-\frac{1}{2}}\sum_{p=1}^{k-1}L\|nu^p - (n-1)u^p\|_H\tau \leq \delta^{-\frac{1}{2}}L\sum_{p=1}^{k-1}\frac{\bar{M}}{L}\frac{(L\delta^{-\frac{1}{2}}t_p)^n}{n!}\tau \leq \frac{\bar{M}}{L}\frac{(L\delta^{-\frac{1}{2}}t_k)^{n+1}}{(n+1)!},$$

$$\left\|\frac{(n+1)u^1 - nu^0}{\tau} - \frac{nu^1 - nu^0}{\tau}\right\|_H = 0,$$

$$\left\|\frac{(n+1)u^k - (n+1)u^{k-1}}{\tau} - \frac{nu^k - nu^{k-1}}{\tau}\right\|_H \leq \sum_{p=1}^{k-1}\frac{1}{2}\left[\|R\|_{H \to H}^{k-p} + \|\widetilde{R}\|_{H \to H}^{k-p}\right]$$

$$\times \left\|f(t_p, nu^p, \frac{\varphi_{p-N} - \varphi_{p-N-1}}{\tau}, \varphi_{p-N}) - f(t_p, (n-1)u^p, \frac{\varphi_{p-N} - \varphi_{p-N-1}}{\tau}, \varphi_{p-N})\right\|_H \tau$$

$$\leq \sum_{p=1}^{k-1}L\|nu^p - (n-1)u^p\|_H\tau \leq \sum_{p=1}^{k-1}L\frac{\bar{M}}{L\delta^{-\frac{1}{2}}}\frac{(L\delta^{-\frac{1}{2}}t_p)^n}{n!}\tau \leq \frac{\bar{M}}{L\delta^{-\frac{1}{2}}}\frac{(L\delta^{-\frac{1}{2}}t_k)^{n+1}}{(n+1)!}$$

for any $k = 2, \cdots, N$. Using the triangle inequality, we get

$$\|(n+1)u^k\|_H \leq \delta^{-1}\left[M + \widetilde{M}\right] + \delta^{-\frac{1}{2}}\bar{M}t_k + \frac{\bar{M}}{L}\frac{\left(\delta^{-\frac{1}{2}}Lt_k\right)^2}{2!} + \cdots + \frac{\bar{M}}{L}\frac{(L\delta^{-\frac{1}{2}}t_k)^{n+1}}{(n+1)!},$$

$$\left\|\frac{(n+1)u^k - (n+1)u^{k-1}}{\tau}\right\|_H \leq \delta^{-\frac{1}{2}}\left[M + \widetilde{M}\right] + \bar{M}t_k$$

$$+\frac{\bar{M}}{L\delta^{-\frac{1}{2}}}\frac{\left(\delta^{-\frac{1}{2}}Lt_k\right)^2}{2!}+\cdots+\frac{\bar{M}}{L\delta^{-\frac{1}{2}}}\frac{(L\delta^{-\frac{1}{2}}t_k)^{n+1}}{(n+1)!}$$

for any $k = 1, \cdots, N$. Therefore, for any $n, n \geq 1$, we have that

$$\|(n+1)u^k - nu^k\|_H \leq \frac{\bar{M}}{L}\frac{(L\delta^{-\frac{1}{2}}t_k)^{n+1}}{(n+1)!},$$

$$\left\|\frac{(n+1)u^k - (n+1)u^{k-1}}{\tau} - \frac{nu^k - nu^{k-1}}{\tau}\right\|_H \leq \frac{\bar{M}}{L\delta^{-\frac{1}{2}}}\frac{(L\delta^{-\frac{1}{2}}t_k)^{n+1}}{(n+1)!}$$

and

$$\|(n+1)u^k\|_H \leq \delta^{-1}\left[M + \tilde{M}\right] + \delta^{-\frac{1}{2}}\bar{M}t_k + \frac{\bar{M}}{L}\frac{\left(\delta^{-\frac{1}{2}}Lt_k\right)^2}{2!} + \cdots + \frac{\bar{M}}{L}\frac{(L\delta^{-\frac{1}{2}}t_k)^{n+1}}{(n+1)!},$$

$$\left\|\frac{(n+1)u^k - (n+1)u^{k-1}}{\tau}\right\|_H \leq \delta^{-\frac{1}{2}}\left[M + \tilde{M}\right] + \bar{M}t_k$$

$$+\frac{\bar{M}}{L\delta^{-\frac{1}{2}}}\frac{\left(\delta^{-\frac{1}{2}}Lt_k\right)^2}{2!} + \cdots + \frac{\bar{M}}{L\delta^{-\frac{1}{2}}}\frac{(L\delta^{-\frac{1}{2}}t_k)^{n+1}}{(n+1)!}$$

for any $k = 1, \cdots, N$ by mathematical induction. In a similar manner, for any n, we can obtain

$$\|A^{\frac{1}{2}}(n+1)u^k - A^{\frac{1}{2}}nu^k\|_H \leq \frac{\bar{M}}{L\delta^{-\frac{1}{2}}}\frac{(L\delta^{-\frac{1}{2}}t_k)^{n+1}}{(n+1)!}$$

and

$$\|(n+1)u^k\|_H \leq \delta^{-\frac{1}{2}}\left[M + \tilde{M}\right] + \bar{M}t_k + \frac{\bar{M}}{L\delta^{-\frac{1}{2}}}\frac{(L\delta^{-\frac{1}{2}}t)^2}{2!} + \cdots + \frac{\bar{M}}{L\delta^{-\frac{1}{2}}}\frac{(L\delta^{-\frac{1}{2}}t_k)^{n+1}}{(n+1)!}.$$

From that and formula (34) it follows that

$$\|u^k\|_H \leq \|0u^k\|_H + \sum_{i=0}^{\infty}\|(i+1)u^k - iu^k\|_H$$

$$\leq \delta^{-1}\left[M + \tilde{M}\right] + \sum_{i=0}^{\infty}\frac{\bar{M}}{L}\frac{(\delta^{-\frac{1}{2}}Lt_k)^{i+1}}{(i+1)!} \leq \delta^{-1}\left[M + \tilde{M}\right] + \frac{\bar{M}}{L}e^{\delta^{-\frac{1}{2}}Lt_k}, \quad 1 \leq k \leq N,$$

$$\left\|\frac{u^k - u^{k-1}}{\tau}\right\|_H \leq \left\|\frac{0u^k - 0u^{k-1}}{\tau}\right\|_H + \sum_{i=0}^{\infty}\left\|\frac{(i+1)u^k - (i+1)u^{k-1}}{\tau} - \frac{iu^k - iu^{k-1}}{\tau}\right\|_H$$

$$\leq \delta^{-\frac{1}{2}}\left[M + \tilde{M}\right] + \sum_{i=0}^{\infty}\frac{\bar{M}}{L\delta^{-\frac{1}{2}}}\frac{(L\delta^{-\frac{1}{2}}t_k)^{i+1}}{(i+1)!} \leq \delta^{-\frac{1}{2}}\left[M + \tilde{M}\right] + \frac{\bar{M}}{L\delta^{-\frac{1}{2}}}e^{\delta^{-\frac{1}{2}}Lt_k}, \quad 1 \leq k \leq N,$$

$$\|A^{\frac{1}{2}}u^k\|_H \leq \|A^{\frac{1}{2}}0u^k\|_H + \sum_{i=0}^{\infty}\|A^{\frac{1}{2}}(i+1)u^k - A^{\frac{1}{2}}iu^k\|_H$$

$$\leq \delta^{-\frac{1}{2}}\left[M + \tilde{M}\right] + \sum_{i=0}^{\infty}\frac{\bar{M}}{L\delta^{-\frac{1}{2}}}\frac{(L\delta^{-\frac{1}{2}}t_k)^{i+1}}{(i+1)!} \leq \delta^{-\frac{1}{2}}\left[M + \tilde{M}\right] + \frac{\bar{M}}{L\delta^{-\frac{1}{2}}}e^{\delta^{-\frac{1}{2}}Lt_k}, \quad 1 \leq k \leq N$$

which proves the existence of a bounded solution of difference scheme (28) in $[0, w]_\tau \times H$ of uniformly with respect to τ.

Step 2. Uniformly boundedness of solution of difference scheme (28) on $[mw, (m+1)w]_\tau$. We consider solution of difference scheme (28) in $mN \leq k \leq (m+1)N$, $m = 1, 2, \ldots$. Then, according to the method of recursive approximation (31), we get

$$u^k = 0u^k + \sum_{i=0}^{\infty} \left[(i+1)u^k - iu^k\right], \tag{37}$$

where

$$\begin{aligned}
0u^{mN+1} &= u^{mN} + \tau R\widetilde{R}\frac{u^{mN} - u^{mN-1}}{\tau}, \\
0u^k &= \frac{R^{k-mN-1} + \widetilde{R}^{k-mN-1}}{2}u^{mN} \\
&\quad + \tau\left(R - \widetilde{R}\right)^{-1}\left(R^{k-mN} - \widetilde{R}^{k-mN}\right)R\widetilde{R}\frac{u^{mN} - u^{mN-1}}{\tau}, \\
&\quad 2 + mN \leq k \leq (m+1)N, \ m = 1, 2, \cdots,
\end{aligned} \tag{38}$$

$$\begin{aligned}
ju^{nN+1} &= u^{mN} + \tau R\widetilde{R}\frac{u^{mN} - u^{mN-1}}{\tau}, \ j = 1, 2, \cdots, \\
ju^k &= \frac{R^{k-mN-1} + \widetilde{R}^{k-mN-1}}{2}u^{mN} + \tau\left(R - \widetilde{R}\right)^{-1}\left(R^{k-mN} - \widetilde{R}^{k-mN}\right)R\widetilde{R}\frac{u^{mN} - u^{mN-1}}{\tau} \\
&\quad + \tau\left(R - \widetilde{R}\right)^{-1}\sum_{p=mN+1}^{k-1} R\widetilde{R}\left(R^{k-p} - \widetilde{R}^{k-p}\right)f(t_p, (j-1)u^p, \frac{u^{p-N} - u^{p-N-1}}{\tau}, u^{p-N})\tau, \\
&\quad 2 + mN \leq k \leq (m+1)N, \ m = 1, 2, \cdots, \ j = 1, 2, \cdots.
\end{aligned} \tag{39}$$

Assume that difference scheme (28) in $[(m-1)N, mN]_\tau \times H$ of uniformly with respect to τ and

$$\|A^{1/2}u^k\|_H \leq M_{m-1}, \quad \left\|\frac{u^k - u^{k-1}}{\tau}\right\|_H \leq \widetilde{M}_{m-1}. \tag{40}$$

Applying formula (38), the triangle inequality and estimates (32), (33), and (40), we get

$$\|0u^{mN+1}\|_H \leq \|A^{-\frac{1}{2}}\|_{H \to H}\left[\left\|A^{\frac{1}{2}}u^{mN}\right\|_H + \left\|\tau A^{\frac{1}{2}}R\widetilde{R}\right\|_{H \to H}\left\|\frac{u^{mN} - u^{mN-1}}{\tau}\right\|_H\right] \leq \delta^{-\frac{1}{2}}\left[M_{m-1} + \widetilde{M}_{m-1}\right],$$

$$\left\|\frac{0u^{mN+1} - u^{mN}}{\tau}\right\|_H \leq \|R\widetilde{R}\|_{H \to H}\left\|\frac{u^{mN} - u^{mN-1}}{\tau}\right\|_H \leq \widetilde{M}_{m-1},$$

$$\|0u^k\|_H \leq \|A^{-\frac{1}{2}}\|_{H \to H}\left[\frac{1}{2}\left[\|R\|_{H \to H}^{k-1} + \|\widetilde{R}\|_{H \to H}^{k-1}\right]\left\|A^{\frac{1}{2}}u^{mN}\right\|_H + \frac{1}{2}\left[\|R\|_{H \to H}^k + \|\widetilde{R}\|_{H \to H}^k\right]\right.$$
$$\left. \times \left\|\frac{u^{mN} - u^{mN-1}}{\tau}\right\|_H\right] \leq \delta^{-\frac{1}{2}}\left[\left\|A^{\frac{1}{2}}u^{mN}\right\|_H + \left\|\frac{u^{mN} - u^{mN-1}}{\tau}\right\|_H\right] \leq \delta^{-\frac{1}{2}}\left[M_{m-1} + \widetilde{M}_{m-1}\right],$$

$$\left\|\frac{0u^k - 0u^{k-1}}{\tau}\right\|_H \leq \left[\frac{1}{2}\left[\|R\|_{H \to H}^{k-1} + \|\widetilde{R}\|_{H \to H}^{k-1}\right]\left\|A^{\frac{1}{2}}u^{mN}\right\|_H + \frac{1}{2}\left[\|R\|_{H \to H}^k + \|\widetilde{R}\|_{H \to H}^k\right]\right.$$
$$\left. \times \left\|\frac{u^{mN} - u^{mN-1}}{\tau}\right\|_H\right] \leq \|Au^{mN}\|_H + \left\|A^{\frac{1}{2}}\frac{u^{mN} - u^{mN-1}}{\tau}\right\|_H \leq M_{m-1} + \widetilde{M}_{m-1}$$

for any $2 + mN \leq k \leq (m+1)N$. Applying formula (39), and estimates (6), (32) and (33), we get

$$\|1u^{mN+1} - 0u^{mN+1}\|_H = 0,$$

$$\|1u^k - 0u^k\|_H \leq \|A^{-\frac{1}{2}}\|_{H \to H}\sum_{p=mN+1}^{k-1}\frac{1}{2}\left[\|R\|_{H \to H}^{k-p} + \|\widetilde{R}\|_{H \to H}^{k-p}\right]$$

$$\times \left\|f(t_p, (j-1)u^p, \frac{u^{p-N} - u^{p-N-1}}{\tau}, u^{p-N})\right\|_H \tau \leq \delta^{-\frac{1}{2}}\widetilde{M}(t_k - mN),$$

$$\left\|\frac{1u^{mN+1}-u^{mN}}{\tau}-\frac{0u^{mN+1}-u^{mN}}{\tau}\right\|_H = 0,$$

$$\left\|\frac{1u^k-1u^{k-1}}{\tau}-\frac{0u^k-0u^{k-1}}{\tau}\right\|_H$$

$$\leq \sum_{p=mN+1}^{k-1}\frac{1}{2}\left[\|R\|_{H\to H}^{k-p}+\|\widetilde{R}\|_{H\to H}^{k-p}\right]\|f(t_p,1u^p,\frac{u^{p-N}-u^{p-N-1}}{\tau},u^{p-N})\|_H \tau \leq \bar{M}(t_k-mN)$$

for any $2+mN \leq k \leq (m+1)N$. Using the triangle inequality, we get

$$\|1u^k\|_H \leq \delta^{-\frac{1}{2}}\left[M+\widetilde{M}\right]+\delta^{-\frac{1}{2}}\bar{M}(t_k-mN),$$

$$\left\|\frac{1u^k-1u^{k-1}}{\tau}\right\|_H \leq M_{m-1}+\widetilde{M}_{m-1}+\bar{M}(t_k-mN)$$

for any $1+mN \leq k \leq (m+1)N$. Applying formula (39), and estimates (7), (32), and (33), we get

$$\|2u^1-1u^1\|_H = 0,$$

$$\|2u^k-1u^k\|_H \leq \|A^{-\frac{1}{2}}\|_{H\to H}\sum_{p=mN+1}^{k-1}\frac{1}{2}\left[\|R\|_{H\to H}^{k-p}+\|\widetilde{R}\|_{H\to H}^{k-p}\right]$$

$$\times \left\|f(t_p,1u^p,\frac{u^{p-N}-u^{p-N-1}}{\tau},u^{p-N})-f(t_p,0u^p,\frac{u^{p-N}-u^{p-N-1}}{\tau},u^{p-N})\right\|_H \tau$$

$$\leq \delta^{-\frac{1}{2}}\sum_{p=mN+1}^{k-1}L\|1u^p-0u^p\|_H \tau \leq \delta^{-\frac{1}{2}}\sum_{p=mN+1}^{k-1}L\delta^{-\frac{1}{2}}\bar{M}t_{p-mN}\tau \leq \delta^{-1}L\bar{M}\frac{(t_k-mN)^2}{2!},$$

$$\left\|\frac{2u^1-2u^0}{\tau}-\frac{1u^1-1u^0}{\tau}\right\|_H = 0,$$

$$\left\|\frac{2u^k-2u^{k-1}}{\tau}-\frac{1u^k-1u^{k-1}}{\tau}\right\|_H \leq \sum_{p=mN+1}^{k-1}\frac{1}{2}\left[\|R\|_{H\to H}^{k-p}+\|\widetilde{R}\|_{H\to H}^{k-p}\right]$$

$$\times \left\|f(t_p,1u^p,\frac{u^{p-N}-u^{p-N-1}}{\tau},u^{p-N})-f(t_p,0u^p,\frac{u^{p-N}-u^{p-N-1}}{\tau},u^{p-N})\right\|_H \tau$$

$$\leq \sum_{p=mN+1}^{k-1}L\|1u^p-0u^p\|_H \tau \leq \sum_{p=mN+1}^{k-1}L\delta^{-\frac{1}{2}}\bar{M}t_{p-mN}\tau \leq \frac{\bar{M}}{L\delta^{-\frac{1}{2}}}\frac{\left(\delta^{-\frac{1}{2}}L(t_k-mN)\right)^2}{2!}$$

for any $2+mN \leq k \leq (m+1)N$. Using the triangle inequality, we get

$$\|2u^k\|_H \leq \delta^{-\frac{1}{2}}\left[M+\widetilde{M}\right]+\delta^{-\frac{1}{2}}\bar{M}(t_k-mN)+\frac{\bar{M}}{L}\frac{\left(\delta^{-\frac{1}{2}}L(t_k-mN)\right)^2}{2!},$$

$$\left\|\frac{2u^k-2u^{k-1}}{\tau}\right\|_H \leq M_{m-1}+\widetilde{M}_{m-1}+\bar{M}(t_k-mN)+\frac{\bar{M}}{L\delta^{-\frac{1}{2}}}\frac{\left(\delta^{-\frac{1}{2}}L(t_k-mN)\right)^2}{2!}$$

for any $1+mN \leq k \leq (m+1)N$. Let

$$\|nu^k-(n-1)u^k\|_H \leq \frac{\bar{M}}{L}\frac{(L\delta^{-\frac{1}{2}}(t_k-mN))^n}{n!},$$

$$\left\|\frac{nu^k - nu^{k-1}}{\tau} - \frac{(n-1)u^k - (n-1)u^{k-1}}{\tau}\right\|_H \leq \frac{\bar{M}}{L\delta^{-\frac{1}{2}}} \frac{(L\delta^{-\frac{1}{2}}(t_k - mN))^n}{n!},$$

$$\|nu^k\|_H \leq \delta^{-1}\left[M_{m-1} + \widetilde{M}_{m-1}\right] + \delta^{-\frac{1}{2}}\bar{M}(t_k - mN)$$

$$+ \bar{M}\frac{\left(\delta^{-\frac{1}{2}}L(t_k - mN)\right)^2}{2!} + \cdots + \frac{\bar{M}}{L}\frac{(L\delta^{-\frac{1}{2}}(t_k - mN))^n}{n!},$$

$$\left\|\frac{nu^k - nu^{k-1}}{\tau}\right\|_H \leq \delta^{-\frac{1}{2}}\left[M_{m-1} + \widetilde{M}_{m-1}\right] + \bar{M}(t_k - mN)$$

$$+ \frac{\bar{M}}{L\delta^{-\frac{1}{2}}}\frac{\left(\delta^{-\frac{1}{2}}L(t_k - mN)\right)^2}{2!} + \cdots + \frac{\bar{M}}{L\delta^{-\frac{1}{2}}}\frac{(L\delta^{-\frac{1}{2}}(t_k - mN))^n}{n!}$$

for any $1 + mN \leq k \leq (m+1)N$. Applying formula (39), and estimates (7), (32), and (33), we get

$$\|(n+1)u^1 - nu^1\|_H = 0,$$

$$\|(n+1)u^k - nu^k\|_H \leq \|A^{-\frac{1}{2}}\|_{H \to H} \sum_{p=mN+1}^{k-1} \frac{1}{2}\left[\|R\|_{H \to H}^{k-p} + \|\widetilde{R}\|_{H \to H}^{k-p}\right]$$

$$\times \left\|f(t_p, nu^p, \frac{u^{p-N} - u^{p-N-1}}{\tau}, u^{p-N})) - f(t_p, (n-1)u^p, \frac{u^{p-N} - u^{p-N-1}}{\tau}, u^{p-N})\right\|_H \tau$$

$$\leq \delta^{-\frac{1}{2}}\sum_{p=mN+1}^{k-1} L\|nu^p - (n-1)u^p\|_H \tau \leq \delta^{-\frac{1}{2}}\sum_{p=mN+1}^{k-1} L\frac{\bar{M}}{L}\frac{(L\delta^{-\frac{1}{2}}t_{p-mN})^n}{n!}\tau$$

$$\leq \frac{\bar{M}}{L}\frac{(L\delta^{-\frac{1}{2}}(t_k - mN))^{n+1}}{(n+1)!},$$

$$\left\|\frac{(n+1)u^1 - nu^0}{\tau} - \frac{nu^1 - nu^0}{\tau}\right\|_H = 0,$$

$$\left\|\frac{(n+1)u^k - (n+1)u^{k-1}}{\tau} - \frac{nu^k - nu^{k-1}}{\tau}\right\|_H \leq \sum_{p=1}^{k-1}\frac{1}{2}\left[\|R\|_{H \to H}^{k-p} + \|\widetilde{R}\|_{H \to H}^{k-p}\right]$$

$$\times \left\|f(t_p, nu^p, \frac{u^{p-N} - u^{p-N-1}}{\tau}, u^{p-N})) - f(t_p, (n-1)u^p, \frac{u^{p-N} - u^{p-N-1}}{\tau}, u^{p-N})\right\|_H \tau$$

$$\leq \sum_{pmN+1}^{k-1} L\|nu^p - (n-1)u^p\|_H \tau \leq \sum_{p=mN+1}^{k-1} L\frac{\bar{M}}{L\delta^{-\frac{1}{2}}}\frac{(L\delta^{-\frac{1}{2}}t_{p-mN})^n}{n!}\tau$$

$$\leq \frac{\bar{M}}{L\delta^{-\frac{1}{2}}}\frac{(L\delta^{-\frac{1}{2}}(t_k - mN))^{n+1}}{(n+1)!}$$

for any $2 + mN \leq k \leq (m+1)N$. Using the triangle inequality, we get

$$\|(n+1)u^k\|_H \leq \delta^{-\frac{1}{2}}\left[M_{m-1} + \widetilde{M}_{m-1}\right] + \delta^{-\frac{1}{2}}\bar{M}(t_k - mN)$$

$$+ \bar{M}\frac{\left(\delta^{-\frac{1}{2}}L(t_k - mN)\right)^2}{2!} + \cdots + \frac{\bar{M}}{L}\frac{(L\delta^{-\frac{1}{2}}(t_k - mN))^{n+1}}{(n+1)!},$$

$$\left\|\frac{(n+1)u^k - (n+1)u^{k-1}}{\tau}\right\|_H \leq M_{m-1} + \widetilde{M}_{m-1} + \bar{M}(t_k - mN)$$

$$+\frac{\bar{M}}{L\delta^{-\frac{1}{2}}}\frac{\left(\delta^{-\frac{1}{2}}L\left(t_k-mN\right)\right)^2}{2!}+\cdots+\frac{\bar{M}}{L\delta^{-\frac{1}{2}}}\frac{(L\delta^{-\frac{1}{2}}\left(t_k-mN\right))^{n+1}}{(n+1)!}$$

for any $1+mN \leq k \leq (m+1)N$. Therefore, for any $n, n \geq 1$, we have that

$$\|(n+1)u^k - nu^k\|_H \leq \frac{\bar{M}}{L}\frac{(L\delta^{-\frac{1}{2}}\left(t_k-mN\right))^{n+1}}{(n+1)!},$$

$$\left\|\frac{(n+1)u^k-(n+1)u^{k-1}}{\tau}-\frac{nu^k-nu^{k-1}}{\tau}\right\|_H \leq \frac{\bar{M}}{L\delta^{-\frac{1}{2}}}\frac{(L\delta^{-\frac{1}{2}}\left(t_k-mN\right))^{n+1}}{(n+1)!}$$

and

$$\|(n+1)u^k\|_H \leq \delta^{-\frac{1}{2}}\left[M_{m-1}+\tilde{M}_{m-1}\right]+\delta^{-\frac{1}{2}}\bar{M}\left(t_k-mN\right)$$

$$+\frac{\bar{M}}{L}\frac{\left(\delta^{-\frac{1}{2}}L\left(t_k-mN\right)\right)^2}{2!}+\cdots+\frac{\bar{M}}{L}\frac{(L\delta^{-\frac{1}{2}}\left(t_k-mN\right))^{n+1}}{(n+1)!},$$

$$\left\|\frac{(n+1)u^k-(n+1)u^{k-1}}{\tau}\right\|_H \leq M_{m-1}+\tilde{M}_{m-1}+\bar{M}\left(t_k-mN\right)$$

$$+\frac{\bar{M}}{L\delta^{-\frac{1}{2}}}\frac{\left(\delta^{-\frac{1}{2}}L\left(t_k-mN\right)\right)^2}{2!}+\cdots+\frac{\bar{M}}{L\delta^{-\frac{1}{2}}}\frac{(L\delta^{-\frac{1}{2}}\left(t_k-mN\right))^{n+1}}{(n+1)!}$$

for any $1+mN \leq k \leq (m+1)N$ by mathematical induction. From that and formula (37) it follows that

$$\|u^k\|_H \leq \|0u^k\|_H + \sum_{i=0}^{\infty}\|(i+1)u^k - iu^k\|_H$$

$$\leq \delta^{-\frac{1}{2}}\left[M_{m-1}+\tilde{M}_{m-1}\right]+\sum_{i=0}^{\infty}\frac{\bar{M}}{L}\frac{(\delta^{-\frac{1}{2}}L\left(t_k-mN\right))^{i+1}}{(i+1)!}$$

$$\leq \delta^{-\frac{1}{2}}\left[M_{m-1}+\tilde{M}_{m-1}\right]+\frac{\bar{M}}{L}e^{\delta^{-\frac{1}{2}}L(t_k-mN)},$$

$$\left\|\frac{u^k-u^{k-1}}{\tau}\right\|_H \leq \left\|\frac{0u^k-0u^{k-1}}{\tau}\right\|_H + \sum_{i=0}^{\infty}\left\|\frac{(i+1)u^k-(i+1)u^{k-1}}{\tau}-\frac{iu^k-iu^{k-1}}{\tau}\right\|_H$$

$$\leq M_{m-1}+\tilde{M}_{m-1}+\sum_{i=0}^{\infty}\frac{\bar{M}}{L\delta^{-\frac{1}{2}}}\frac{(L\delta^{-\frac{1}{2}}\left(t_k-mN\right))^{i+1}}{(i+1)!}$$

$$\leq M_{m-1}+\tilde{M}_{m-1}+\frac{\bar{M}}{L\delta^{-\frac{1}{2}}}e^{\delta^{-\frac{1}{2}}L(t_k-mN)},$$

$$\left\|A^{\frac{1}{2}}u^k\right\|_H \leq \left\|A^{\frac{1}{2}}0u^k\right\|_H + \sum_{i=0}^{\infty}\left\|A^{\frac{1}{2}}(i+1)u^k - A^{\frac{1}{2}}iu^k\right\|_H \leq M_{m-1}+\tilde{M}_{m-1}$$

$$+\sum_{i=0}^{\infty}\frac{\bar{M}}{L\delta^{-\frac{1}{2}}}\frac{(L\delta^{-\frac{1}{2}}\left(t_k-mN\right))^{i+1}}{(i+1)!} \leq M_{m-1}+\tilde{M}_{m-1}+\frac{\bar{M}}{L\delta^{-\frac{1}{2}}}e^{\delta^{-\frac{1}{2}}L(t_k-mN)}$$

for any $1+mN \leq k \leq (m+1)N$ which proves the existence of a bounded solution of difference scheme (28) in $[m\omega, (m+1)w]_\tau \times H$ of uniformly with respect to τ.

Step 3. Uniqueness of solution of difference scheme (28). We prove uniqueness of the uniformly bounded solution of problem (28). Suppose that there is a bounded solution $v^\tau = \{v^k\}_{k=0}^\infty$ of problem (28) and $v^\tau \neq u^\tau$. Denoting $z^\tau = v^\tau - u^\tau$ and using Equation (28), we get

$$\begin{cases} \frac{z^{k+1}-2z^k+z^{k-1}}{\tau^2} + Az^{k+1} \\ = f(t_k, v^k, \frac{v^{k-N}-v^{k-N-1}}{\tau}, v^{k-N}) - f(t_k, u^k, \frac{u^{k-N}-u^{k-N-1}}{\tau}, u^{k-N}), \ k \geq 1, \\ (I + \tau^2 A)\frac{z^{k+1}-z^k}{\tau} = \frac{z^k-z^{k-1}}{\tau}, \ k = mN, \ m = 1, ..., \\ z^k = 0, \ -N \leq k \leq 0 \end{cases}$$

for z^τ. We consider the interval $1 \leq k \leq N$. Since $v^{k-N} = u^{k-N} = \varphi(t_{k-N})$, we have that

$$\begin{cases} \frac{z^{k+1}-2z^k+z^{k-1}}{\tau^2} + Az^{k+1} \\ = f(t_k, v^k, \frac{\varphi_{k-N}-\varphi_{k-N-1}}{\tau}, \varphi_{k-N}) - f(t_k, u^k, \frac{\varphi_{k-N}-\varphi_{k-N-1}}{\tau}, \varphi_{k-N}), \ 1 \leq k \leq N, \\ (I + \tau^2 A)z^1 = 0, z^0 = 0. \end{cases}$$

Therefore,

$$\|z^k\|_H \leq \|A^{-\frac{1}{2}}\|_{H \to H} \sum_{p=1}^{k-1} \frac{1}{2} \left[\|R\|_{H \to H}^{k-p} + \|\widetilde{R}\|_{H \to H}^{k-p} \right]$$

$$\times \left\| f(t_k, v^k, \frac{\varphi_{k-N} - \varphi_{k-N-1}}{\tau}, \varphi_{k-N}) - f(t_k, u^k, \frac{\varphi_{k-N} - \varphi_{k-N-1}}{\tau}, \varphi_{k-N}) \right\|_H \tau.$$

Applying estimates estimates (7), (32), and (33), we get

$$\|z^k\|_H \leq \delta^{-\frac{1}{2}} L \sum_{m=1}^{k-1} \|z^m\|_H \tau.$$

Using the discrete analogy of the integral inequality, we get

$$\|z^k\|_H \leq 0.$$

From that it follows that $z^k = 0, 1 \leq k \leq N$ which proves the uniqueness of a bounded solution of problem (28) in $[0, w]_\tau \times H$ of uniformly with respect to τ. Using the same method and mathematical induction, we can prove the uniqueness of a bounded solution of problem (28) in $[mw, (m+1)w]_\tau \times H$ of uniformly with respect to τ. Theorem 8 is proved. □

5. Applications of Theorem 8

First, we consider the initial value problem (17) for the one dimensional semininear hyperbolic differential equation with time delay term and with nonlocal conditions.

The discretization of problem (17) is provided in two steps. To the differential operator A generated by problem (17), we assign the difference operator A_h^x by the formula

$$A_h^x \varphi_h(x) = \{-(a(x)\varphi_{\bar{x}})_{x,r} + \delta \varphi_r(x)\}_1^{K-1}, \tag{41}$$

acting in the space of grid functions $\varphi_h(x) = \{\varphi_r\}_0^K$ satisfying the conditions $\varphi_0 = \varphi_K$, $\varphi_1 - \varphi_0 = \varphi_K - \varphi_{K-1}$. It is known that A_h^x is a self-adjoint positive definite operator in $L_{2h} = L_2([0,l]_h)$ [51]. With the help of A_h^x, we arrive at the initial value problem

$$\begin{cases} \frac{d^2 u^h(t,x)}{dt^2} + A_h^x u^h(t,x) = f^h(t,x,u^h(t,x), u_t^h(t-w,x), u^h(t-w,x)), \\ 0 < t < \infty, \ x \in [0,l]_h, \\ u^h(t,x) = \varphi^h(t,x), \ -w \le t \le 0, \ x \in [0,l]_h. \end{cases} \quad (42)$$

In the second step, we replace problem (42) by first order of accuracy difference scheme (28)

$$\begin{cases} \frac{u_{k+1}^h(x) - 2u_k^h(x) + u_{k-1}^h(x)}{\tau^2} + A_h^x u_{k+1}^h(x) = f(t_k, x, u_k^h(x), \frac{u_{k-N}^h(x) - u_{k-N-1}^h(x)}{\tau}, u_{k-N}^h(x)), \\ t_k = k\tau, \ 1 \le k < \infty, \ N\tau = w, \\ (I + \tau^2 A_h^x) \frac{u_{k+1}^h(x) - u_k^h(x)}{\tau} = \frac{u_k^h(x) - u_{k-1}^h(x)}{\tau}, k = mN, m = 0, 1, ..., \\ u_k^h(x) = \varphi^h(t_k, x), \ t_k = k\tau, \ -N \le k \le 0, \ x \in [0,l]_h. \end{cases} \quad (43)$$

Theorem 9. *Suppose that assumptions of Theorem 2 hold. Then, there exists a unique solution $\{u_k^h\}_{k=0}^\infty$ of difference scheme (43) which is bounded in $[m\omega, (m+1)w]_\tau \times L_{2h}$, $m = 0, 1, \cdots$ of uniformly with respect to τ and h.*

Proof. Difference scheme (43) can be written in abstract form (28) in a Hilbert space $L_{2h} = L_2([0,l]_h)$ with self-adjoint positive definite operator $A_h = A_h^x$ by formula (41). Here, $f(t_k, x, u_k^h(x), \frac{u_{k-N}^h(x) - u_{k-N-1}^h(x)}{\tau}, u_{k-N}^h(x))$ and $u_k^h = u_k^h(x)$ are abstract mesh functions defined on $[0,l]_h$ with the values in $H = L_{2h}$. Therefore, the proof of Theorem 9 is based on Theorem 8 and symmetry properties of the difference operator A_h^x. □

Second, we study the initial nonlocal boundary value problem (22) for one dimensional semilinear delay hyperbolic equations type with involution. The discretization of problem (22) is provided in two steps. To the differential operator A generated by problem (22), we assign the difference operator A_h^x by the formula

$$A_h^x \varphi^h(x) = \{-(a(x)\varphi_{\bar{x}}(x))_{x,r} - \beta(a(-x)\varphi_{\bar{x}}(-x))_{x,r} + \delta\varphi^r(x)\}_{-M+1}^{M-1}, \quad (44)$$

acting in the space of grid functions $\varphi^h(x) = \{\varphi^r\}_{-M}^M$ satisfying the conditions $\varphi^{-M} = \varphi^M = 0$. It is known that A_h^x is a self-adjoint positive definite operator in $L_{2h} = L_2([-l,l]_h)$ [52]. With the help of A_h^x, we arrive at the initial value problem

$$\begin{cases} \frac{d^2 u^h(t,x)}{dt^2} + A_h^x u^h(t,x) = f^h(t,x,u^h(t,x), u_t^h(t-w,x), u^h(t-w,x)), \\ 0 < t < \infty, \ x \in [-l,l]_h, \\ u^h(t,x) = \varphi^h(t,x), \ -w \le t \le 0, \ x \in [-l,l]_h. \end{cases} \quad (45)$$

In the second step, we replace problem (45) by first order of accuracy difference scheme (28)

$$\begin{cases} \frac{u^h_{k+1}(x) - 2u^h_k(x) + u^h_{k-1}(x)}{\tau^2} + A^x_h u^h_{k+1}(x) = f(t_k, x, u^h_k(x), \frac{u^h_{k-N}(x) - u^h_{k-N-1}(x)}{\tau}, u^h_{k-N}(x)), \\ t_k = k\tau, \ 1 \leq k < \infty, \ N\tau = \omega, \\ (I + \tau^2 A^x_h) \frac{u^h_{k+1}(x) - u^h_k(x)}{\tau} = \frac{u^h_k(x) - u^h_{k-1}(x)}{\tau}, \ k = mN, \ m = 0, 1, \ldots, \\ u^h_k(x) = \varphi^h(t_k, x), \ t_k = k\tau, \ -N \leq k \leq 0, \ x \in [-l, l]_h. \end{cases} \quad (46)$$

Theorem 10. *Suppose that assumptions of Theorem 3 hold. Then, there exists a unique solution $\left\{u^h_k\right\}_{k=0}^{\infty}$ of difference scheme (46) which is bounded in $[m\omega, (m+1)\omega]_\tau \times L_{2h}$, $m = 0, 1, \cdots$ uniformly with respect to τ and h.*

Proof. Difference scheme (46) can be written in abstract form (28) in a Hilbert space $L_{2h} = L_2([-l, l]_h)$ with self-adjoint positive definite operator $A_h = A^x_h$ by formula (44). Here, $f(t_k, x, u^h_k(x), \frac{u^h_{k-N}(x) - u^h_{k-N-1}(x)}{\tau}, u^h_{k-N}(x))$ and $u^h_k = u^h_k(x)$ are abstract mesh functions defined on $[-l, l]_h$ with the values in $H = L_{2h}$. Therefore, the proof of Theorem 10 is based on Theorem 8 and symmetry properties of the difference operator A^x_h. □

Third, we study the initial boundary value problem (23) for multidimensional semilinear delay hyperbolic equations.

The discretization of problem (23) is provided in two steps. In the first step, here and in future, we define the grid space

$$\overline{\Omega}_h = \{x = x_r = (h_1 j_1, \cdots, h_m j_m), \ j = (j_1, \cdots, j_m), \ 0 \leq j_r \leq N_r,$$
$$N_r h_r = 1, \ r = 1, \cdots, m\}, \ \Omega_h = \overline{\Omega}_h \cap \Omega, \ S_h = \overline{\Omega}_h \cap S.$$

We introduce the Banach space $L_{2h} = L_2(\overline{\Omega}_h)$ of the grid functions $\varphi_h(x) = \{\varphi(h_1 r_1, \ldots, h_m r_m)\}$ defined on $\overline{\Omega}_h$, equipped with the norm

$$\|\varphi_h\|_{L_{2h}} = \left(\sum_{x \in \overline{\Omega}_h} |\varphi_h(x)|^2 h_1 \cdots h_m\right)^{1/2}$$

to the differential operator A generated by problem (23), we assign the difference operator A^x_h by the formula

$$A^x_h u_h(x) = -\sum_{r=1}^{m} (a_r(x) u_{\overline{x}_r, h})_{x_r, j_r} \quad (47)$$

acting in the space of grid functions $u_h(x)$, satisfying the conditions $u_h(x) = 0 (\forall x \in S_h)$. It is known that A^x_h is a self-adjoint positive definite operator in L_{2h}. With the help of A^x_h, we arrive at the initial value problem

$$\begin{cases} \frac{d^2 u_h(t,x)}{dt^2} + A^x_h u_h(t, x) = f^h(t, x, u^h(t, x), u^h_t(t - w, x), u^h(t - w, x)), \\ 0 < t < \infty, \ x \in \Omega_h, \\ u_h(t, x) = \varphi_h(t, x), \ -w \leq t \leq 0, \ x \in \overline{\Omega}_h. \end{cases} \quad (48)$$

In the second step, we replace problem (48) by first order of accuracy difference scheme (28)

$$\begin{cases} \frac{u_h^{k+1}(x) - 2u_h^k(x) + u_h^{k-1}(x)}{\tau^2} + A_h^x u_h^{k+1}(x) = f(t_k, x, u_k^h(x), \frac{u_{k-N}^h(x) - u_{k-N-1}^h(x)}{\tau}, u_{k-N}^h(x)), \\ t_k = k\tau, \ 1 \leq k < \infty, \ N\tau = \omega, \ x \in \Omega_h, \\ (I + \tau^2 A_h^x) \frac{u_h^{k+1}(x) - u_h^k(x)}{\tau} = \frac{u_h^k(x) - u_h^{k-1}(x)}{\tau}, \ k = mN, \ m = 0, 1, ..., \\ u_k^h(x) = \varphi_h(t_k, x), \ t_k = k\tau, \ -N \leq k \leq 0, \ x \in \overline{\Omega}_h. \end{cases} \quad (49)$$

Theorem 11. *Suppose that assumptions of Theorem 4 hold. Then, there exists a unique solution $\{u_k^h\}_{k=0}^\infty$ of difference scheme (49) which is bounded in $[m\omega, (m+1)\omega]_\tau \times L_{2h}$, $m = 0, 1, \cdots$ uniformly with respect to τ and h.*

Proof. Difference scheme (49) can be written in abstract form (28) in a Hilbert space $L_{2h} = L_2(\overline{\Omega}_h)$ with self-adjoint positive definite operator $A_h = A_h^x$ by formula (47). Here, $f(t_k, x, u_k^h(x), \frac{u_{k-N}^h(x) - u_{k-N-1}^h(x)}{\tau}, u_{k-N}^h(x))$ and $u_k^h = u_k^h(x)$ are abstract mesh functions defined on $\overline{\Omega}_h$ with the values in $H = L_{2h}$. Therefore, the proof of Theorems 11 is based on the abstract Theorem 8 and symmetry properties of the difference operator A_h^x defined by formula (47) and the following theorem on coercivity inequality for the solution of the elliptic problem in L_{2h} [53]. □

Theorem 12. *For the solutions of the elliptic difference problem*

$$\begin{cases} A_h^x u^h(x) = \omega^h(x), \ x \in \Omega_h, \\ u^h(x) = 0, \ x \in S_h, \end{cases}$$

the coercivity inequality

$$\sum_{r=1}^n \left\| u^h_{x_r \overline{x}_r} \right\|_{L_{2h}} \leq M_1 \|\omega^h\|_{L_{2h}}$$

is satisfied, where M_1 does not depend on h and ω^h.

Fourth, we study the initial boundary value problem (26) for multidimensional semilinear delay hyperbolic equations. The discretization of problem (23) is provided in two steps. To the differential operator A generated by problem (26), we assign the difference operator A_h^x by the formula

$$A_h^x u^h(x) = -\sum_{r=1}^m \left(a_r(x) u_{\overline{x}_r}^h \right)_{x_r, j_r} + \delta u^h(x) \qquad (50)$$

acting in the space of grid functions $u^h(x)$, satisfying the conditions $D^h u^h(x) = 0$ ($\forall x \in S_h$). Here D^h is the approximation of operator $\frac{\partial}{\partial \overrightarrow{p}}$. With the help of A_h^x, we arrive at the initial value problem (48). In the second step, we replace problem (48) by first order of accuracy difference scheme (28), we get Equation (49).

Theorem 13. *Suppose that assumptions of Theorem 6 hold. Then, there exists a unique solution $\left\{u_k^h\right\}_{k=0}^{\infty}$ of difference scheme (49) which is bounded in $[m\omega, (m+1)\,w]_\tau \times L_{2h}$, $m = 0, 1, \cdots$ uniformly with respect to τ and h.*

Proof. Difference scheme (49) can be written in abstract form (28) in a Hilbert space $L_{2h} = L_2(\overline{\Omega}_h)$ with self-adjoint positive definite operator $A_h = A_h^x$ by formula (50). Here, $f(t_k, x, u_k^h(x), \frac{u_{k-N}^h(x) - u_{k-N-1}^h(x)}{\tau}, u_{k-N}^h(x))$ and $u_k^h = u_k^h(x)$ are abstract mesh functions defined on $\overline{\Omega}_h$ with the values in $H = L_{2h}$. Therefore, the proof of Theorems 13 is based on the abstract Theorem 8 and symmetry properties of the difference operator A_h^x defined by formula (50) and the following theorem on coercivity inequality for the solution of the elliptic problem in L_{2h} [53]. □

Theorem 14. *For the solutions of the elliptic difference problem*

$$\begin{cases} A_h^x u^h(x) = \omega^h(x), \ x \in \Omega_h, \\ D^h u^h(x) = 0, \ x \in S_h, \end{cases}$$

the coercivity inequality

$$\sum_{r=1}^n \left\| u^h_{x_r \overline{x}_r} \right\|_{L_{2h}} \leq M_2 \|\omega^h\|_{L_{2h}}$$

is satisfied, where M_2 does not depend on h and ω^h.

6. Numerical Experiments

In general, it is not able to get the exact solution of semilinear hyperbolic problems. Therefore, numerical results for the solution of difference schemes for one and two dimensional semilinear hyperbolic equations with time delay are presented. These results fit with the theoretical results perfectly.

6.1. One Dimensional Case

For the numerical experiment, we consider the mixed problem

$$\begin{cases} \frac{\partial^2 u(t,x)}{\partial t^2} - \frac{\partial^2 u(t,x)}{\partial x^2} = 2e^{-t} \sin x + \cos\left(u(t,x)\,u(t-1,x)\right) \\ \quad - \cos\left(e^{-t} \sin x\, u(t-1,x)\right), 0 < t < \infty, \ 0 < x < 2\pi, \\ u(t,x) = e^{-t} \sin x, \ 0 \leq x \leq 2\pi, \ -1 \leq t \leq 0, \\ u(t,0) = u(t,2\pi), \ u_x(t,0) = u_x(t,2\pi), \ t \geq 0 \end{cases} \tag{51}$$

for the semilinear delay one dimensional hyperbolic differential equation with nonlocal boundary conditions. The exact solution of problem (51) is $u(t,x) = e^{-t}\sin x$. We will consider the following iterative difference scheme of first order of approximation in t for the numerical solution of problem (51)

$$\begin{cases} \frac{{}_ju_n^{k+1}-2({}_ju_n^k)+{}_ju_n^{k-1}}{\tau^2} - \frac{{}_ju_{n+1}^{k+1}-2({}_ju_n^{k+1})+{}_ju_{n-1}^{k+1}}{h^2} = 2e^{-t_k}\sin x_n \\ \quad + \cos\left(\left({}_{j-1}u_n^k\right)u_n^{k-N}\right) - \cos\left(e^{-t_k}\sin x_n \left(u_n^{k-N}\right)\right), \\ t_k = k\tau, \ x_n = nh, \ 1 \leq k < \infty, \ 1 \leq n \leq M-1, \ N\tau = 1, \ Mh = 2\pi, \\ \\ \frac{{}_ju_n^{k+1}-{}_ju_n^k}{\tau} - \frac{\tau}{h^2}\left({}_ju_{n+1}^{k+1} - {}_ju_{n+1}^k - 2\left({}_ju_n^{k+1} - {}_ju_n^k\right) + {}_ju_{n-1}^{k+1} - {}_ju_{n-1}^k\right) \\ = \frac{u_n^k - u_n^{k-1}}{\tau}, \ k = mN+1, m = 0,1,...,k \geq 1, \\ u_n^k = e^{-t_k}\sin x_n, \ t_k = k\tau, \ x_n = nh, \ 0 \leq n \leq M, \ -N \leq k \leq 0, \\ {}_ju_0^k = {}_ju_M^k, \ {}_ju_1^k - {}_ju_0^k = {}_ju_M^k - {}_ju_{M-1}^k, \ 0 \leq k < \infty, \ j = 1,2,... \end{cases} \quad (52)$$

for the semilinear delay hyperbolic equation. Here and in future j denotes the iteration index and an initial guess ${}_0u_n^k$, $k \geq 1$, $0 \leq n \leq M$ is to be made.

For solving difference scheme (52), the numerical steps are given below. For $0 \leq k < N$, $0 \leq n \leq M$ the algorithm is as follows :

1. $j=1$.
2. ${}_{j-1}u_n^k$ is known.
3. ${}_ju_n^k$ is calculated.
4. If the max absolute error between ${}_{j-1}u_n^k$ and ${}_ju_n^k$ is greater than the given tolerance value, take $j = j+1$ and go to step 2. Otherwise, terminate the iteration process and take ${}_ju_n^k$ as the result of the given problem.

We write (52) in matrix form

$$A\left({}_ju^{k+1}\right) + B\left({}_ju^k\right) + C\left({}_ju^{k-1}\right) = R\varphi({}_{j-1}u^k, u^{k-N}), \ 1 \leq k < \infty,$$

$$u^k = \{e^{-t_k}\sin x_n\}_{n=0}^M, \ -N \leq k \leq 0, \quad (53)$$

$$\frac{{}_ju^{k+1} - {}_ju^k}{\tau} - \left\{\frac{\tau}{h^2}\left({}_ju_{n+1}^{k+1} - {}_ju_{n+1}^k - 2\left({}_ju_n^{k+1} - {}_ju_n^k\right) + {}_ju_{n-1}^{k+1} - {}_ju_{n-1}^k\right)\right\}_{n=1}^{M-1}$$
$$= \frac{u^k - u^{k-1}}{\tau}, \ k = mN+1, k \geq 1. \quad (54)$$

Here R, A, B, and C are $(M+1) \times (M+1)$ matrices given below:

$$R = \begin{bmatrix} 1 & 0 & 0 & 0 & 0 & . & 0 & 0 \\ 0 & 1 & 0 & 0 & 0 & . & 0 & 0 \\ 0 & 0 & 1 & 0 & 0 & . & 0 & 0 \\ . & . & . & . & . & . & . & . \\ 0 & 0 & 0 & 0 & 0 & . & 1 & 0 \\ 0 & 0 & 0 & 0 & 0 & . & 0 & 1 \end{bmatrix},$$

$$A = \begin{bmatrix} 1 & 0 & 0 & 0 & 0 & . & 0 & -1 \\ a & b & a & 0 & 0 & . & 0 & 0 \\ 0 & a & b & a & 0 & . & 0 & 0 \\ 0 & 0 & a & b & a & . & 0 & 0 \\ 0 & 0 & 0 & a & b & . & 0 & 0 \\ . & . & & & & . & & . \\ 0 & 0 & 0 & 0 & 0 & . & b & a \\ 1 & -1 & 0 & 0 & 0 & . & -1 & 1 \end{bmatrix},$$

$$B = \begin{bmatrix} 0 & 0 & 0 & 0 & 0 & . & 0 & 0 \\ 0 & c & 0 & 0 & 0 & . & 0 & 0 \\ 0 & 0 & c & 0 & 0 & . & 0 & 0 \\ 0 & 0 & 0 & c & 0 & . & 0 & 0 \\ . & . & & & & . & & . \\ 0 & 0 & 0 & 0 & 0 & . & c & 0 \\ 0 & 0 & 0 & 0 & 0 & . & 0 & 0 \end{bmatrix},$$

$$C = \begin{bmatrix} 0 & 0 & 0 & 0 & 0 & . & 0 & 0 \\ 0 & d & 0 & 0 & 0 & . & 0 & 0 \\ 0 & 0 & d & 0 & 0 & . & 0 & 0 \\ 0 & 0 & 0 & d & 0 & . & 0 & 0 \\ . & . & & & & . & & . \\ 0 & 0 & 0 & 0 & 0 & . & d & 0 \\ 0 & 0 & 0 & 0 & 0 & . & 0 & 0 \end{bmatrix},$$

where

$$a = -\frac{1}{h^2}, \quad b = \frac{1}{\tau^2} + \frac{2}{h^2}, \quad c = -\frac{2}{\tau^2}, \quad d = \frac{1}{\tau^2}.$$

Finally, here $\varphi(_{j-1}u^k, u^{k-N})$ and $_j u^s$, $s = k$, $k \pm 1$ are $(M+1) \times 1$ column vectors as

$$\varphi(_{j-1}u^k, u^{k-N}) = \begin{bmatrix} 0 \\ _j\varphi_1^k \\ . \\ _j\varphi_{M-1}^k \\ 0 \end{bmatrix}, \quad _j u^s = \begin{bmatrix} _j u_0^s \\ _j u_1^s \\ . \\ _j u_{M-1}^s \\ _j u_M^s \end{bmatrix},$$

$$_j\varphi_n^k = 2e^{-t_k}\sin x_n + \cos\left((_{j-1}u_n^k)(u_n^{k-N})\right) - \cos\left(e^{-t_k}\sin x_n u_n^{k-N}\right)$$

for $1 \leq k \leq N-1$, $1 \leq n \leq M-1$.

So, we have the initial value problem for the second order difference equation with respect to k with matrix coefficients. From Equations (53) and (54) it follows that

$$_j u^{k+1} = -A^{-1}\left(B_j u^k - C_j u^{k-1} + A^{-1} R\varphi^k(_{j-1}u^k, u^{k-N})\right),$$

$$_j u^k = \{e^{-t_k}\sin x_n\}_{n=0}^M, \quad -N \leq k \leq 0,$$

$$_j u^{k+1} = \psi\left(u^k, u^{k-1}\right), \quad k = mN+1, m = 0, 1, ..., k \geq 1. \tag{55}$$

Here, $\psi\left(u^k, u^{k-1}\right)$ is $(M+1) \times 1$ column vector defined by formula (54).

In computations the initial guess is chosen as $_0u^k = \{\sin x_n\}_{n=0}^{M}$ and when the maximum errors between two consecutive results of iterative difference scheme (52) become less than 10^{-8}, the iterative process is terminated. We present numerical experimental results for different values of N and M and u_n^k represent the numerical solutions of difference scheme (52) at (t_k, x_n). The table of numerical results is constructed for $N = M = 30, 60, 120$ in $t \in [0,1], t \in [1,2], t \in [2,3]$, respectively and the errors are computed by the following formula

$$mE_M^N = \max_{mN+1 \leq k \leq (m+1)N, 0 \leq n \leq M} \left| u(t_k, x_n) -_j u_n^k \right|.$$

As can be seen from tables, these numerical experiments support the theoretical statements. The number of iterations and maximum errors are decreasing with the increase of grid points.

In Table 1, as we increase values of M and N each time starting from $M = N = 30$ by a factor of 2 the errors in the first order of accuracy difference scheme decrease approximately by a factor of $1/2$. The errors presented in Table 1 indicate the first order of accuracy of the difference scheme.

Table 1. The errors (52) (Number of the iteration = j).

$mE_M^N/(N,M)$	(30,30)	(60,60)	(120,120)
in $t \in [0,1]$	$4.1195 \times 10^{-3}, j = 6$	$2.0322 \times 10^{-3}, j = 6$	$1.0098 \times 10^{-3}, j = 6$
in $t \in [1,2]$	$2.3014 \times 10^{-3}, j = 6$	$1.1297 \times 10^{-3}, j = 6$	$5.6051 \times 10^{-4}, j = 2$
in $t \in [2,3]$	$1.0245 \times 10^{-3}, j = 6$	$5.0161 \times 10^{-4}, j = 6$	$2.4864 \times 10^{-4}, j = 6$

6.2. Two-Dimensional Case

For the numerical experiment, we consider the mixed boundary value problem

$$\begin{cases} \frac{\partial^2 u(t,x,y)}{\partial t^2} - \frac{\partial^2 u(t,x,y)}{\partial x^2} - \frac{\partial^2 u(t,x,y)}{\partial y^2} = 2e^{-t} \sin x \sin y + \cos(u(t,x,y)u(t-1,x,y)) \\ - \cos(e^{-t} \sin x \sin y\, u(t-1,x,y)),\ 0 < t < \infty,\ 0 < x,y < \pi, \\ u(t,x,y) = e^{-t} \sin x \sin y,\ 0 \leq x,y \leq \pi,\ -1 \leq t \leq 0, \\ u(t,0,y) = u(t,\pi,y) = 0,\ 0 \leq y \leq \pi,\ t \geq 0, \\ u(t,x,0) = u(t,x,\pi) = 0,\ 0 \leq x \leq \pi,\ t \geq 0 \end{cases} \quad (56)$$

for the semilinear two dimensional delay hyperbolic equation. The exact solution of problem (56) is $u(t,x) = e^{-t}\sin x \sin y$. We will consider the following iterative difference scheme of first order of approximation in t for the numerical solution of the initial-boundary value problem (56)

$$\begin{cases} \frac{_ju_{n,i}^{k+1} - 2(_ju_{n,i}^k) + _ju_{n,i}^{k-1}}{\tau^2} - \frac{_ju_{n+1,i}^{k+1} - 2(_ju_{n,i}^{k+1}) + _ju_{n-1,i}^{k+1}}{h^2} - \frac{_ju_{n,i+1}^{k+1} - 2(_ju_{n,i}^{k+1}) + _ju_{n,i-1}^{k+1}}{h^2} \\ = 2e^{-t_k}\sin x_n \sin x_i + \cos\left((_{j-1}u_{n,i}^k)u_{n,i}^{k-N}\right) - \cos\left(e^{-t_k}\sin x_n \sin x_i u_{n,i}^{k-N}\right), \\ t_k = k\tau, \ x_n = nh, \ 1 \leq k < \infty, \ 1 \leq n,i \leq M-1, \ N\tau = 1, \ Mh = \pi, \\ u_{n,i}^k = e^{-t_k}\sin x_n \sin x_i, \ t_k = k\tau, \ x_n = nh, \ 0,i \leq M, \ -N \leq k \leq 0, \\ \frac{_ju_{n,i}^{k+1} - _ju_{n,i}^k}{\tau} - \frac{\tau}{h^2}\left(_ju_{n+1,i}^{k+1} - _ju_{n+1,i}^k - 2\left(_ju_{n,i}^{k+1} - _ju_{n,i}^k\right) + _ju_{n,i-1}^{k+1} - _ju_{n,i-1}^k\right) \\ - \frac{\tau}{h^2}\left(_ju_{n,i+1}^{k+1} - _ju_{n,i+1}^k - 2\left(_ju_{n,i}^{k+1} - _ju_{n,i}^k\right) + _ju_{n-1,i}^1 - _ju_{n-1,i}^k\right) \\ = \frac{u_{n,i}^k - u_{n,i}^{k-1}}{\tau}, \ k = mN+1, m = 0,1,...,k \geq 1, \\ _ju_{0,i}^k = _ju_{M,i}^k = 0, \ 0 \leq i \leq M, _ju_{n,0}^k = _ju_{n,M}^k = 0, \ 0 \leq n \leq M, \\ 0 \leq k < \infty, \ j = 1,2,... \end{cases} \quad (57)$$

for the semilinear delay hyperbolic equation. Here and in the future j denotes the iteration index and an initial guess $_0u_{n,i}^k$, $k \geq 1$, $0 \leq n,i \leq M$ is to be made. For solving difference scheme (57), the numerical steps are given below. For $0 \leq k < N$, $0 \leq n,i \leq M$ the algorithm is as follows:

1. $j = 1$.
2. $_{j-1}u_{n,i}^k$ is known.
3. $_ju_{n,i}^k$ is calculated.
4. If the max absolute error between $_{j-1}u_{n,i}^k$ and $_ju_{n,i}^k$ is greater than the given tolerance value, take $j = j+1$ and go to step 2. Otherwise, terminate the iteration process and take $_ju_{n,i}^k$ as the result of the given problem.

We write Equation (57) in matrix form

$$A\left(_ju^{k+1}\right) + B\left(_ju^k\right) + C\left(_ju^{k-1}\right) = R\varphi(_{j-1}u^k, u^{k-N}), \ 1 \leq k < \infty,$$

$$u^k = \{e^{-t_k}\sin x_n \sin x_i\}_{n,i=0}^M, \ -N \leq k \leq 0, \quad (58)$$

$$\frac{_ju^{k+1} - _ju^k}{\tau} - \left\{\frac{\tau}{h^2}\left(_ju_{n+1,i}^{k+1} - _ju_{n+1,i}^k - 2\left(_ju_{n,i}^{k+1} - _ju_{n,i}^k\right) + _ju_{n,i-1}^{k+1} - _ju_{n,i-1}^k\right)\right\}_{n,i=1}^{M-1}$$

$$- \left\{\frac{\tau}{h^2}\left(_ju_{n,i+1}^{k+1} - _ju_{n,i+1}^k - 2\left(_ju_{n,i}^{k+1} - _ju_{n,i}^k\right) + _ju_{n-1,i}^{k+1} - _ju_{n-1,i}^k\right)\right\}_{n,i=1}^{M-1}$$

$$= \frac{u^k - u^{k-1}}{\tau}, \ k = mN+1, m = 0,1,...,k \geq 1.$$

Here R, A, B, and C are $(M+1) \times (M+1) \times (M+1)$ given matrices and $\varphi(_{j-1}u^k, u^{k-N})$ and $_ju^s$, $s = k$, $k \pm 1$ are given $(M+1) \times (M+1) \times 1$ column vectors. Therefore, we will use the same algorithm as the one dimensional case.

So, we have the initial value problem for the second order difference equation with respect to k with matrix coefficients. From Equations (53) and (54) it follows that

$$_j u^{k+1} = -A^{-1}\left(B_j u^k - C_j u^{k-1} + A^{-1} R \varphi^k(_{j-1}u^k, u^{k-N})\right), \; 1 \leq k < \infty,$$

$$u^k = \{e^{-t_k} \sin x_n \sin x_i\}_{n,i=0}^{M}, \; -N \leq k \leq 0,$$

$$_j u^{k+1} = \psi\left(u^k, u^{k-1}\right), \; k = mN+1, m = 0, 1, ..., k \geq 1. \tag{59}$$

Here, $\psi\left(u^k, u^{k-1}\right)$ is the given $(M+1) \times (M+1) \times 1$ column vector.

In computations the initial guess is chosen as $_0 u^k = \{\sin x_n \sin x_i\}_{n,i=0}^{M}$ and when the maximum errors between two consecutive results of iterative difference scheme (57) become less than 10^{-6}, the iterative process is terminated. We present numerical results for different values of N and M and $u_{n,i}^k$ represent the numerical solutions of this difference scheme at (t_k, x_n, x_i). The table is constructed for $N = M = 20, 40, 80$ in $t \in [0,1]$, $t \in [1,2]$, $t \in [2,3]$, respectively and the errors are computed by the following formula

$$mE_{M,M}^N = \max_{mN+1 \leq k \leq (m+1)N, 0 \leq n, i \leq M} \left| u(t_k, x_n, x_i) - {_j u_{n,i}^k} \right|.$$

As can be seen from table, these numerical experiments support the theoretical statements. The number of iterations and maximum errors are decreasing with the increase of grid points.

In Table 2, as we increase values of M and N each time starting from $M = N = 30$ by a factor of 2 the errors in the first order of accuracy difference scheme decrease approximately by a factor of $1/2$. The errors presented in tables indicate the the time convergence order is one. This result fits with the theoretical results perfectly.

Table 2. The errors of difference scheme (57) (Number of the iteration = j).

$mE_{M,M}^N/(N,M,M)$	(20,20,20)	(40,40,40)	(80,80,80)
in $t \in [0,1]$	1.2517×10^{-2}, $j=4$	0.6554×10^{-2}, $j=4$	0.3348×10^{-2}, $j=4$
in $t \in [1,2]$	1.2517×10^{-2}, $j=3$	0.6556×10^{-2}, $j=3$	0.3348×10^{-2}, $j=3$
in $t \in [2,3]$	2.3678×10^{-3}, $j=2$	1.4306×10^{-3}, $j=2$	0.7870×10^{-3}, $j=2$

6.3. Conclusions and Our Future Plans

1. In the present paper, the main theorem on the existence and uniqueness of a bounded solution of the initial value problem for a semilinear hyperbolic equation with time delay in a Hilbert space with the self adjoint positive definite operator is established. In applications, the existence and uniqueness of a bounded solution of four problems for semilinear hyperbolic equations with time delay are obtained. A first order of accuracy difference scheme for the numerical solution of the abstract problem is presented. The theorem on the existence and uniqueness of an uniformly bounded solution of this difference scheme with respect to τ is established. In applications, the existence and uniqueness of a uniformly bounded solutions with respect to time and space stepsizes of difference schemes for four semilinear partial differential equations with time delay are obtained. Numerical results for the solution of difference schemes for one and two dimensional semilinear delay hyperbolic equation are presented.

2. We are interested in studying uniformly boundedness of solutions of high order of accuracy difference schemes uniformly with respect to time stepsize of approximate solutions of this initial-value problem, in which bounded solutions were established without any assumptions in respect to the

grid steps τ and h. We have not been able to establish such type of results for the solution of the very well-known second order difference scheme

$$\begin{cases} \frac{u^{k+1}-2u^k+u^{k-1}}{\tau^2} + Au^k = f(t_k, u^k, \frac{u^{k-N}-u^{k-N-1}}{\tau}, u^{k-N}), \\ t_k = k\tau, \ 1 \leq k < \infty, \ N\tau = \omega, \\ (I+\tau^2 A)\frac{u^{k+1}-u^k}{\tau} = \frac{2u^k-3u^{k-1}+u^{k-2}}{\tau}, \ k = mN, \ m = 1, ..., \\ u^k = \varphi(t_k), \ t_k = k\tau, \ -N \leq k \leq 0. \end{cases}$$

Note that absolute stable two-step difference schemes of the high order of approximation for hyperbolic partial differential equations were presented and investigated in papers [11,54]. Applying methods of the present paper and papers [11,54] we can establish the similar stability and convergence results of this paper for the solution of the absolute stable two-step difference schemes of high order of approximation for semilinear delay hyperbolic equations.

3. Investigate the uniform to-step difference schemes and asymptotic formulas for the solution of initial value perturbation problem

$$\begin{cases} \varepsilon^2 u''(t) + Au(t) = f(t, u(t), u_t(t-w), u(t-w)), \ t > 0, \\ u(t) = \varphi(t), \ -w \leq t \leq 0 \end{cases}$$

for a semilinear delay hyperbolic equation in a Hilbert space H with the self adjoint positive definite operator A and with $\varepsilon \in (0, \infty)$ parameter multiplying the highest order derivative term.

In [31], the uniform difference schemes and asymptotic formulas for the solution of initial value perturbation problem for a linear hyperbolic equation in a Hilbert space with the self adjoint positive definite operator and with $\varepsilon \in (0, \infty)$ parameter multiplying the highest order derivative term were presented and investigated.

4. Investigate the initial value problem

$$\begin{cases} u_{tt}(t)dt + Au(t)dt = f(t, u(t), u_t(t-w), u(t-\omega))dw_t, \\ w_t = \sqrt{t}\xi, \ \xi \in N(0,1), \ t > 0, \\ u(t) = 0, \ -\omega \leq t \leq 0 \end{cases}$$

for a semilinear stochastic hyperbolic equation with time delay in a Hilbert space H with the self adjoint positive definite operator A. Here, w_t is a standard Wiener process given on the probability space (Q, F, P).

Note that absolute stable difference schemes for stochastic linear hyperbolic equations in Hilbert spaces were presented and investigated in [30].

Finally, in paper [55], a Lie algebra approach is applied to solve an SIS model where infection rate and recovery rate are time-varying. The method presented here has been used widely in chemical and physical sciences but not in epidemic applications due to insufficient symmetries.

Author Contributions: Investigation, A.A. and D.A.

Funding: This research was funded by "Russian Foundation for Basic Research (RFBR) grant number 16–01–00450."

Acknowledgments: The publication has been prepared with the support of the "RUDN University Program 5–100" and dedicated in memory of Pavel Evseevich Sobolevskii. The authors are grateful to Francisco Rodríguez and reviewers of this paper for the useful comments and relevant references.

Conflicts of Interest: The authors declare no conflict of interest.

References

1. Ardito, A.; Ricciardi, P. Existence and regularity for linear delay partial differential equations. *Nonlinear Anal.* **1980**, *4*, 411–414. [CrossRef]
2. Arino, A. *Delay Differential Equations and Applications*; Springer: Berlin/Heidelberg, Germany, 2006.
3. Bhalekar, S.; Patade, J. Analytic solutions of nonlinear with proportional delays. *Appl. Comput. Math.* **2016**, *15*, 331–345.
4. Blasio, G.D. Delay differential equations with unbounded operators on delay terms. *Nonlinear Anal.-Theory Appl.* **2003**, *52*, 1–18. [CrossRef]
5. Kurulay, G.; Ozbay, H. Design of first order controllers for a flexible robot arm with time delay. *Appl. Comput.* **2017**, *16*, 48–58.
6. Skubachevskii, A.L. On the problem of attainment of equilibrium for control-system with delay. *Dokl. Akad.* **1994**, *335*, 157–160.
7. Vlasov, V.V.; Rautian, N.A. *Spectral Analysis of Functional Differential Equations*; MAKS Press: Moscow, Russia, 2016. (In Russian)
8. Shang, Y. On the delayed scaled consensus problems. *Appl. Sci.* **2017**, *7*, 713. [CrossRef]
9. Atay, F.M. On the duality between consensus problems and Markov processes, with application to delay systems. *Markov Process. Relat. Field* **2016**, *22*, 537–553.
10. Lu, X. Combined iterative methods for numerical solutions of parabolic problems with time delays. *Appl. Math. Comput.* **1998**, *89*, 213–224. [CrossRef]
11. Ashyralyev, A.; Sobolevskii, P.E. A note on the difference schemes for hyperbolic equations. *Abstr. Appl. Anal.* **2001**, *6*, 63–70. [CrossRef]
12. Agirseven, D. Approximate solutions of delay parabolic equations with the Dirichlet condition. *Abstr. Appl. Anal.* **2012**, *2012*, 682752. [CrossRef]
13. Ashyralyev, A.; Agirseven, D. Stability of parabolic equations with unbounded operators acting on delay terms. *Electron. J. Differ. Equ.* **2014**, *2014*, 1–13.
14. Ashyralyev, A.; Agirseven, D. On source identification problem for a delay parabolic equation. *Nonlinear Anal. Model. Control* **2014**, *19*, 335–349. [CrossRef]
15. Ashyralyev, A.; Agirseven, D. Stability of delay parabolic difference equations. *Filomat* **2014**, *28*, 995–1006. [CrossRef]
16. Ashyralyev, A.; Agirseven, D. Well-posedness of delay parabolic equations with unbounded operators acting on delay terms. *Bound. Value Probl.* **2014**, *126*. [CrossRef]
17. Ashyralyev, A.; Agirseven, D. Well-posedness of delay parabolic difference equations. *Adv. Differ. Equ.* **2014**, *2014*. [CrossRef]
18. Ashyralyev, A.; Agirseven, D. On convergence of difference schemes for delay parabolic equations. *Comput. Math. Appl.* **2013**, *66*, 1232–1244. [CrossRef]
19. Ashyralyev, A.; Agirseven, D.; Ceylan, B. Bounded solutions of delay nonlinear evolutionary equations. *J. Comput. Appl. Math.* **2017**, *318*, 69–78. [CrossRef]
20. HenrÃ-quez, H.R.; Cuevas, C.; Caicedo, A. Almost periodic solutions of partial differential equations with delay. *Adv. Differ. Equ.* **2015**, *2015*. [CrossRef]
21. Hao, Z.P.; Fan, K.; Cao, W.R.; Sun, Z.Z. A finite difference scheme for semilinear space-fractional diffusion equations with time delay. *Appl. Math. Comput.* **2016**, *275*, 238–254. [CrossRef]
22. Liang, H. Convergence and asymptotic stability of Galerkin methods for linear parabolic equations with delays. *Appl. Math. Comput.* **2015**, *264*, 160–178. [CrossRef]
23. Bhrawy, A.H.; Abdelkawy, M.A.; Mallawi, F. An accurate Chebyshev pseudospectral scheme for multi-dimensional parabolic problems with time delays. *Bound. Value Probl.* **2015**, *2015*. [CrossRef]
24. Ismailov, Z.I.; Guler, B.O.; Ipek, P. Solvable time-delay differential operators for first order and their spectrums. *Hacet. J. Math. Stat.* **2016**, *45*, 755–764.
25. Piriadarshani, D.; Sengadir, T. Existence of solutions and semi-discretization for PDE with infinite delay. *Differ. Equ. Appl.* **2015**, *7*, 313–331. [CrossRef]

26. Castro, M.A.; Rodriguez, F.; Cabrera, J.; Martin, J.A. Difference schemes for time-dependent heat conduction models with delay. *Int. J. Comput. Math.* **2014**, *91*, 53–61. [CrossRef]
27. Fattorini, H.O. *Second Order Linear Differential Equations in Banach Spaces*; Elsevier Science Publishing Company: Amsterdam, The Netherlands, 1985.
28. Goldstein, J.A. *Semigroups of Linear Operators and Applications*; Oxford Mathematical Monographs; The Clarendon Press Oxford University Press: New York, NY, USA, 1985.
29. Krein, S.G. *Linear Differential Equations in Banach Space*; American Mathematical Society: Providence, RI, USA, 1971.
30. Ashyralyev, A.; Akat, M. An approximation of stochastic hyperbolic equations: Case with Wiener process. *Math. Appl. Sci.* **2013**, *36*, 1095–1106. [CrossRef]
31. Ashyralyev, A.; Fattorini, H.O. On uniform difference-schemes for 2nd-order singular pertubation problems in Banach spaces. *SIAM J. Math. Anal.* **1992**, *23*, 29–54. [CrossRef]
32. Vasilev, V.V.; Krein, S.G.; Piskarev, S. *Operator Semigroups, Cosine Operator Functions, and Linear Differential Equations*; Itogi Nauki i Tekhniki Academy of Science USSR: Moscow, Russia, 1990.
33. Shang, Y. Non-hyperbolicity of random graphs with given expected degrees. *Stoch. Models* **2013**, *29*, 451–462. [CrossRef]
34. Shang, Y. Lack of Gromov-hyperbolicity in small-world networks. *Cent. Eur. J. Math.* **2012**, *10*. [CrossRef]
35. Ashyralyyeva, M.A.; Ashyraliyev, M. On the numerical solution of identification hyperbolic-parabolic problems with the Neumann boundary condition. *Bull. Karaganda Univ.-Math.* **2018**, *91*, 69–74. [CrossRef]
36. Ashyralyyeva, M.; Ashyraliyev, M. Numerical Solutions of Source Identification Problem for Hyperbolic-Parabolic Equations. In *AIP Conference Proceedings*; AIP Publishing: Melville, NY, USA, 2018; Volume 1997, p. 020048.
37. Zilal, D.; Ashyraliyev, M. Difference schemes for the semilinear integral-differential equation of the hyperbolic type. *Filomat* **2018**, *32*, 1009–1018.
38. Mohanty, R.K. An unconditionally stable finite difference formula for a linear second order one space dimensional hyperbolic equation with variable coefficients. *Appl. Math. Comput.* **2005**, *165*, 229–236. [CrossRef]
39. Mohanty, R.K. An operator splitting technique for an unconditionally stable difference method for a linear three space dimensional hyperbolic equation with variable coefficients. *Appl. Math. Comput.* **2005**, *165*, 549–557. [CrossRef]
40. Mohanty, R.K. An operator splitting method for an unconditionally stable difference scheme for a linear hyperbolic equation with variable coefficients in two space dimensions. *Appl. Math. Comput.* **2004**, *152*, 799–806. [CrossRef]
41. Piskarev, S. Stability of difference schemes in Cauchy problems with almost periodic solutions. *Differentsial' nye Uravneniya* **1994**, *20*, 689–695. (In Russian)
42. Piskarev, S. *Principles of Discretization Methods III*; Report ak -3410; Acoustic Institute, Academy of Science USSR: Moscow, Russia, 1986. (In Russian)
43. Sobolevskii, P.E.; Chebotaryeva, L.M. Approximate solution by method of lines of the Cauchy problem for an abstract hyperbolic equations. *Vyssh. Uchebn. Zav. Mat.* **1977**, *5*, 103–116. (In Russian)
44. Ashyralyev, A.; Pastor, J.; Piskarev, S.; Yurtsever, H.A. Second order equations in functional spaces: Qualitative and discrete well-posedness. *Abstr. Appl. Anal.* **2015**, *2015*, 1–63. [CrossRef]
45. Poorkarimi, H.; Wiener, J. Bounded solutions of non-linear hyperbolic equations with delay. In *Proceedings of the VII International Conference on Non-Linear Analysis*, Arlington, TX, USA, 28 July–1 August 1986; Volume 1, pp. 471–478.
46. Poorkarimi, H.; Wiener, J.; Shah, S.M. On the exponential growth of solutions to non-linear hyperbolic equations. *Int. J. Math. Sci.* **1989**, *12*, 539–546.
47. Shah, S.M.; Poorkarimi, H.; Wiener, J. Bounded solutions of retarded nonlinear hyperbolic equations. *Bull. Allahabad Math. Soc.* **1986**, *1*, 1–14.
48. Wiener, J. *Generalized Solutions of Functional Differential Equations*; World Scientific: Singapore, 1993.
49. Ashyralyev, A.; Agirseven, D. Bounded solutions of nonlinear hyperbolic equations with time delay. *Electron. J. Differ. Equ.* **2018**, *2018*, 1–15.
50. Ashyralyev, A.; Sobolevskii, P.E. *New Difference Schemes for Partial Differential Equations*; Birkhäuser Verlag: Basel, Switzerland, 2004.

51. Ashyralyev, A. A survey of results in the theory of fractional spaces generated by positive operators. *TWMS J. Pure Appl. Math.* **2015**, *6*, 129–157.
52. Ashyralyev, A.; Sarsanbi, A. Well-posedness of a parabolic equation with involution. *Numer. Funct. Anal. Optim.* **2017**, *38*, 1295–1305. [CrossRef]
53. Sobolevskii, P.E. *Difference Methods for the Approximate Solution of Differential Equations*; Voronezh State University Press: Voronezh, Russia, 1975. (In Russian)
54. Ashyralyev, A.; Sobolevskii, P.E. Two new approaches for construction of the high order of accuracy difference schemes for hyperbolic differential equations. *Discret. Dyn. Nat. Soc.* **2005**, *2005*, 183–213. [CrossRef]
55. Shang, Y. A Lie algebra approach to susceptible-infected-epidemics. *Electron. J. Differ. Equ.* **2012**, *233*, 1–7.

© 2019 by the authors. Licensee MDPI, Basel, Switzerland. This article is an open access article distributed under the terms and conditions of the Creative Commons Attribution (CC BY) license (http://creativecommons.org/licenses/by/4.0/).

Article

Exact and Nonstandard Finite Difference Schemes for Coupled Linear Delay Differential Systems

María Ángeles Castro [1], Miguel Antonio García [1], José Antonio Martín [1] and Francisco Rodríguez [1,2,*]

[1] Department of Applied Mathematics, University of Alicante, Apdo. 99, 03080 Alicante, Spain; ma.castro@ua.es (M.Á.C.); miguel.garcia@ua.es (M.A.G.); jose.martin@ua.es (J.A.M.)
[2] Multidisciplinary Institute for Environmental Studies (IMEM), University of Alicante, Apdo. 99, 03080 Alicante, Spain
* Correspondence: f.rodriguez@ua.es

Received: 5 September 2019; Accepted: 27 October 2019; Published: 3 November 2019

Abstract: In recent works, exact and nonstandard finite difference schemes for scalar first order linear delay differential equations have been proposed. The aim of the present work is to extend these previous results to systems of coupled delay differential equations $X'(t) = AX(t) + BX(t-\tau)$, where X is a vector, and A and B are commuting real matrices, in general not simultaneously diagonalizable. Based on a constructive expression for the exact solution of the vector equation, an exact scheme is obtained, and different nonstandard numerical schemes of increasing order are proposed. Dynamic consistency properties of the new nonstandard schemes are illustrated with numerical examples, and proved for a class of methods.

Keywords: delay systems; nonstandard numerical methods; dynamic consistency

1. Introduction

Due to the presence of time lags in the dynamics of most real systems, delay differential equations (DDE) have become basic instruments in the mathematical modelling of a wide range of problems in science and engineering, such as in population biology, physiology, epidemiology, economics, and control problems (see, e.g., [1–5], and references therein), and special methods have been developed to compute numerical solutions for DDE [6]. In the case of differential problems without delay, exact schemes have been defined for different particular problems, and the use of nonstandard finite difference (NSFD) numerical schemes has gained increasing interest in the last years [7–9]. The NSFD numerical schemes can be competitive in terms of accuracy while providing dynamically consistent solutions, i.e., they can provide numerical discrete solutions that inherit the structural properties defining the dynamical behaviour of the original continuous equation [10]. The possibility of defining NSFD schemes that reproduce the qualitative behaviour of the continuous solutions has made them specially attractive for population and epidemiology models (e.g., [11–15]), and they have also been proposed for some problems with delay [16–21]. However, for DDE models the construction of exact schemes, and consequently of NSFD methods derived from them, has not been much developed.

In [22], a NSFD method was proposed for the scalar first order linear delay problem

$$x'(t) = \alpha x(t) + \beta x(t-\tau), \quad t > 0, \qquad (1)$$
$$x(t) = f(t), \quad -\tau \leq t \leq 0, \qquad (2)$$

where $\alpha, \beta \in \mathbb{R}$, $\tau > 0$, and $f : [-\tau, 0] \to \mathbb{R}$ is a continuous function. The method of [22] was exact in the initial time interval $0 \leq t \leq \tau$, and then switched to a NSFD method of second order at most. More recently [23], an exact scheme for problem (1)–(2) was constructed, valid in the whole domain of definition, and a family of increasing order NSFD schemes was defined. The NSFD methods presented in [22,23] were shown to be consistent with different dynamical properties of the continuous problem (1)–(2).

In the present work, we consider the coupled linear delay system

$$X'(t) = AX(t) + BX(t-\tau), \quad t > 0, \tag{3}$$

satisfying the initial condition

$$X(t) = F(t), \quad -\tau \leq t \leq 0, \tag{4}$$

where $X(t)$ and $F(t)$ are d-dimensional real vector functions, and A and B are $d \times d$ commuting real matrices, in general not simultaneously diagonalizable.

The usefulness of nonstandard schemes for scalar linear delay problems and their possible advantages over alternative numerical methods have been discussed in [22,23]. Particularly, the family of schemes proposed in [23] allows the computation of numerical solutions for scalar linear delay problems with the required degree of accuracy and with comparatively low computational complexity. Moreover, the numerical approximations obtained with these nonstandard schemes reproduce dynamical properties of the exact continuous solutions, such as asymptotic stability, positivity, and oscillation behaviour.

The aim, and main contribution, of the present work is to make available, for a wide class of coupled linear delay differential systems, NSFD methods that possess analogous advantages to those in the scalar setting, exhibiting similar properties in terms of accuracy and dynamic consistency. It is to be remarked that for a class of the new NSFD schemes proposed in this work, the $\mathcal{F}_M$ schemes as defined in Theorem 3, it is rigorously proved that they preserve delay dependent stability. This is a property that usual alternative methods, such as natural Runge-Kutta methods, do not possess, and that is challenging to prove for numerical methods for linear delay systems [6] (p. 356).

There are two main difficulties when dealing with problem (3)–(4), compared with the corresponding scalar problem (1)–(2). Firstly, the obtention of an exact constructive solution that would allow deriving an exact scheme. Secondly, once the new NSFD schemes are defined, the process of proving dynamical properties, which is much more complex than in the scalar case. To overcome these difficulties, the key point is to assume commutativity of the coupled coefficient matrices, a property also considered in other problems involving delay systems [24]. With this assumption, a compact expression for the exact solution of problem (3)–(4), analogous to the scalar case, can be obtained. Also, for commuting matrices, a common Schur basis exists and both matrix coefficients in (3), A and B, can be simultaneously reduced to triangular form, which facilitates analyzing the dynamical properties of the new proposed NSFD schemes.

This paper is structured as follows. In the next section, based on a constructive expression for the exact solution of the initial value vector problem (3)–(4), an exact scheme that is valid in the whole domain of definition is obtained. In Section 3, a family of new nonstandard schemes of increasing order of accuracy are proposed. Next, in Section 4, dynamic consistency properties of the new nonstandard schemes are illustrated with numerical examples and proved for a class of methods. In the final section, the results are summarized and discussed.

2. Exact Numerical Scheme

In our next theorem we present an explicit expression for the solution of problem (3)–(4), derived by using the method of steps [25] (pp. 45–47) and an integral convolution [26] (p. 67), in a similar way as was done in [27] for the scalar problem (1)–(2).

Theorem 1. *The exact solution of (3)–(4) is given by $X(t) = F(t)$, for $-\tau \leq t \leq 0$, and, for $(m-1)\tau < t \leq m\tau$ and $m \geq 1$,*

$$X(t) = \sum_{k=0}^{m-1} \frac{B^k(t-k\tau)^k}{k!} e^{A(t-k\tau)} F(0) + \sum_{k=0}^{m-2} \frac{B^{k+1}}{k!} \int_{-\tau}^{0} (t-(k+1)\tau-s)^k e^{A(t-(k+1)\tau-s)} F(s) ds$$

$$+ \frac{B^m}{(m-1)!} \int_{-\tau}^{t-m\tau} (t-m\tau-s)^{m-1} e^{A(t-m\tau-s)} F(s) ds, \quad (5)$$

where the second summation is assumed to be empty for $m = 1$.

Proof. For $m = 1$, one has $X(t) = e^{At} F(0) + B \int_{-\tau}^{t-\tau} e^{A(t-\tau-s)} F(s) ds$, so that $X(0) = F(0)$ and $X'(t) = AX(t) + BF(t-\tau) = AX(t) + BX(t-\tau)$. For $m > 1$, it is also immediate to check that $X'(t) = AX(t) + BX(t-\tau)$, and that the expressions given by (5) for two consecutive intervals agree at the connecting points $t = m\tau$. Thus, $X(t)$ is continuous for $t > -\tau$, with continuous derivative for $t > 0$, and satisfies (3)–(4). □

From the exact solution given in Theorem 1, an exact numerical difference scheme can be obtained, in a similar way as done in [23] for the scalar case, as shown in the next theorem.

Theorem 2. *Let $h > 0$ such that $Nh = \tau$, for some integer $N \geq 1$. Writing $t_n \equiv nh$ and $X_n \equiv X(t_n)$, for $n \geq -N$, the numerical solution given by $X_n = F(t_n)$, for $-N \leq n \leq 0$, and, for $(m-1)\tau \leq nh < m\tau$ and $m \geq 1$ by*

$$X_{n+1} = e^{Ah} \sum_{k=0}^{m-1} \frac{B^k h^k}{k!} X_{n-kN} + \frac{B^m}{(m-1)!} \int_{t_n-m\tau}^{t_n-m\tau+h} (t_n-m\tau+h-s)^{m-1} e^{A(t_n-m\tau+h-s)} F(s) ds, \quad (6)$$

defines an exact numerical scheme for problem (3)–(4).

Proof. Write $X(t) = E_1(t) + E_2(t) + E_3(t)$, corresponding to the three terms in expression (5). Then, expanding the binomial terms and rearranging and renaming indices, one has

$$E_1(t_{n+1}) = E_1(t_n + h) = \sum_{k=0}^{m-1} \frac{B^k(t_n - k\tau + h)^k}{k!} e^{A(t_n - k\tau + h)} F(0)$$

$$= e^{Ah} \sum_{k=0}^{m-1} \sum_{r=0}^{k} \frac{B^r h^r}{r!} \frac{B^{k-r}(t_n - r\tau - (k-r)\tau)^{k-r}}{(k-r)!} e^{A(t_n - r\tau - (k-r)\tau)} F(0)$$

$$= e^{Ah} \sum_{k=0}^{m-1} \frac{B^k h^k}{k!} \sum_{r=0}^{m-1-k} \frac{B^r(t_n - k\tau - r\tau)^r}{r!} e^{A(t_n - k\tau - r\tau)} F(0) = e^{Ah} \sum_{k=0}^{m-1} \frac{B^k h^k}{k!} E_1(t_n - kN).$$

In a similar way, one gets

$$E_2(t_{n+1}) = e^{Ah} \sum_{k=0}^{m-2} \frac{B^k h^k}{k!} E_2(t_n - kN) = e^{Ah} \sum_{k=0}^{m-1} \frac{B^k h^k}{k!} E_2(t_n - kN),$$

since $E_2(t_n - (m-1)N) = 0$ for $(m-1)\tau \leq t_n < m\tau$. Also,

$$E_3(t_{n+1}) = e^{Ah} \sum_{k=0}^{m-1} \frac{B^k h^k}{k!} E_3(t_n - kN) + \frac{B^m}{(m-1)!} \int_{t_n-m\tau}^{t_n-m\tau+h} (t_n-m\tau+h-s)^{m-1} e^{A(t_n-m\tau+h-s)} F(s) ds,$$

so that expression (6) is recovered. □

Remark 1. The expressions given in Theorems 1 and 2 are also valid when $A = 0$, i.e., for the particular case of the pure delay problem

$$X'(t) = BX(t - \tau), t > 0, \qquad X(t) = F(t), -\tau \leq t \leq 0. \tag{7}$$

If A and B are diagonal, or in the case where they are simultaneously diagonalizable after the usual change of variables, problem (3)–(4) consists of d independent scalar problems, and the expressions given by Theorems 1 and 2 for each component of $X(t)$ coincide with those given in [23] for the corresponding scalar problems.

Example 1. Figure 1 presents a numerical example of application of the results of this section, showing the continuous solution given by Theorem 1 (lines) and the exact numerical solution of Theorem 2 with $N = 5$ (points), for the problem (3)–(4) with parameters $\tau = 1$ and

$$A = \begin{pmatrix} -3/2 & 1 \\ -2 & 3/2 \end{pmatrix}, \quad B = \begin{pmatrix} 5/4 & -1 \\ 2 & -7/4 \end{pmatrix}, \quad F(t) = \begin{pmatrix} 2(t+1) \\ (t+1)^2 \end{pmatrix}.$$

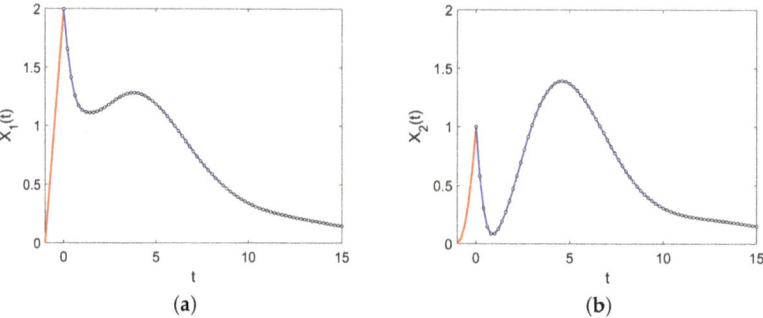

Figure 1. Exact solutions (lines) and numerical solutions provided by the exact scheme (points) for the two components of Example 1. (**a**) First component, $X_1(t)$. (**b**) Second component, $X_2(t)$.

3. Nonstandard Finite Difference Methods of Increasing Orders

The exact numerical solution given by Theorem 2 has the drawback of the integral term in (6), as an exact expression could be obtained for certain initial functions $F(t)$, but in general a numerical approximation would be needed. A class of methods could be derived by approximating this integral term, either by using some numerical integration algorithm or by approximating the initial function with some family of functions that allowed the explicit computation of the integral. Instead, as proposed in [23] for the scalar problem, a family of nonstandard methods of increasing orders can be derived by computing the exact solution in the first M intervals and then discarding the integral term, as shown in the next theorem. We define two classes of methods of order M, $\mathcal{F}_M$ and $\mathcal{T}_M$ methods, depending on whether the full sum in (6) is kept or a truncated sum is used.

Theorem 3. Let $N \geq 1$ and $h = \tau/N$. For a given $M \geq 1$, assume that the values of X_n, for $n = -N \ldots MN$, are computed using the exact scheme of Theorem 2. Define F_M and T_M schemes to compute successive values for any $m > M$ by the expressions

$$\mathcal{F}_M := \quad X_{n+1} = e^{Ah} \sum_{k=0}^{m-1} \frac{B^k h^k}{k!} X_{n-kN}, \quad (m-1)\tau \leq nh < m\tau, \tag{8}$$

$$\mathcal{T}_M := \quad X_{n+1} = e^{Ah} \sum_{k=0}^{M} \frac{B^k h^k}{k!} X_{n-kN}, \quad (m-1)\tau \leq nh < m\tau. \tag{9}$$

Then, both numerical schemes, $\mathcal{F}_M$ and $\mathcal{T}_M$, have local error $O(h^{M+1})$ and order M.

Proof. Let $\|\ \|$ be any vector norm and a compatible norm for matrices, and consider the scheme $\mathcal{T}_M$. Assume that $\|X(t_k) - X_k\| = O(h^{M+1})$ for $k \leq n$, which is the case for $nh \leq M\tau$. Then, for $m \geq M+1$ and $(m-1)\tau \leq nh < m\tau$, using (6), one gets

$$\|X(t_{n+1}) - X_{n+1}\| \leq \|e^{Ah}\| \sum_{k=0}^{M} \frac{\|B\|^k h^k}{k!} \|X(t_{n-kN}) - X_{n-kN}\|$$
$$+ \frac{\|B\|^m}{(m-1)!} \int_{t_n-m\tau}^{t_n-m\tau+h} (t_n - m\tau + h - s)^{m-1} \|e^{A(t_n-m\tau+h-s)}\| \|F(s)\| ds. \tag{10}$$

Let $M = \|B\|$, and $M_1, M_2 > 0$ such that $\|e^{As}\| < M_1$ and $\|F(s)\| < M_2$ for $s \in [0,h]$. Then, by the induction hypothesis, the first term in (10) is $O(h^{M+1})$ and the second term is bounded by

$$\frac{M^m M_1 M_2}{(m-1)!} \int_{t_n-m\tau}^{t_n-m\tau+h} (t_n - m\tau + h - s)^{m-1} ds < \frac{M^m M_1 M_2}{(m-1)!} h^m \leq O(h^{M+1}).$$

Similar arguments result in the same bounds holding for the scheme $\mathcal{F}_M$. □

Remark 2. *The results of Theorem 3 also hold if the values for X_n in the first intervals are computed using any numerical method of order at least $O(h^{M+1})$, instead of using the exact scheme. Although both types of schemes, $\mathcal{F}_M$ and $\mathcal{T}_M$, have the same order, more general dynamic consistency properties can be proved for the class of $\mathcal{F}_M$ schemes, as shown in the next section.*

The error analysis of the methods provided by Theorem 3 is illustrated in the next two figures. Errors of numerical solutions for Example 1, computed using $\mathcal{T}_M$ schemes of three different orders, are shown in Figure 2 (top). The corresponding errors relative to the expected order, i.e., errors divided by h^M, are shown in Figure 2 (bottom), with results in agreement with the expected orders given by Theorem 3.

Figure 3 presents the errors in relation with the size of the mesh for numerical solutions of Example 1 computed using $\mathcal{T}_3$, the truncated method with $M = 3$. Errors overlap when divided by h^3, clearly showing that the method is of order three, as established in Theorem 3.

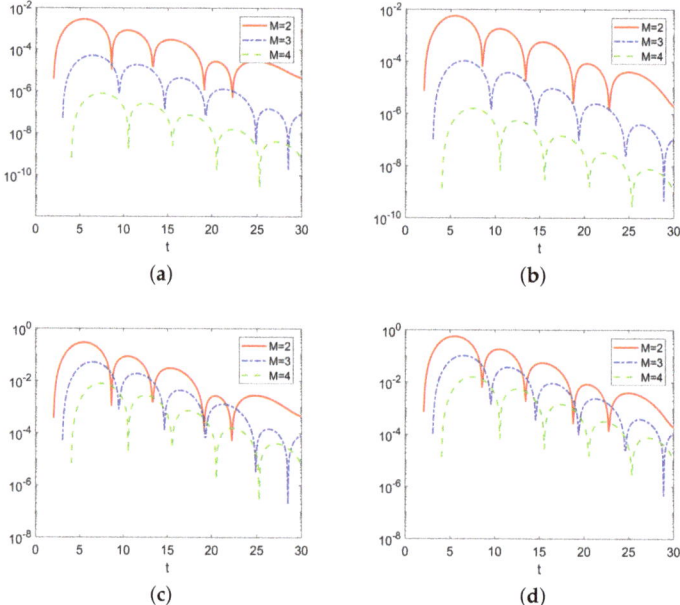

Figure 2. Absolute errors (log-scale) of numerical solutions for Example 1, computed using three different $\mathcal{T}_M$ schemes of increasing orders, with $h = 0.1$. (**a**,**b**) Absolute errors for the first and second component, respectively. (**c**,**d**) Errors divided by h^M.

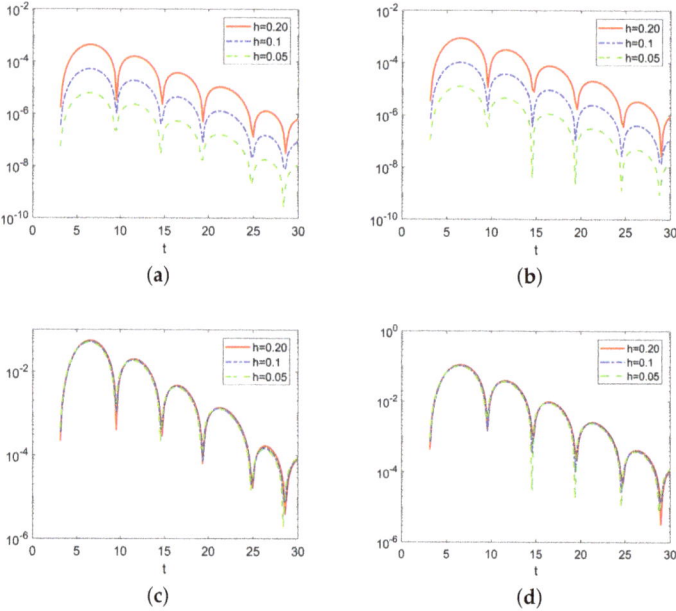

Figure 3. Errors (log-scale) of numerical solutions for Example 1, computed with the method $\mathcal{T}_3$ using three different mesh sizes. (**a**,**b**) Absolute errors for the first and second component, respectively. (**c**,**d**) Errors divided by h^3.

4. Dynamic Consistency Properties

In this section we analyse the consistency between dynamic properties of the numerical solutions resulting from applying the $\mathcal{F}_M$ and $\mathcal{T}_M$ schemes defined in Theorem 3 and the continuous solutions of problem (3)–(4).

4.1. Asymptotic Stability

First we will show that the $\mathcal{F}_M$ schemes defined in Theorem 3 preserve delay-dependent stability, i.e., that they are $\tau(0)$-stable [28].

It is well known that for the trivial solution of (3)–(4) to be asymptotically stable it is necessary and sufficient that all the roots λ_i of the characteristic equation

$$\det(\lambda I - A - e^{-\lambda \tau} B) = 0, \tag{11}$$

where I is the $d \times d$ identity matrix, have negative real parts, $\Re(\lambda_i) < 0$. This condition, involving a transcendental equation with an infinite number of roots, is difficult to verify in general. However, when A and B commute, there is a common Schur basis for them, and they can be simultaneously reduced to triangular form, with elements in the diagonal corresponding to the eigenvalues of each matrix [29]. Thus, in this case, condition (11) is equivalent to

$$\prod_{i=1}^{d} (\lambda - \alpha_i - e^{-\lambda \tau} \beta_i) = 0, \tag{12}$$

where (α_i, β_i) are pairs of eigenvalues of A and B, as they appear in the i diagonal position in the common triangular form. Hence, writing (α, β) for any of these pairs, it follows that if the trivial solution of (3)–(4) is asymptotically stable then

$$\lambda - \alpha - e^{-\lambda \tau} \beta = 0 \tag{13}$$

implies $\Re(\lambda) < 0$.

Consider now the difference equations system (8) defining the $\mathcal{F}_M$ scheme. For any n such that $(m-1)\tau \le nh = n\tau/N < m\tau$, the integer part of n/N is $[n/N] = m - 1$. Thus, we can write (8) in the form of a Volterra difference system of convolution type,

$$X_{n+1} = \sum_{j=0}^{n} B_j X_{n-j}, \tag{14}$$

by setting $B_j = 0$, the d-dimensional zero matrix, when $j \ne kN$, and

$$B_j = e^{Ah} \frac{B^{j/N} h^{j/N}}{(j/N)!} \tag{15}$$

when $j = kN$, for integer k. Thus, using the Z-transform method, it holds that the system (14) is asymptotically stable if all roots of the characteristic equation

$$\det(zI - \tilde{B}(z)) = 0, \tag{16}$$

satisfy $|z| < 1$ [30] (Theorem 5.21), where $\tilde{B}(z)$ is the Z transform of B. In this case,

$$\tilde{B}(z) = \sum_{j=0}^{\infty} B_j z^{-j} = e^{Ah} \sum_{k=0}^{\infty} \frac{B^k h^k}{k!} z^{-kN} = e^{Ah} e^{Bh/z^N}. \tag{17}$$

Now we have the basis to prove our next theorem.

Theorem 4 ($\tau(0)$-stability). *Consider problem (3)–(4) and the $\mathcal{F}_M$ schemes defined in Theorem 3. If the trivial solution of (3)–(4) is asymptotically stable then the numerical solutions computed using $\mathcal{F}_M$ schemes are also asymptotically stable.*

Proof. From the common triangular decompositions of A and B, it follows that every root of (16) must satisfy, for some pair of ordered eigenvalues (α, β),

$$z - e^{\alpha h} e^{\beta h / z^N} = 0 \implies \ln(z) - \alpha h - \beta h / z^N = 0. \tag{18}$$

Writing $\ln(z) = \lambda \tau / N$, so that $z^{-N} = \exp(-\lambda \tau)$, one gets from (18)

$$\lambda \tau / N - \alpha h - \beta h \exp(-\lambda \tau) = 0, \tag{19}$$

which is equivalent to (13), since $h = \tau/N$. Hence, if the trivial solution of (3)–(4) is asymptotically stable it must hold that $\Re(\lambda) < 0$, and therefore $|z| = \exp(\Re(\lambda \tau / N)) < 1$. □

Remark 3. *For the class of $\mathcal{T}_M$ schemes, a general and unconditional result similar to Theorem 4 is not to be expected, as shown by considering the simple case where $A = 0$, $M = 1$, and $N = 1$, so that the $\mathcal{T}_1$ scheme reduces to*

$$X_{n+1} = X_n + Bh X_{n-1}. \tag{20}$$

If B has a real eigenvalue β, the trivial solution of the pure delay problem (7) is asymptotically stable if $|\beta| < \pi/2$, while the asymptotic stability of (20) requires the more stringent condition $|\beta| < 1$ [31].

Delay Independent Stability

Our next theorem shows that the class of $\mathcal{T}_M$ schemes do preserve absolute or delay independent stability, i.e., that they are P-stable [6] (p. 296). This is also trivially the case for $\mathcal{F}_M$ schemes, as P-stability is a weaker condition than $\tau(0)$-stability.

Theorem 5 (P-stability). *Consider problem (3)–(4) and the $\mathcal{T}_M$ schemes defined in Theorem 3. If the trivial solution of (3)–(4) is asymptotically stable for all values of τ, then the numerical solutions computed using $\mathcal{T}_M$ schemes are also asymptotically stable.*

Proof. Using the common triangular forms of A and B, and considering a pair of ordered eigenvalues (α, β), a necessary condition for the trivial solution of (3)–(4) to be delay-independent asymptotically stable is [31,32]

$$\Re(\alpha) + |\beta| < 0. \tag{21}$$

The solution of the difference system (9) defining the $\mathcal{T}_M$ scheme is asymptotically stable if all roots of the characteristic equation

$$\det\left(z^{MN+1} I - e^{Ah} \sum_{k=0}^{M} \frac{B^k h^k}{k!} z^{(M-k)N}\right) = 0 \tag{22}$$

are inside the unit disc. A nonzero z is a root of (22) if for a pair (α, β) it holds that

$$z - e^{\alpha h} \sum_{k=0}^{M} \frac{\beta^k h^k}{k!} z^{-kN} = 0. \tag{23}$$

Thus, if condition (21) hold and we assume that there is a root with $|z| \geq 1$, we would get a contradiction, since, from (23),

$$|z| \leq e^{\Re(\alpha h)} \sum_{k=0}^{M} \frac{|\beta|^k h^k}{k!} |z|^{-kN} < e^{(\Re(\alpha)+|\beta|)h} < 1. \qquad (24)$$

□

The stability analysis provided by Theorems 4 and 5 assures that, for a fixed delay, the region of asymptotic stability for (3)–(4) is contained in the region of asymptotic stability of $\mathcal{F}_M$ schemes, while for $\mathcal{T}_M$ schemes it can only be assured that the region of asymptotic stability of (3)–(4) for all delays is contained in the corresponding region for the numerical solution. However, $\mathcal{T}_M$ schemes usually perform much better than can be guaranteed, as shown in the next example.

Example 2. *Figure 4 shows the numerical solutions computed with the method $\mathcal{T}_2$, with $N = 5$, for the pure delay problem (7) with parameters*

$$B = \begin{pmatrix} -0.435 & 0.0325 \\ 0.13 & -0.435 \end{pmatrix}, \quad F(t) = \begin{pmatrix} \cos(\pi t) \\ (t+1)^2 \end{pmatrix},$$

and two different values of delay, $\tau = 3$ and $\tau = 3.3$.

Matrix B has real eigenvalues, $\lambda_1 = -0.37$ and $\lambda_2 = -0.5$. Hence, in this case the trivial solution of (7) is asymptotically stable if all the eigenvalues β of B satisfy $|\beta|\tau < \pi/2$ [6] (p. 289), i.e., for $\tau < \pi$. As shown in Figure 4, the numerical solutions present the correct behaviour, even for values of τ close to the limit of stability. For both components, the solution approach zero as t increases for $\tau = 3$, inside the region of stability (Figure 4, top), while they diverge for $\tau = 3.3$, outside the region of stability (Figure 4, bottom).

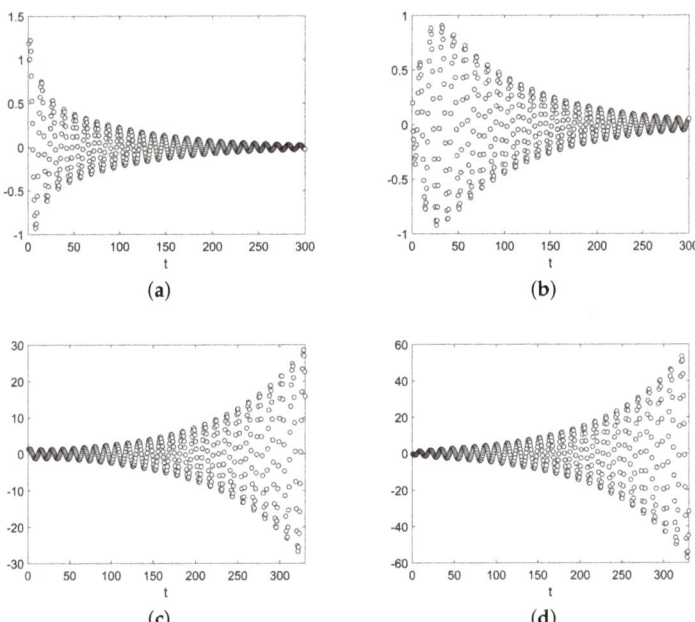

Figure 4. Numerical solutions computed with the method $\mathcal{T}_2$ for Example 2 with two different values of delay, showing stable and unstable behaviours. (**a**,**b**) First and second component, respectively, with delay $\tau = 3$. (**c**,**d**) First and second component, respectively, with delay $\tau = 3.3$.

4.2. Oscillation and Positivity

Our next theorem shows that $\mathcal{F}_M$ schemes also preserve the oscillation properties of exact solutions for problem (3)–(4).

We recall that a solution of (3) is said to oscillate if every component of the solution has arbitrary large zeros; otherwise it is called non-oscillatory [33] (Definition 5.0.1). It is known that every solution of the delay differential system (3) oscillates if and only if the characteristic Equation (11) has no real roots [33] (Theorem 5.1.1).

Theorem 6 (Oscillation). *If every solution of (3)–(4) oscillates, then the numerical solutions computed using $\mathcal{F}_M$ schemes also oscillate.*

We will use the result of the following lemma, whose proof is similar to that of Theorem 7.1.1 in [33].

Lemma 1. *Consider the linear system of difference Equation (8) defining the $\mathcal{F}_M$ scheme. Every solution of (8) oscillates if and only if the characteristic Equation (16) has no positive roots.*

Proof of Theorem 6. Using common triangular decompositions of A and B, if every solution of (3)–(4) oscillates then, for any pair of ordered eigenvalues (α, β), Equation (13), or equivalently Equation (19), has no real roots. If we assume that there is a non-oscillatory solution of (8) we get a contradiction, since, from Lemma 1, there would be a positive z satisfying

$$z - e^{\alpha h} e^{\beta h / z^N} = 0, \qquad (25)$$

and writing $z = \exp(\lambda h)$, we would get Equation (19) with λ a real root. $\square$

Remark 4. *For the class of $\mathcal{T}_M$ schemes, a general result similar to Theorem 6 seems difficult, although particular cases could be dealt with, as shown in our next proposition.*

Proposition 1. *If every solution of the pure delay problem (7) oscillates, then the numerical solutions computed using the $\mathcal{T}_1$ scheme also oscillate.*

Proof. For the pure delay problem (7), an equivalent condition for every solution to oscillate is that B has no real eigenvalues in the interval $[-1/e\tau, +\infty)$ [33] (Theorem 5.2.2). The characteristic equation for the system of difference equations (9) defining the $\mathcal{T}_1$ scheme, i.e., Equation (22) with $A = 0$ and $M = 1$, reads

$$\det\left((z^{N+1} - z^N)I - Bh\right) = 0, \qquad (26)$$

and every solution oscillates if (26) has no positive roots [33] (Theorem 7.1.1). But z is a root of (26) if for an eigenvalue β of B it holds that

$$z^{N+1} - z^N = \beta h. \qquad (27)$$

Thus, if every solution of (7) oscillates, so that any possible real eigenvalue β of B satisfies $\beta \tau < -1/e$, and we assume that there is a positive root of (27), we get a contradiction. From (27), if z is positive, then β is real and $z^N(z-1) = \beta \tau / N < 0$. Hence, it follows that $z < 1$ and

$$N z^N (1 - z) = -\beta \tau > e^{-1}.$$

But for $0 < z < 1$, the maximum value of $N z^N (1 - z)$ is attained when $z = N/(N+1)$, so that

$$N z^N (1 - z) \leq \left(\frac{N}{N+1}\right)^{N+1} < e^{-1}.$$

Example 3. *Figure 5 shows the numerical solution computed with the method $\mathcal{T}_2$, with $N = 10$, for the first component of the pure delay problem (7) with parameters $\tau = 1$ and B and $F(t)$ as in Example 2. In this case, every solution oscillates if all the eigenvalues β of B satisfy $|\beta|\tau > 1/e \approx 0.3679$. As shown in Figure 4, the numerical solutions preserve the correct behaviour, even for a value of τ very close to the limit of oscillation.*

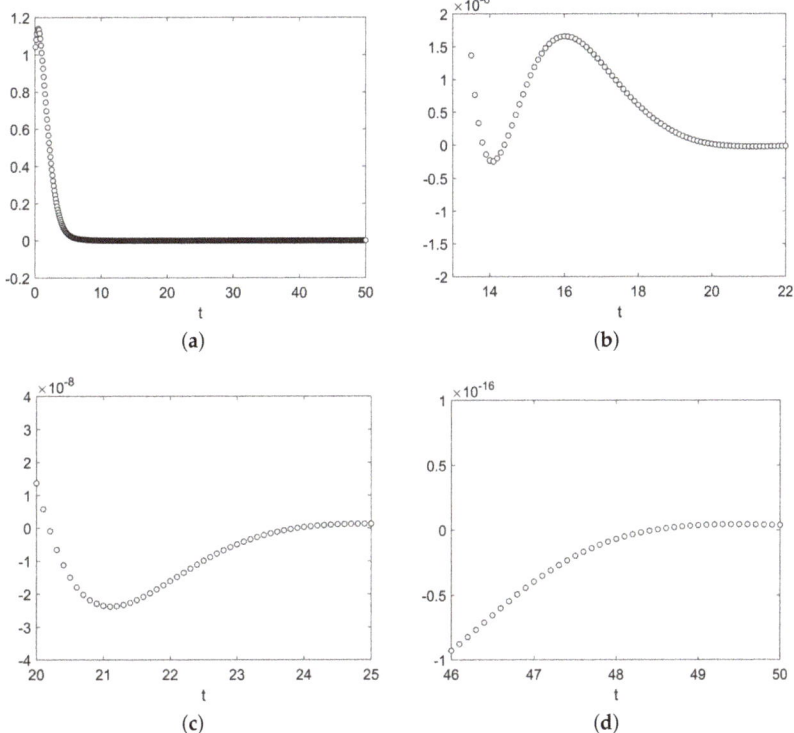

Figure 5. Numerical solution for the first component of Example 1 and zoom-views in different intervals. (**a**) $t \in [0, 50]$. (**b**) $t \in [13, 22]$. (**c**) $t \in [20, 25]$. (**d**) $t \in [46, 50]$.

Positivity

Conditions for the solution of a DDE system to preserve positivity, in the sense that for any component-wise positive initial function $F(t)$ the solution always remains positive, are necessarily very restrictive.

Consider the pure delay problem (7). If $B = (b_{ij}) > 0$ element-wise, i.e., $b_{ij} > 0$, $i,j = 1 \ldots d$, then it is clear from the expression of the exact solution given in (5) that for any component-wise positive initial function $F(t)$ all components of the solution $X(t)$ remain positive for all $t > 0$. In this case, it is also clear from the expressions of $\mathcal{F}_M$ and $\mathcal{T}_M$ schemes given in Theorem 3 that the numerical solutions computed with both methods also remain positive for all $t > 0$.

If B is only non-negative, i.e., $B \geq 0$ element-wise, then the exact as well as the numerical solutions remain non-negative for any non-negative initial function and all $t > 0$. The condition of all elements of B being non-negative is also necessary to preserve positivity, for if there is an element of B, say b_{1r}, negative, then it is possible to find an initial function, component-wise positive, for which some component of $X(t)$ becomes negative, already in the first interval $0 < t < \tau$. To see this, take $F(t)$ with components $F_r(t) = t^2$ and $F_j(t) = \delta t^2$, $j \neq r$, and choose δ such that

$$0 < \delta < -b_{1r}/|\sum_{j \neq r} b_{1j}|.$$

Taking into account that, from (5), for $0 < t < \tau$ one gets $X(t) = BG(t)$, where the components of $G(t)$ are $G_r(t) = h(t)$ and $G_j(t) = \delta h(t)$, $j \neq r$, with $h(t) = ((t-\tau)^3 - (-\tau)^3)/3$, it follows that the first component of $X(t)$ becomes negative,

$$X_1(t) = b_{1r}h(t) + \delta \sum_{j \neq r} b_{1j}h(t) < \left(b_{1r} - b_{1r} \frac{\sum_{j \neq r} b_{1j}}{|\sum_{j \neq r} b_{1j}|} \right) h(t) < 0,$$

since $h(t) > 0$ for $t \in (0, \tau)$.

For the general linear problem (3)–(4), if $B > 0$ and also $A > 0$ element-wise, then it is also immediate that positivity is preserved both in the exact solution and in the numerical solutions computed using the $\mathcal{F}_M$ and $\mathcal{T}_M$ schemes. For $B \geq 0$, non-negativity of the solutions, both exact and numerical, is preserved if A is Metzler, i.e., with non-negative off-diagonal elements, as then $\exp(At)$ is non-negative for any $t > 0$.

5. Conclusions

Despite the growing interest in NSFD methods, including their application to some problems with delay, the scheme presented in Theorem 2 is possibly the first example of an exact scheme for a system of delay differential equations, generalising to systems of linear DDE with commuting matrix coefficients the results presented in [23] for scalar linear DDE problems.

The families of $\mathcal{F}_M$ and $\mathcal{T}_M$ schemes defined in Theorem 3 allow the computation of numerical solutions for problem (3)–(4) with high accuracy and low computational costs for extended time intervals, showing good dynamic consistency properties. In particular, $\mathcal{F}_M$ schemes have been proved to preserve delay-dependent asymptotic stability of the continuous solution, i.e., they are $\tau(0)$-stable difference methods, while $\mathcal{T}_M$ schemes have been proved to preserve delay-independent asymptotic stability, i.e., they are P-stable methods. Also, $\mathcal{F}_M$ schemes preserve the oscillation behaviour of the exact solution, which has also been proved for the $\mathcal{T}_1$ scheme when applied to the pure delay problem (7). Both types of scheme also provide numerical solutions that remain positive, or non-negative, when the original problem satisfy conditions assuring the corresponding property.

Several problems and lines of research are open from the results presented in this work. Proving dynamic consistency properties similar to those of $\mathcal{F}_M$ schemes for some particular $\mathcal{T}_M$ schemes, either in general or when applied to some type of problems or under certain conditions, could deserve further attention, as $\mathcal{T}_M$ schemes offer the same accuracy than $\mathcal{F}_M$ schemes with reduced computational needs. Applying the new schemes to low order systems, e.g., with coefficients being 2×2 or 3×3 matrices, might allow to express the systems of difference equations defining the schemes in the more usual form of a NSFD method, with derivatives for each component being approximated by the corresponding increments divided by a scalar function $\varphi(h) = h + O(h^2)$, as has been done for some examples of systems without delay [34,35]. This could also be the case when considering problems where the matrix coefficients A and B posses some special structure.

Author Contributions: Conceptualization, M.Á.C. and F.R.; methodology and formal analysis, all authors; writing—original draft preparation, F.R.; writing—review and editing, all authors.

Funding: This research was funded by Ministerio de Economía y Competitividad grant number CGL2017-89804-R.

Conflicts of Interest: The authors declare no conflict of interest. The funders had no role in the design of the study; in the collection, analyses, or interpretation of data; in the writing of the manuscript, or in the decision to publish the results.

References

1. Kuang, Y. *Delay Differential Equations. With Applications in Population Dynamics*; Academic Press: San Diego, CA, USA, 1993.
2. Kolmanovskii, V.; Myshkis, A. *Introduction to the Theory and Applications of Functional Differential Equations*; Kluwer Academic Publishers: Dordrecht, The Netherlands, 1999.
3. Smith, H. *An Introduction to Delay Differential Equations with Applications to the Life Sciences*; Springer: New York, NY, USA, 2011.
4. Wu, M.; He, Y.; She, J.-H. *Stability Analysis and Robust Control of Time-Delay Systems*; Science Press: Beijing, China; Springer: Heidelberg, Germany, 2010.
5. Chalishajar, D.N.; Kumar, A. Total controllability of the second order semi-linear differential equation with infinite delay and non-instantaneous impulses. *Math. Comput. Appl.* **2018**, *23*, 32. [CrossRef]
6. Bellen, A.; Zennaro, M. *Numerical Methods for Delay Differential Equations*; Oxford University Press: Oxford, UK, 2003.
7. Mickens, R.E. *Nonstandard Finite Difference Models of Differential Equations*; World Scientific: Singapore, 1994.
8. Mickens, R.E. (Ed.) *Advances on Applications of Nonstandard Finite Difference Schemes*; World Scientific: Singapore, 2005.
9. Patidar, K.C. Nonstandard finite difference methods: Recent trends and further developments. *J. Differ. Equ. Appl.* **2016** *22*, 817–849. [CrossRef]
10. Stuart, A.M.; Humphries, A.R. *Dynamical Systems and Numerical Analysis*; Cambridge University Press: Cambridge, UK, 1998.
11. Jódar, L.; Villanueva, R.J.; Arenas, A.J.; González, G.C. Nonstandard numerical methods for a mathematical model for influenza disease. *Math. Comput. Simul.* **2008**, *79*, 622–633. [CrossRef]
12. Arenas, A.J.; Moraño, J.A.; Cortés, J.C. Non-standard numerical method for a mathematical model of RSV epidemiological transmission. *Comput. Math. Appl.* **2008**, *56*, 670–678. [CrossRef]
13. Arenas, A.J.; González-Parra, G.; Chen-Charpentier, B.M. A nonstandard numerical scheme of predictor–corrector type for epidemic models. *Comput. Math. Appl.* **2010**, *59*, 3740–3749. [CrossRef]
14. Mickens, R.E.; Washington, T.M. NSFD discretizations of interacting population models satisfying conservation laws. *Comput. Math. Appl.* **2013**, *66*, 2307–2316. [CrossRef]
15. Guerrero, F.; González-Parra, G.; Arenas, A.J. A nonstandard finite difference numerical scheme applied to a mathematical model of the prevalence of smoking in Spain: A case study. *Comput. Appl. Math.* **2014**, *33*, 13–25. [CrossRef]
16. Sekiguchi, M.; Ishiwata, E. Global dynamics of a discretized SIRS epidemic model with time delay. *J. Math. Anal. Appl.* **2010**, *371*, 195–202. [CrossRef]
17. Wang, Y. Dynamics of a nonstandard finite-difference scheme for delay differential equations with unimodal feedback. *Commun. Nonlinear Sci. Numer. Simul.* **2012**, *17*, 3967–3978. [CrossRef]
18. Su, H.; Li, W.; Ding, X. Numerical dynamics of a nonstandard finite difference method for a class of delay differential equations. *J. Math. Anal. Appl.* **2013**, *400*, 25–34. [CrossRef]
19. Xu, J.; Geng, Y. Stability preserving NSFD scheme for a delayed viral infection model with cell-to-cell transmission and general nonlinear incidence. *J. Differ. Equ. Appl.* **2017**, *23*, 893–916. [CrossRef]
20. Zhou, J.; Yang, Y. Global dynamics of a discrete viral infection model with time delay, virus-to-cell and cell-to-cell transmissions. *J. Differ. Equ. Appl.* **2017**, *23*, 1853–1868. [CrossRef]
21. Dang, Q.A.; Hoang, M.T. Lyapunov direct method for investigating stability of nonstandard finite difference schemes for metapopulation models. *J. Differ. Equ. Appl.* **2018**, *24*, 15–47. [CrossRef]
22. Garba, S.M.; Gumel, A.B.; Hassan, A.S.; Lubuma, J.M-S. Switching from exact scheme to nonstandard finite difference scheme for linear delay differential equation. *Appl. Math. Comput.* **2015**, *258*, 388–403. [CrossRef]

23. García, M.A.; Castro, M.A.; Martín, J.A.; Rodríguez, F. Exact and nonstandard numerical schemes for linear delay differential models. *Appl. Math. Comput.* **2018**, *338*, 337–345. [CrossRef]
24. Pospíšil, M. Representation and stability of solutions of systems of functional differential equations with multiple delays. *Electron. J. Qual. Theory* **2012**, *54*, 1–30. [CrossRef]
25. Bellman, R.; Cooke, K.L. *Differential-Difference Equations*; Academic Press: New York, NY, USA, 1963.
26. El'sgol'ts, L.E.; Norkin, S.B. *Introduction to the Theory and Application of Differential Equations with Deviating Arguments*; Academic Press: New York, NY, USA, 1973.
27. Martín, J.A.; Rodríguez, F.; Company, R. Analytic solution of mixed problems for the generalized diffusion equation with delay. *Math. Comput. Model.* **2004**, *40*, 361–369. [CrossRef]
28. Guglielmi, N. On the asymptotic stability properties of Runge–Kutta methods for delay differential equations. *Numer. Math.* **1997**, *77*, 467–485. [CrossRef]
29. Horn, R.A.; Johnson, C.R. *Matrix Analysis*; Cambridge University Press: Cambridge, UK, 1985.
30. Elaydi, S.N. *An Introduction to Difference Equations*; Springer: New York, NY, USA, 1996.
31. Čermák, J.; Jánský, J.; Nechvátal, L. Exact versus discretized stability regions for a linear delay differential equation. *Appl. Math. Comput.* **2019**, *347*, 712–722. [CrossRef]
32. Nishiguchi, J. On parameter dependence of exponential stability of equilibrium solutions in differential equations with a single constant delay. *Discrete Contin. Dyn. Syst.* **2016**, *36*, 5657–5679. [CrossRef]
33. Györi, I.; Ladas, G. *Oscillation Theory of Delay Differential Equations*; Clarendon Press: Oxford, UK, 1991.
34. Roeger, L.-I.W. Exact finite-difference schemes for two-dimensional linear systems with constant coefficients. *J. Comput. Appl. Math.* **2008**, *219*, 102–109. [CrossRef]
35. Quang, A.D.; Tuan, H.M. Exact finite difference schemes for three-dimensional linear systems with constant coefficients. *J. Vietnam J. Math.* **2018**, *46*, 471–492. [CrossRef]

© 2019 by the authors. Licensee MDPI, Basel, Switzerland. This article is an open access article distributed under the terms and conditions of the Creative Commons Attribution (CC BY) license (http://creativecommons.org/licenses/by/4.0/).

MDPI
St. Alban-Anlage 66
4052 Basel
Switzerland
Tel. +41 61 683 77 34
Fax +41 61 302 89 18
www.mdpi.com

Mathematics Editorial Office
E-mail: mathematics@mdpi.com
www.mdpi.com/journal/mathematics

www.ingramcontent.com/pod-product-compliance
Lightning Source LLC
LaVergne TN
LVHW070439100526
838202LV00014B/1625